Dictionary of Computer Vision and Image Processing

R.B. Fisher
University of Edinburgh
K. Dawson-Howe
Trinity College Dublin
A. Fitzgibbon
Oxford University
C. Robertson
Heriot-Watt University
E. Trucco
Heriot-Watt University

John Wiley & Sons, Ltd

Copyright © 2005 John Wiley & Sons Ltd, The Atrium, Southern Gate, Chichester, West Sussex PO19 8SQ, England

Telephone (+44) 1243 779777

Email (for orders and customer service enquiries): cs-books@wiley.co.uk
Visit our Home Page on www.wiley.com

Other Wiley Editorial Offices

John Wiley & Sons Inc., 111 River Street, Hoboken, NJ 07030, USA

Jossey-Bass, 989 Market Street, San Francisco, CA 94103-1741, USA

Wiley-VCH Verlag GmbH, Boschstr. 12, D-69469 Weinheim, Germany

John Wiley & Sons Australia Ltd, 42 McDougall Street, Milton, Queensland 4064, Australia

John Wiley & Sons (Asia) Pte Ltd, 2 Clementi Loop #02-01, Jin Xing Distripark, Singapore 129809

John Wiley & Sons Canada Ltd, 22 Worcester Road, Etobicoke, Ontario, Canada M9W 1L1

Wiley also publishes its books in a variety of electronic formats. Some content that appears in print may not be available in electronic books.

British Library Cataloguing in Publication Data

A catalogue record for this book is available from the British Library

ISBN-13 978-0-470-01526-1 (PB)
ISBN-10 0-470-01526-8 (PB)

Typeset in 9/9.5pt Garamond by Integra Software Services Pvt. Ltd, Pondicherry, India
Printed and bound in Great Britain by Antony Rowe Ltd, Chippenham, Wiltshire
This book is printed on acid-free paper responsibly manufactured from sustainable forestry in which at least two trees are planted for each one used for paper production.

From Bob to Rosemary, Mies, Hannah, Phoebe and Lars

From AWF to Liz, to my parents, and again to D.

To Karen and Aidan. Thanks pips!

From Ken to Jane, William and Susie

From Manuel to Emily, Francesca and Alistair

Contents

Preface

This dictionary arose out of a continuing interest in the resources needed by beginning students and researchers in the fields of image processing, computer vision and machine vision (however you choose to define these overlapping fields). As instructors and mentors, we often found confusion about what various terms and concepts mean for the beginner. To support these learners, we have tried to define the key concepts that a competent generalist should know about these fields. The results are definitions for more than 2500 terms.

This is a dictionary, not an encyclopedia, so the definitions are necessarily brief and are not intended to replace a proper textbook explanation of the term. We have tried to capture the essentials of the terms, with short examples or mathematical precision where feasible or necessary for clarity. Further information about many of the terms can be found in the references that follow this preface. These are mostly general textbooks, each providing a broad view of a portion of the field. Some of the concepts are also quite recent and, although commonly used in research publications, have not yet appeared in mainstream textbooks. Thus this book is also a useful source for recent terminology and concepts.

Certainly some concepts are missing, but we have scanned both textbooks and the research literature to find the central and commonly used terms. Many additional terms also arose as part of the definition process itself.

Although the dictionary was intended for beginning and intermediate students and researchers, as we developed the dictionary it was clear that we also had some confusions and vague understandings of the concepts. It also surprised us that some terms had multiple usages. To improve quality and coverage, each definition was reviewed during development by at least two people besides its author. We hope that this has caught any errors and vagueness, as well as reproduced the alternative meanings. Each of the co-authors is quite experienced in the topics covered here, but it was still educational to learn more about our field in the process of compiling the dictionary.

We hope that you find using the dictionary equally valuable.

While we have strived for perfection, we recognize that we might have made some errors or been insufficiently precise. Hence, there is a web site where errata and other materials can be found: http://homepages.inf.ed.ac.uk/rbf/CVDICT/ If you spot an error, please email us: rbf@inf.ed.ac.uk

To help the reader, terms appearing elsewhere in the dictionary are underlined. We have tried to be reasonably thorough about this, but some terms, such as 2D, 3D, light, camera, image, pixel and color were so commonly used that we decided to not cross-reference these.

We have tried to be consistent with the mathematical notation: italics for scalars (s), arrowed italics for points and vectors ($\vec{v}$), and bold font letters for matrices ($\mathbf{M}$).

The contents of the dictionary have been much improved with the advice and suggestions from the dictionary's international panel of experts:

Andrew Blake
Aaron Bobick
Chris Brown
Stefan Carlsson
Henrik Christensen
Roberto Cipolla
James L. Crowley
Patrick Flynn
Vaclav Hlavac
Anil Jain
Avinash Kak
Ales Leonardis
Song-De Ma
Gerard Medioni
Joe L. Mundy
Shmuel Peleg
Maria Petrou
Keith Price
Azriel Rosenfeld
Amnon Shashua
Yoshiaki Shirai
Milan Sonka
Chris Taylor
Demetri Terzopoulos
John Tsotsos
Shimon Ullman
Andrew Zisserman
Steven Zucker

References

1. D. Ballard, C. Brown. *Computer Vision*. Prentice Hall, 1982.

   A classic textbook presenting an overview of techniques in the early days of computer vision. Still a source of very useful information.

2. D. Forsyth, J. Ponce. *Computer Vision – A Modern Approach*. Prentice Hall, 2003.

   A recent, comprehensive book covering both 2D (image processing) and 3D material.

3. R. M. Haralick, L. G. Shapiro. *Computer and Robot Vision*. Addison-Wesley Longman Publishing, 1992.

   A well-known, extensive collection of algorithms and techniques, with mathematics worked out in detail. Mostly image processing, but some 3D vision as well.

4. E. Hecht. *Optics*. Addison-Wesley, 1987.

   A key resource for information on light, geometrical optics, distortion, polarization, Fourier optics, etc.

5. B. K. P. Horn. *Robot Vision*. MIT Press, 1986.

   A classic textbook in computer vision. Especially famous for the treatment of optic flow. Dated nowadays, but still very interesting and useful.

6. A. Jain. *Fundamentals of Digital Image Processing*. Prentice Hall Intl, 1989.

   A little dated, but still a thorough introduction to the key topics in 2D image processing and analysis. Particularly useful is the information on various whole image transforms, such as the 2D Fourier transform.

7. R. Jain, R. Kasturi, B. Schunck. *Machine Vision*. McGraw-Hill, 1995.

A good balance of image processing and 3D vision, including typically 3D topics like model-based matching. Reader-friendly presentation, also graphically.

8. V. S. Nalwa. *A Guided Tour of Computer Vision*. Addison-Wesley, 1993.

   A discursive, compact presentation of computer vision at the beginning of the 90s. Good to get an overview of the field as it was, and quickly.

9. M. Petrou and P. Bosdogianni. *Image Processing: The Fundamentals*. Wiley Interscience, 1999.

   A student-oriented textbook on image processing, focusing on enhancement, compression, restoration and pre-processing for image understanding.

10. M. Sonka, V. Hlavac, R. Boyle. *Image Processing, Analysis, and Machine Vision.* Chapman & Hall, 1993.

    A well-known, exhaustive textbook covering much image processing and a good amount of 3D vision alike, so that algorithms are sometimes only sketched. A very good reference book.

11. G. Stockman, L. Shapiro. *Computer Vision*. Prentice Hall. 2001.

    A thorough and broad 2D and 3D computer vision book, suitable for use as a course textbook and for reference.

12. E. Trucco, A. Verri. *Introductory Techniques for 3-D Computer Vision*. Prentice Hall, 1998.

    This book gives algorithms and theory for many central 3D algorithms and topics, and includes supporting detail from 2D and image processing where appropriate.

13. S. E. Umbaugh. *Computer Vision and Image Processing*. Prentice Hall, 1998.

    A compact book on image processing, coming with its own development kit and examples on a CD.

0

1D: One dimensional, usually in reference to some structure. Examples include: 1) a signal $x(t)$ that is a function of time t, 2) the dimensionality of a single property value or 3) one degree of freedom in shape variation or motion.

2D: Two dimensional. A space describable using any pair of orthogonal basis vectors consisting of two elements.

2D coordinate system: A system associating uniquely 2 real numbers to any point of a plane. First, two intersecting lines (axes) are chosen on the plane, usually perpendicular to each other. The point of intersection is the origin of the system. Second, metric units are established on each axis (often the same for both axes) to associate numbers to points. The coordinates P_x and P_y of a point, P, are obtained by projecting P onto each axis in a direction parallel to the other axis, and reading the numbers at the intersections:

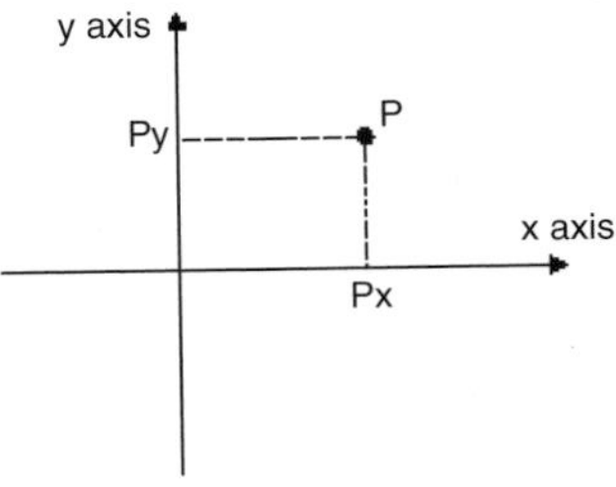

2D Fourier transform: A special case of the general Fourier transform often used to find structures in images.

2D image: A matrix of data representing samples taken at discrete intervals. The data may be from a variety of sources and sampled in a variety of ways. In computer vision applications the image values are often encoded color or monochrome intensity samples taken by digital cameras but may also be

Dictionary of Computer Vision and Image Processing R.B. Fisher, K. Dawson-Howe, A. Fitzgibbon, C. Robertson and E. Trucco

range data. Some typical intensity values are:

Image values

2D input device: A device for sampling light intensity from the real world into a 2D matrix of measurements. The most popular two dimensional imaging device is the charge-coupled device (CCD) camera. Other common devices are flatbed scanners and X-ray scanners.

2D point: A point in a 2D space, that is, characterized by two coordinates; most often, a point on a plane, for instance an image point in pixel coordinates. Notice, however, that two coordinates do not necessarily imply a plane: a point on a 3D surface can be expressed either in 3D coordinates or by two coordinates given a surface parameterization (see surface patch).

2D point feature: Localized structures in a 2D image, such as interest points, corners and line meeting points (X, Y and T shaped for example). One detector for these features is the SUSAN corner finder.

2D pose estimation: A fundamental open problem in computer vision where the correspondence between two sets of 2D points is found. The problem is defined as follows: Given two sets of points $\{\vec{x}_j\}$ and $\{\vec{y}_k\}$, find the Euclidean transformation $\{\mathbf{R}, \vec{t}\}$ (the pose) and the match matrix $\{\mathbf{M}_{jk}\}$ (the correspondences) that best relates them. A large number of techniques has been used to address this problem, for example tree-pruning methods, the Hough transform and geometric hashing. A special case of 3D pose estimation.

2D projection: A transformation mapping higher dimensional space onto two dimensional space. The simplest method is to simply discard higher dimensional coordinates, although generally a viewing position is used and the projection is performed.

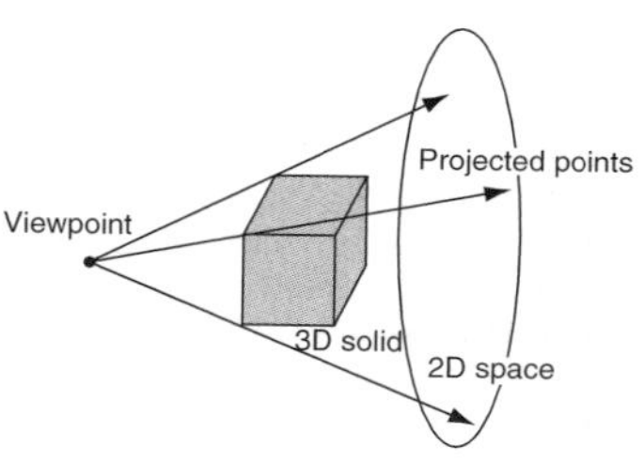

For example, the main steps for a computer graphics projection are as follows: apply normalizing transform to 3D point world coordinates; clip against canonical view volume; project onto projection plane; transform into viewport in 2D device coordinates for display. Commonly used projections functions are parallel projection or perspective projection.

2.5D image: A range image obtained by scanning from a single viewpoint. This allows the data to be represented in a single image array, where each pixel value encodes the distance to the observed scene. The reason this is not called a 3D image is to make explicit the fact that the back sides of the scene objects are not represented.

2.5D sketch: Central structure of Marr's theory of vision. An intermediate description of a scene indicating the visible surfaces and their arrangement with respect to the viewer. It is built from several different elements: the contour, texture and shading information coming from the primal sketch, stereo information and motion. The description is theorized to be a kind of buffer where partial resolution of the objects takes place. The name $2\frac{1}{2}$D sketch stems from the fact that although local changes in depth and discontinuities are well resolved, and the absolute distance to all scene points may remain unknown.

3D: Three dimensional. A space describable using any triple of mutually orthogonal basis vectors consisting of three elements.

3D coordinate system: Same as 2D coordinate system, but in three dimensions.

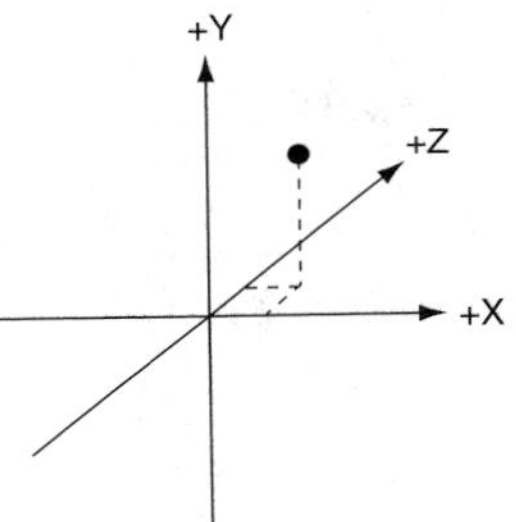

3D data: Data described in all three spatial dimensions. See also range data, CAT and NMR. An example of a 3D data set is:

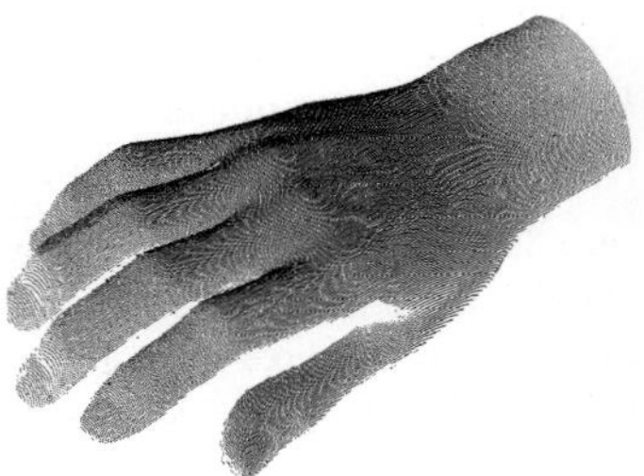

3D data acquisition: Sampling data in all three spatial dimensions. There are a variety of ways to perform this sampling, for example using structured light triangulation.

3D image: See range image.

3D interpretation: A 3D model, *e.g.*, a solid object, that explains an image or a set of image data. For instance, a certain configuration of image lines can be explained as the perspective projection of a polyhedron; in simpler words, the image lines are the images of some of the polyhedron's lines. See also image interpretation.

3D model: A description of a 3D object that primarily describes its shape. Models of this sort are regularly used as exemplars in model based recognition and 3D computer graphics.

3D moments: A special case of moment where the data comes from a set of 3D points.

3D object: A subset of $\mathbb{R}^3$. In computer vision, often taken to mean a volume in $\mathbb{R}^3$ that is bounded by a surface. Any solid object around you is an example: table, chairs, books, cups, and you yourself.

3D point: An infinitesimal volume of 3D space.

3D point feature: A point feature on a 3D object or in a 3D environment. For instance, a corner in 3D space.

3D pose estimation: 3D pose estimation is the process of determining the transformation (translation and rotation) of an object in one coordinate frame with respect to another coordinate frame. Generally, only rigid objects are considered, models of those object exist *a priori*, and we wish to determine the position of that object in an image on the basis of matched features. This is a fundamental open problem in computer vision where the correspondence between two sets of 3D points is found. The problem is defined as follows: Given two sets of points $\{\vec{x}_j\}$ and $\{\vec{y}_k\}$, find the parameters of an Euclidean transformation $\{\mathbf{R}, \vec{t}\}$ (the pose)and the match matrix $\{\mathbf{M}_{jk}\}$ (the correspondences) that best relates them. Assuming the points correspond, they should match exactly under this transformation.

3D reconstruction: A general term referring to the computation of a 3D model from 2D images.

3D skeleton: See skeleton

3D stratigraphy: A modeling and visualization tool used to display different underground layers. Often used for visualizations of archaeological sites or for detecting different rock and soil structures in geological surveying.

3D structure recovery: See 3D reconstruction.

3D texture: The appearance of texture on a 3D surface when imaged, for instance, the fact that the density of texels varies with distance due to perspective effects. 3D surface properties (*e.g.*, shape, distances, orientation) can be estimated from such effects. See also shape from texture, texture orientation.

3D vision: A branch of computer vision dealing with characterizing data composed of 3D measurements. For example, this may involve segmentation of the data into individual surfaces that are then used to identify the data as one of several models. Reverse engineering is a specialism inside 3D vision.

4 connectedness: A type of image connectedness in which each rectangular pixel is considered to be connected to the four neighboring pixels that share a common crack edge. See also 8 connectedness. This figure shows the four pixels connected to the central pixel (*):

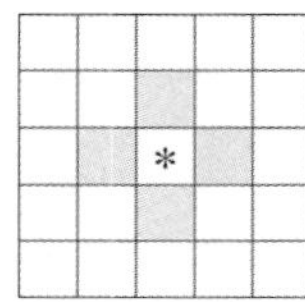

and the four groups of pixels joined by 4 connectedness:

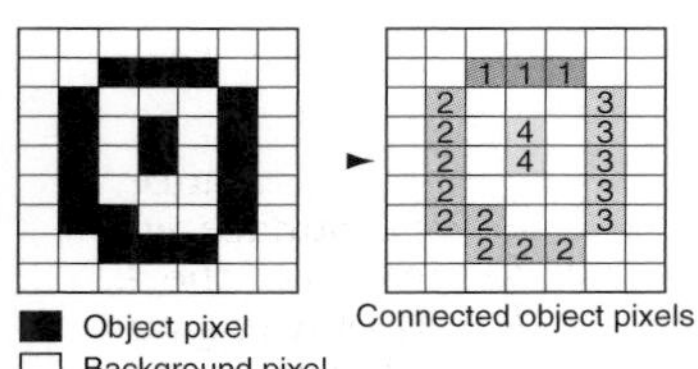

8 connectedness: A type of image connectedness in which each rectangular pixel is considered to be connected to all eight neighboring pixels.

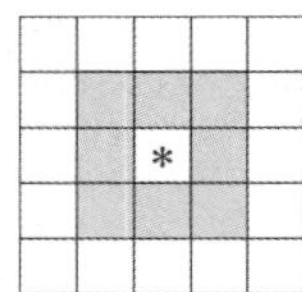

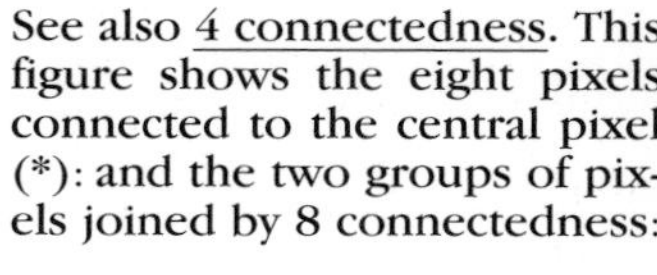
See also 4 connectedness. This figure shows the eight pixels connected to the central pixel (*): and the two groups of pixels joined by 8 connectedness:

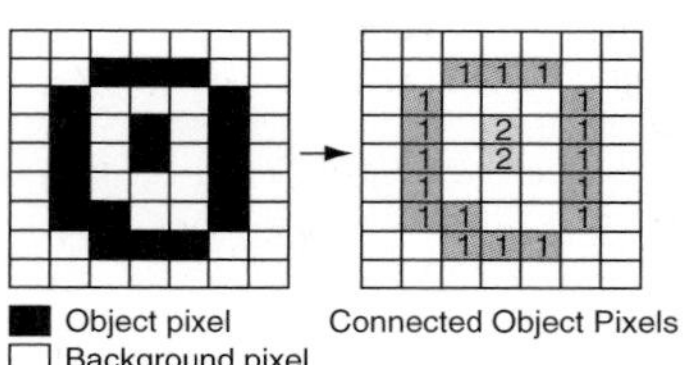

A

A*: A search technique that performs best-first searching based on an evaluation function that combines the cost so far and the estimated cost to the goal.

a posteriori **probability**: Literally, "after" probability. It is the probability $p(s|e)$ that some situation s holds after some evidence e has been observed. This contrasts with the *a priori* probability $p(s)$ that is the probability of s before any evidence is observed. Bayes, rule is often used to compute the *a posteriori* probability from the *a priori* probability and the evidence.

a priori **probability**: Suppose that there is a set Q of equally likely outcomes for a given action. If a particular event E could occur of any one of a subset S of these outcomes, then the *a priori* or theoretical probability of E is defined by

$$P(E) = \frac{size(S)}{size(Q)}$$

aberration: Problem exhibited by a lens or a mirror whereby unexpected results are obtained. There are two types of aberration commonly encountered: chromatic aberration, where different frequencies of light focus at different positions – and spherical aberration, where light passing through the edges of a lens (or mirror) focuses at slightly different positions.

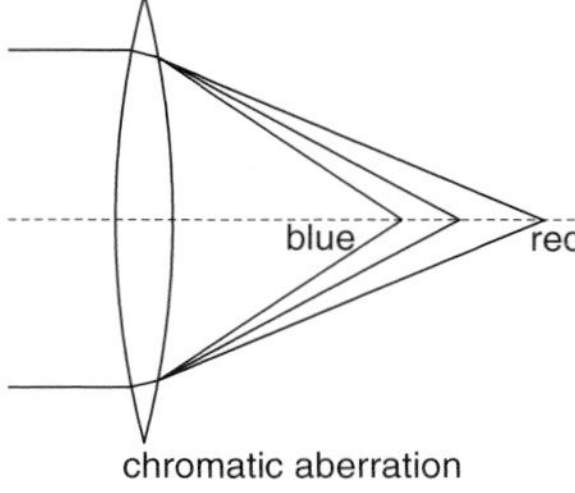

chromatic aberration

absolute conic: The conic in 3D projective space that is the intersection of the unit (or any) sphere with the plane at infinity. It consists only of complex points. Its importance in computer vision is due to its role in the problem of autocalibration: the image of the absolute conic

Dictionary of Computer Vision and Image Processing R.B. Fisher, K. Dawson-Howe, A. Fitzgibbon, C. Robertson and E. Trucco

(IAC), a 2D conic, is represented by a 3×3 matrix ω that is the inverse of the matrix $\mathbf{K}\mathbf{K}^\top$, where $\mathbf{K}$ is the matrix of the internal camera calibration parameters. Thus, identifying ω allows the camera calibration to be computed.

absolute coordinates: Generally used in contrast to *local* or *relative* coordinates. A coordinate system that is referenced to some external datum. For example, a pixel in a satellite image might be at (100,200) in image coordinates, but at (51:48:05N, 8:17:54W) in georeferenced absolute coordinates.

absolute orientation: In photogrammetry, the problem of registering two corresponding sets of 3D points. Used to register a photogrammetric reconstruction to some absolute coordinate system. Often expressed as the problem of determining the rotation $\mathbf{R}$, translation $\vec{t}$ and scale s that best transforms a set of *model* points $\{\vec{m}_1, \ldots, \vec{m}_n\}$ to corresponding data points $\{\vec{d}_1, \ldots, \vec{d}_n\}$ by minimizing the least-squares error

$$\epsilon(R, \vec{t}, s) = \sum_{i=1}^{n} \|\vec{d}_i - s(\mathbf{R}\vec{m}_i + \vec{t})\|^2$$

to which a solution may be found in terms of the singular value decomposition.

absolute point: A 3D point defining the origin of a coordinate system.

absolute quadric: The symmetric 4×4 rank 3 matrix $\Omega = \begin{pmatrix} \mathbf{I}_3 & \vec{0}_3 \\ \vec{0}_3^\top & 0 \end{pmatrix}$. Like the absolute conic, it is defined to be invariant under Euclidean transformations, is rescaled under similarities, takes the form $\Omega = \begin{pmatrix} \mathbf{A}^\top\mathbf{A} & \vec{0}_3 \\ \vec{0}_3^\top & 0 \end{pmatrix}$ under affine transforms and becomes an arbitrary 4×4 rank 3 matrix under projective transforms.

absorption: Attenuation of light caused by passing through an optical system or being incident on an object surface.

accumulation method: A method of accumulating evidence in histogram form, then searching for peaks, which correspond to hypotheses. See also Hough transform, generalized Hough transform.

accumulative difference: A means of detecting motion in image sequences. Each frame in the sequence is compared to a reference frame (after registration if necessary) to produce a difference image. Thresholding the difference image gives a binary motion mask. A counter for each pixel location in the accumulative image is incremented every time the difference between the reference image and the current image exceeds some threshold. Used for change detection.

accuracy: The error of a value away from the true value. Contrast this with precision.

acoustic sonar: Sound Navigation And Ranging. A device that is used primarily for the detection and location of objects (*e.g.*, underwater or in air, as in mobile robotics, or internal to a human body, as in medical ultrasound) by reflecting and intercepting acoustic waves. It operates with acoustic waves in an analogous way to that of radar, using both the time of flight and Doppler effects, giving the radial component of relative position and velocity.

ACRONYM: A vision system developed by Brooks that attempted to recognize three dimensional objects from two dimensional images, using generalized cylinder primitives to represent both stored model and objects extracted from the image.

active appearance model: A generalization of the widely used active shape model approach that includes all of the information in the image region covered by the target object, rather than just that near modeled edges. The active appearance model has a statistical model of the shape and gray-level appearance of the object of interest. This statistical model generalizes to cover most valid examples. Matching to an image involves finding model parameters that minimize the difference between the image and a synthesized model example, projected into the image.

active blob: A region based approach to the tracking of non-rigid motion in which an active shape model is used. The model is based on an initial region that is divided using Delaunay triangulation and then each patch is tracked from frame to frame (note that the patches can deform).

active contour models: A technique used in model based vision where object boundaries are detected using a deformable curve representation such as a snake. The term active refers to the ability of the snake to deform shape to better match the image data. See also active shape model.

active contour tracking: A technique used in model based vision where object boundaries are tracked in a video sequence using active contour models.

active illumination: A system of lighting where intensity, orientation, or pattern may be continuously controlled and altered. This kind of system may be used to generate structured light.

active learning: Learning about the environment through interaction (*e.g.*, looking at an object from a new viewpoint).

active net: An active shape model that parameterizes a triangulated mesh.

active sensing: 1) A sensing activity carried out in an active or purposive way, for instance where a camera is moved in space to acquire multiple or optimal views of an object. (See also active vision, purposive vision, sensor planning.)

2) A sensing activity implying the projection of a pattern of energy, for instance a laser line, onto the scene. See also laser stripe triangulation, structured light triangulation.

active shape model: Statistical models of the shapes of objects that can deform to fit to a new example of the object. The shapes are constrained by a statistical shape model so that they may vary only in ways seen in a training set. The models are usually formed by using principal component analysis to identify the dominant modes of shape variation in observed examples of the shape. Model shapes are formed by linear combinations of the dominant modes.

active stereo: An alternative approach to traditional binocular stereo. One of the cameras is replaced with a structured light projector, which projects light onto the object of interest. If the camera calibration is known, the triangulation for computing the 3D coordinates of object points simply involves finding the intersection of a ray and known structures in the light field.

active surface: 1) A surface determined using a range sensor; 2) an active shape model that deforms to fit a surface.

active triangulation: Determination of surface depth by triangulation between a light source at a known position and a camera that observes the effects of the illuminant on the scene. Light stripe ranging is one form of active triangulation. A variant is to use a single scanning laser beam to illuminate the scene and use a stereo pair of cameras to compute depth.

active vision: An approach to computer vision in which the camera or sensor is moved in a controlled manner, so as to simplify the nature of a problem. For example, rotating a camera with constant angular velocity while maintaining fixation at a point allows absolute calculation of scene point depth, instead of only relative depth that depends on the camera speed. (See also kinetic depth.)

active volume: The volume of interest in a machine vision application.

activity analysis: Analyzing the behavior of people or objects in a video sequence, for the purpose of identifying the immediate actions occurring or the long term sequence of actions. For example, detecting potential intruders in a restricted area.

acuity: The ability of a vision system to discriminate (or resolve) between closely arranged visual stimuli. This can be measure using a grating, *i.e.*, a pattern of parallel black and white stripes of equal widths. Once the bars become too close, the grating becomes indistinguishable from a uniform image of the same average intensity as the bars. Under optimal lighting, the minimum spacing that a person can resolve is 0.5 min of arc.

adaptive: The property of an algorithm to adjust its parameters to the data at hand in order to optimize performance. Examples include adaptive contrast enhancement, adaptive filtering and adaptive smoothing.

adaptive coding: A scheme for the transmission of signals over unreliable channels, for example a wireless link. Adaptive coding varies the parameters of the encoding to respond to changes in the channel, for example "fading", where the signal-to-noise ratio degrades.

adaptive contrast enhancement: An image processing operation that applies histogram equalization locally across an image.

adaptive edge detection: Edge detection with adaptive thresholding of the gradient magnitude image.

adaptive filtering: In signal processing, any filtering process in which the parameters of the filter change over time, or where the parameters are different at different parts of the signal or image.

adaptive histogram equalization: A localized method of improving image contrast. A histogram is constructed of the gray levels present. These gray levels are re-mapped so that the histogram is approximately flat. It can be made perfectly flat by dithering.

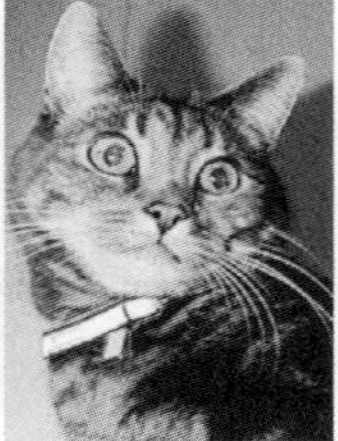

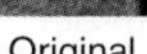

Original

After adaptive histogram equalization

adaptive Hough transform: A Hough transform method that iteratively increases the resolution of the parameter space quantization. It is particularly useful for dealing with high dimensional parameter spaces. Its disadvantage is that sharp peaks in the histogram can be missed.

adaptive meshing: Methods for creating simplified meshes where elements are made smaller in regions of high detail (rapid changes in surface orientation) and larger in regions of low detail, such as planes.

adaptive pyramid: A method of multi-scale processing where small areas of image having some feature in common (say color) are first extracted into a graph representation. This graph is then manipulated, for example by pruning or merging, until the level of desired scale is reached.

adaptive reconstruction: Data driven methods for creating statistically significant data in areas of a 3D data cloud where data may be missing due to sampling problems.

adaptive smoothing: An iterative smoothing algorithm that avoids smoothing over edges. Given an image $I(x, y)$, one iteration of adaptive smoothing proceeds as follows:

1. Compute gradient magnitude image $G(x, y) = |\nabla I(x, y)|$
2. Make weights image $W(x, y) = e^{-\lambda G(x,y)}$
3. Smooth the image

$$S(x, y) = \frac{\sum_{i=-1}^{1} \sum_{j=-1}^{1} A_{xyij}}{\sum_{i=-1}^{1} \sum_{j=-1}^{1} B_{xyij}}$$

where

$$A_{xyij} = I(x+i, y+j)\, W(x+i, y+j)$$

$$B_{xyij} = W(x+i, y+j)$$

adaptive thresholding: An improved image thresholding technique where the threshold value is varied at each pixel. A common technique is to use the average intensity in a neighbourhood to set the threshold.

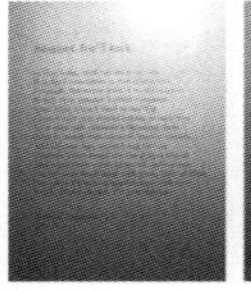

Image, I

Smoothed, S

Thresholded I > S–6

adaptive triangulation: See adaptive meshing.

adaptive visual servoing: See visual servoing.

additive color: The way in which multiple wavelengths of light can be combined to allow other colors to be perceived (*e.g.*, if equal amounts of green and red light are shone on a sheet of white paper the paper will appear to be illuminated with a yellow light source). Contrast this with subtractive color. (See plate section for a color version of this figure.)

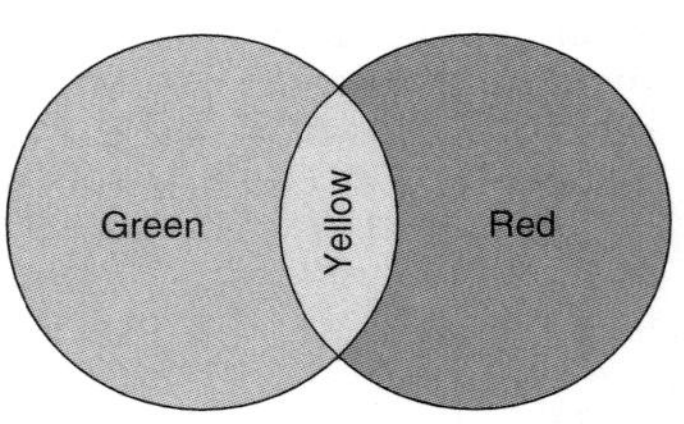

additive noise: Generally image independent noise that is added to it by some external process. The recorded image I at pixel (i, j) is then the sum of the true signal S and the noise N.

$$I_{i,j} = S_{i,j} + N_{i,j}$$

The noise added at each pixel (i, j) could be different.

adjacent: Commonly meaning "next to each other", whether in a physical sense of being connected pixels in an image, image regions sharing some common boundary, nodes in a graph connected by an arc or components in a geometric model sharing some common bounding component, etc. Formally defining "adjacent" can be somewhat heuristic because you may need a way to specify closeness (*e.g.*, on a quantized grid of pixels) or consider how much shared "boundary" is required before two structures are adjacent.

adjacency: See adjacent.

adjacency graph: A graph that shows the adjacency between structures, such as segmented image regions. The nodes of the graph are the structures and an arc implies adjacency of the two structures connected by the arc. This figure shows the graph associated with the segmented image on the left:

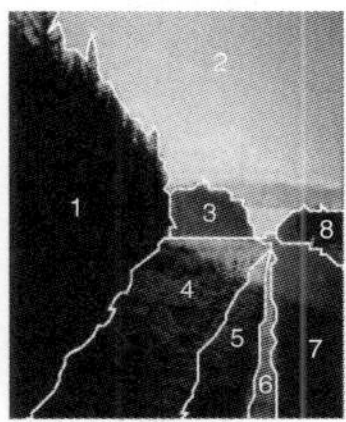

Regions   Adjacency graph

affine: A term first used by Euler. Affine geometry is a study of properties of geometric objects that remain invariant under affine transformations (mappings). These include: parallelness, cross ratio, adjacency.

affine arc length: For a parametric equation of a curve $\vec{f}(u) = (x(u), y(u))$, arc length is not preserved under an affine transformation. The affine length

$$\tau(u) = \int_0^u (\dot{x}\ddot{y} - \ddot{x}\dot{y})^{\frac{1}{3}}$$

is invariant under affine transformations.

affine camera: A special case of the projective camera that is obtained by constraining the 3×4 camera parameter matrix $\mathbf{T}$ such that $T_{3,1} = T_{3,2} = T_{3,3} = 0$ and reducing the camera parameter vector from 11 degrees of freedom to 8.

affine curvature: A measure of curvature based on the affine arc length, τ. For a parametric equation of a curve $\vec{f}(u) = (x(u), y(u))$, its affine curvature, μ, is

$$\mu(\tau) = x''(\tau)y'''(\tau) - x'''(\tau)y''(\tau)$$

affine flow: A method of finding the movement of a surface patch by estimating the affine transformation parameters required to transform the patch from its position in one view to another.

affine fundamental matrix: The fundamental matrix which is obtained from a pair of cameras under affine viewing conditions. It is a 3×3 matrix whose upper left 2×2 submatrix is all zero.

affine invariant: An object or shape property that is not changed (*i.e.*, is invariant) by the application of an affine transformation. See also invariant.

affine length: See affine arc length.

affine moment: Four shape measures derived from second- and third-order moments that remain invariant under affine transformations. They are given by

$$I_1 = \frac{\mu_{20}\mu_{02} - \mu_{11}^2}{\mu_{00}^4}$$

$$I_2 = (\mu_{30}^2\mu_{03}^2 - 6\mu_{30}\mu_{21}\mu_{12}\mu_{03} + 4\mu_{30}\mu_{12}^3 + 4\mu_{21}^3\mu_{03} - 3\mu_{21}^2\mu_{12}^2)/\mu_{00}^{10}$$

$$I_3 = (\mu_{20}(\mu_{21}\mu_{03} - \mu_{12}^2) - \mu_{11}(\mu_{30}\mu_{03} - \mu_{21}\mu_{12}) + \mu_{02}(\mu_{30}\mu_{12} - \mu_{21}^2))/\mu_{00}^7$$

$$I_4 = (\mu_{20}^3\mu_{03}^2 - 6\mu_{20}^2\mu_{11}\mu_{12}\mu_{03} - 6\mu_{20}^2\mu_{02}\mu_{21}\mu_{03} + 9\mu_{20}^2\mu_{02}\mu_{12}^2 + 12\mu_{20}\mu_{11}^2\mu_{21}\mu_{03} + 6\mu_{20}\mu_{11}\mu_{02}\mu_{30}\mu_{03} - 18\mu_{20}\mu_{11}\mu_{02}\mu_{21}\mu_{12} - 8\mu_{11}^3\mu_{30}\mu_{03} - 6\mu_{20}\mu_{02}^2\mu_{30}\mu_{12} + 9\mu_{20}\mu_{02}^2\mu_{21}^2 + 12\mu_{11}^2\mu_{02}\mu_{30}\mu_{12} - 6\mu_{11}\mu_{02}^2\mu_{30}\mu_{21} + \mu_{02}^3\mu_{30}^2)/\mu_{00}^{11}$$

where each μ is the associated central moment.

affine quadrifocal tensor: The form taken by the quadrifocal tensor when specialized to the viewing conditions modeled by the affine camera.

affine reconstruction: A three dimensional reconstruction where the ambiguity in the choice of basis is affine only. Planes that are parallel in the Euclidean basis are parallel in the affine reconstruction. A projective reconstruction can be upgraded to affine by identification of the plane at infinity, often by locating the absolute conic in the reconstruction.

affine stereo: A method of scene reconstruction using two calibrated views of a scene from known view points. It is a simple but very robust approximation to the geometry of stereo vision, to estimate positions, shapes and surface orientations. It can be calibrated very easily by observing just four reference points. Any two views of the same planar surface will be related by an affine transformation that maps one image to the other. This consists of a translation and a tensor, known as the disparity gradient tensor representing the distortion in image shape. If the standard unit vectors X and Y in one image are the projections of some vectors on the object surface and the linear mapping between images is represented by a 2×3 matrix A, then the first two columns of A will be the corresponding vectors in the other image. Since the centroid of the plane will map to both image centroids, it can be used to find the surface orientation

affine transformation: A special set of transformations in Euclidean geometry that preserve some properties of the construct being transformed.

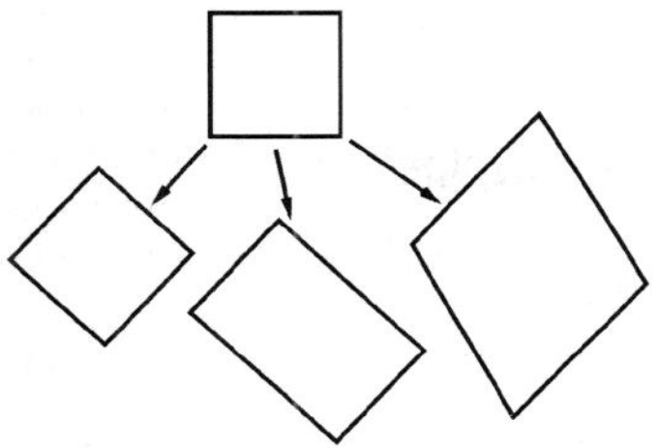

Affine transformations preserve:

- Collinearity of points: if three points belong to the same straight line, their images under affine transformations also belong to the same line and the middle point remains between the other two points.
- Parallel lines remain parallel, concurrent lines remain concurrent (images of intersecting lines intersect).
- The ratio of length of line segments of a given line remains constant.
- The ratio of areas of two triangles remains constant.
- Ellipses remain ellipses and the same is true for parabolas and hyperbolas.
- Barycenters of triangles (and other shapes) map into the corresponding barycenters.

Analytically, affine transformations are represented in the matrix form

$$f(x) = \mathbf{A}x + b$$

where the determinant det(**A**) of the square matrix **A** is not 0. In 2D the matrix is 2×2; in 3D it is 3×3.

affine trifocal tensor: The form taken by the trifocal tensor when specialized to the viewing conditions modeled by the affine camera.

affinely invariant region: Image patches that automatically deform with changing viewpoint in such a way that they cover identical physical parts of a scene. Since such regions can are describable by a set of invariant features they are relatively easy to match between views under changing illumination.

agglomerative clustering: A class of iterative clustering algorithms that begin with a large number of clusters and at each iteration merge pairs (or tuples) of clusters. Stopping the process at a certain number of iterations gives the final set of clusters, or the process can be run until only one cluster remains, and the progress of the algorithm represented as a dendrogram.

albedo: Whiteness. Originally a term used in astronomy to describe reflecting power.

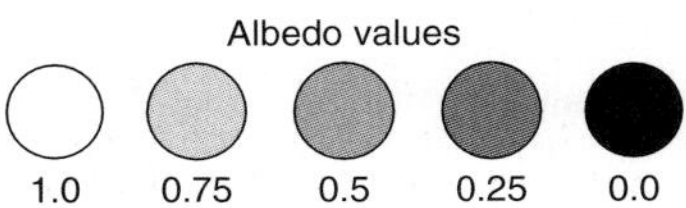

If a body reflects 50% of the light falling on it, it is said to have albedo 0.5.

algebraic distance: A linear distance metric commonly used in computer vision applications because of its simple form and standard matrix based least mean square estimation operations. If a curve or surface is defined implicitly by $f(\vec{x}, \vec{a}) = 0$ (*e.g.*, $\vec{x} \cdot \vec{a} = 0$ for a hyperplane) the algebraic distance of a point $\vec{x}_i$ to the surface is simply $f(\vec{x}_i, \vec{a})$.

aliasing: The erroneous replacement of high spatial frequency (HF) components by low-frequency ones when a signal is sampled. The affected HF components are those that are higher than the Nyquist frequency, or half the sampling frequency. Examples include the slowing of periodic signals by strobe lighting, and corruption of areas of detail in image resizing. If the source signal has no HF components, the effects of aliasing are avoided, so the low pass filtering of a signal to remove HF components prior to sampling is one form of anti-aliasing. The image below is the perspective projection of a checkerboard. The image is obtained by sampling the scene at a set of integer locations. First figure: The spatial frequency increases as the plane recedes, producing aliasing artifacts (jagged lines in the foreground, moiré patterns in the background). Second figure: removing high-frequency components (*i.e.*, smoothing) before downsampling mitigates the effect.

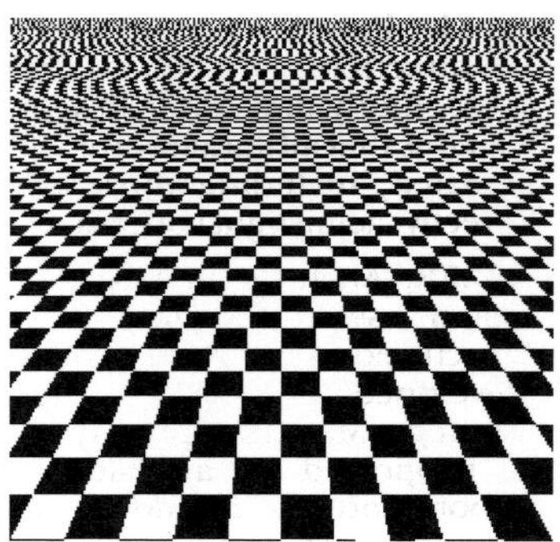

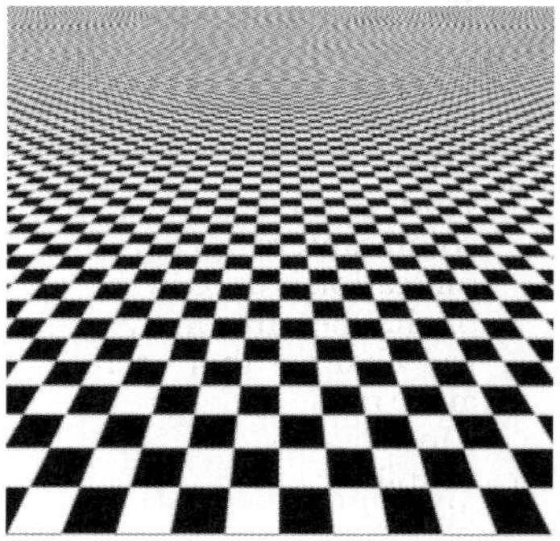

alignment: An approach to geometric model matching by registering a geometric model to the image data.

ALVINN: Autonomous Land Vehicle In a Neural Network. An early attempt, at Carnegie-Mellon University, to learn a complex behaviour (maneuvering a vehicle) by observing humans.

ambient light: Illumination by diffuse reflections from all surfaces within a scene (including the sky, which acts as an external distant surface). In other words, light that comes from all directions, such as the sky on a cloudy day. Ambient light ensures that all surfaces are illuminated, including those not directly facing light sources.

AMBLER: An autonomous active vision system using both structured light and sonar, developed by NASA and Carnegie-Mellon University. It is supported by a 12-legged robot and is intended for planetary exploration.

amplifier noise: Spurious additive noise signal generated by the electronics in a sampling device. The standard model for this type of noise is Gaussian. It is independent of the signal. In color cameras, where more amplification is used in the blue color channel than in the green or red channel there tends to be more noise in the blue channel. In well-designed electronics amplifier noise is generally negligible.

analytic curve finding: A method of detecting parametric curves by first transforming data into a feature space that is then searched for the hypothesized curve parameters. Examples might be line finding using the Hough transform.

anamorphic lens: A lens having one or more cylindrical surfaces. Anamorphic lenses are used in photography to produce images that are compressed in one dimension. Images can later be restored to true form using another reversing anamorphic lens set. This form of lens is used in wide-screen movie photography.

anatomical map: A biological model usable for alignment with or region labeling of a corresponding image dataset. For example, one could use a model of the brain's functional regions to assist in the identification of brain structures in an NMR dataset.

AND operator: A boolean logic operator that combines two input binary images, applying the AND logic

p	q	$p\&q$
0	0	0
0	1	0
1	0	0
1	1	1

at each pair of corresponding pixels. This approach is used to select image regions. The rightmost image below is the result of ANDing the two leftmost images.

angiography: A method for imaging blood vessels by introducing a dye that is opaque when photographed by X-ray. Also the study of images obtained in this way.

angularity ratio: Given two figures, X and Y, $\alpha_i(X)$ and $\beta_j(Y)$ are angles subtending convex parts of the contour of the figure X and $\gamma_k(X)$ are angles subtending plane parts of the contour of figure X, then the angularity ratios are:

$$\sum_i \frac{\alpha_i(X)}{360^\circ}$$

and

$$\frac{\sum_i \beta_j(X)}{\sum_k \gamma_k(X)}$$

anisotropic filtering: Any filtering technique where the filter parameters vary over the image or signal being filtered.

anomalous behavior detection: Special case of surveillance where human movement is analyzed. Used in particular to detect intruders or behavior likely to precede or indicate crime.

antimode: The minimum between two maxima. For example one method of threshold selection is done by determining the antimode in a bimodal histogram.

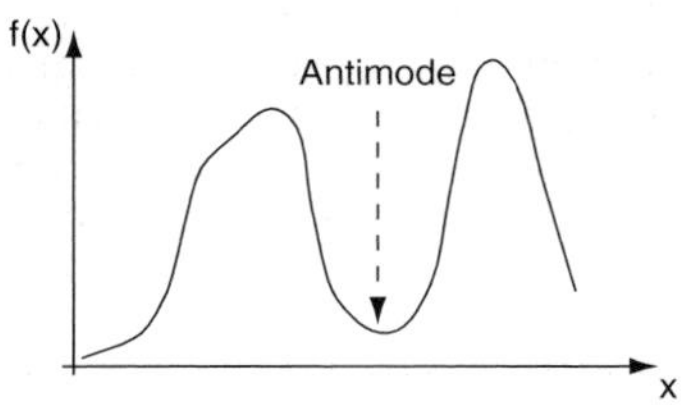

aperture: Opening in the lens diaphragm of a camera through which light is admitted. This device is often arranged so that the amount of light can be controlled accurately. A small aperture reduces the amount of light available, but increases the depth of field. This figure shows nearly closed (left) and nearly open (right) aperture positions:

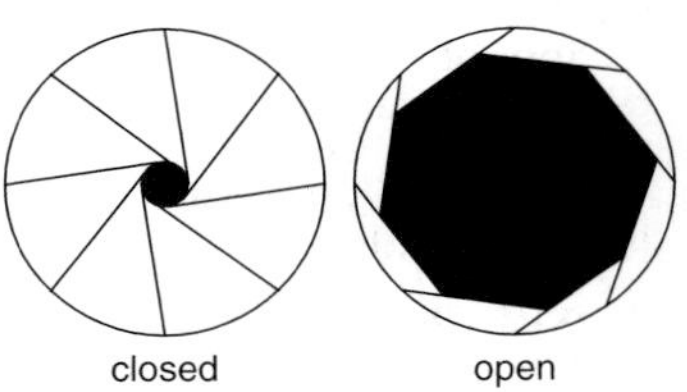

aperture control: Mechanism for varying the size of a camera's aperture.

aperture problem: If a motion sensor has a finite receptive field, it perceives the world through something resembling an aperture, making the motion of a homogeneous contour seem locally ambiguous. Within that aperture, different physical motions are therefore indistinguishable. For example, the two alternative motions of the square below are identical in the circled receptive fields:

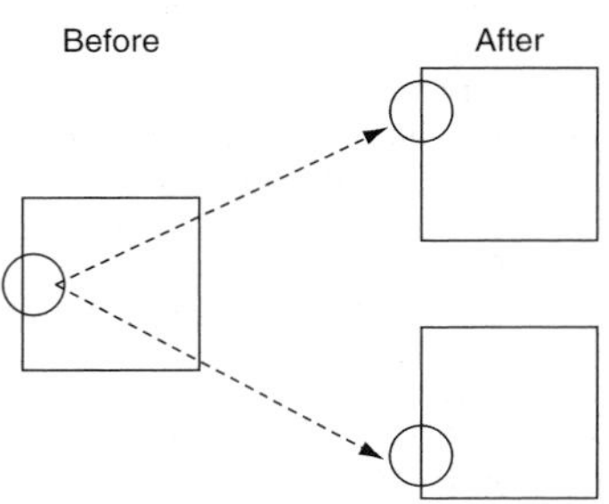

apparent contour: The apparent contour of a surface S in 3D, is the set of critical values of the projection of S on a plane, in other words, the silhouette. If the surface is transparent, the apparent contour can be decomposed into a collection of closed curves with double points and cusps. The convex envelope of an apparent contour is also the boundary of its convex hull.

apparent motion: The 3D motion suggested by the image motion field, but not necessarily matching the real 3D motion. The reason for this mismatch is the motion fields may be ambiguous, that is, may be generated by different 3D motions, or light source movement. Mathematically, there may be multiple solutions to the problem of reconstructing 3D motion from the image motion field. See also visual illusion, motion estimation.

appearance: The way an object looks from a particular viewpoint under particular lighting conditions.

appearance based recognition: Object recognition where the object model encodes the possible appearances of the object (as contrasted with a geometric model that encodes the shape as used in model based recognition). In principle, it is impossible to encode all appearances when occlusions are considered; however, small numbers of appearances can often be adequate, especially if there are not many models in the model base. There are many approaches to appearance based recognition, such as using a principal component model to encode all appearances in a compressed framework, using color histograms to summarize the appearance, or using a set of local appearance descriptors such as Gabor filters extracted at interest points. A common feature of these approaches is learning the models from examples.

appearance based tracking: Methods for object or target recognition in real time, based on image pixel values in each frame rather than derived features. Temporal filtering, such as the Kalman filter, is often used.

appearance change: Changes in an image that are not easily accounted for by motion, such as an object actually changing form.

appearance enhancement transform: Generic term for operations applied to images to change, or enhance, some aspect of them. Examples include brightness adjustment, contrast adjustment, edge sharpening, histogram equalization, saturation adjustment or magnification.

appearance flow: Robust methods for real time object recognition from a sequence of images depicting a moving object. Changes in the images are used rather than the images themselves. It is analogous to processing using optical flow.

appearance model: A representation used for interpreting images that is based on the appearance of the object. These models are usually learned by using multiple views of the objects. See also active appearance model and appearance based recognition.

appearance prediction: Part of the science of appearance engineering, where an object texture is changed so that the viewer experience is predictable.

appearance singularity: An image position where a small change in viewer position can cause a dramatic change in the appearance of the observed scene, such as the appearance or disappearance of image features. This is contrasted with changes occurring when in a generic viewpoint. For example, when viewing the corner of a cube from a distance, a small change in viewpoint still leaves the three surfaces at the corner visible. However, when the viewpoint moves into the infinite plane containing one of the cube faces (a singularity), one or more of the planes disappears.

arc length: If f is a function such that its derivative f' is continuous on some closed interval $[a, b]$ then the arc length of f from $x = a$ to $x = b$ is the integral

$$\int_a^b \sqrt{1+[f'(x)]^2}\mathrm{d}x$$

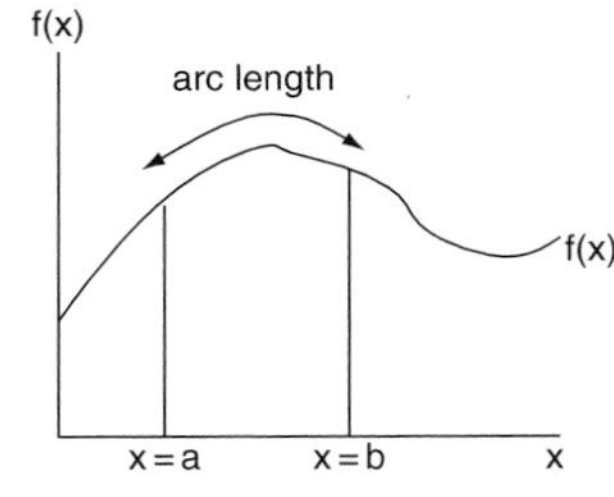

arc of graph: Two nodes in a graph can be connected by an arc. The dashed lines here are the arcs:

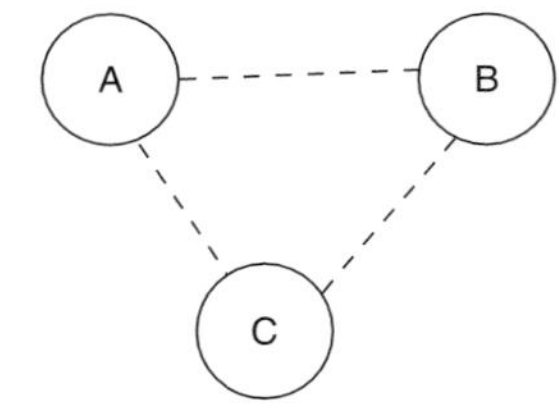

architectural model reconstruction: A generic term for reverse engineering buildings based on collected 3D data as well as libraries of building constraints.

area: The measure of a region or surface's extension in some given units. The units could be image units, such as square pixels, or in scene units, such as square centimeters.

area based: Image operation that is applied to a region of an image, as opposed to pixel based.

array processor: A group of time-synchronized processing elements that perform computations on data distributed across them. Some array processors have elements that communicate only with their immediate neighbors, as in the topology shown below. See also single instruction multiple data.

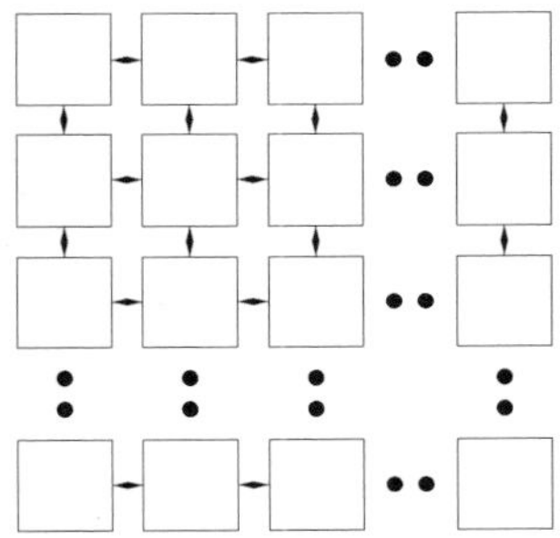

arterial tree segmentation: Generic term for methods used in finding internal pipe-like structures in medical images. Example image types are NMR images, angiograms and X-rays. Example trees are bronchial systems and veins.

articulated object: An object composed by a number of (usually) rigid subparts or components connected by joints, which can be arranged in a number of different configurations. The human body is a typical example.

articulated object model: A representation of an articulated object that includes both its separate parts and their range of movement (typically joint angles) relative to each other.

articulated object segmentation: Methods for acquiring an articulated object from 2D or 3D data.

articulated object tracking: Tracking an articulated object in an image sequence. This includes both the pose of the object and also its shape parameters, such as joint angles.

aspect graph: A graph of the set of views (aspects) of an object, where the arcs of the graph are transitions between two neighboring views (the nodes) and a change between aspects is called a visual event. See also characteristic view. This graph shows some of the aspects of the hippopotamus

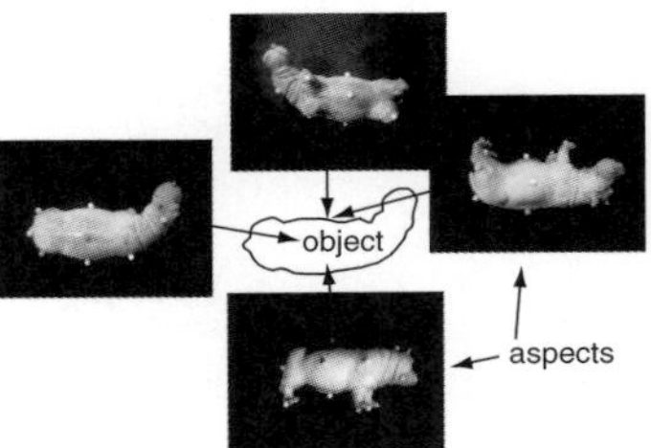

aspect ratio: 1) The ratio of the sides of the bounding box of an object, where the orientation of the box is chosen to maximize this ratio. Since this measure is scale invariant it is a useful metric for object recognition. 2) In a camera, it is the ratio of the horizontal to vertical pixel sizes. 3) In an image, it is the ratio of the image width to height. For example, an image of 640 by 480 pixels has an aspect ratio of 4:3.

aspects: See characteristic view and aspect graph.

association graph: A graph used in structure matching, such as matching a geometric model to a data description. In this graph, each node corresponds to a pairing between a model and a data feature (with the implicit assumption that they are compatible). Arcs in the graph mean that the two connected nodes are pairwise compatible. Finding maximal cliques is one technique for finding good matches. The graph below shows a set of pairings of model features A, B and C with image features a, b, c and d. The maximal clique consisting of A:a, B:b and C:c is one match hypothesis.

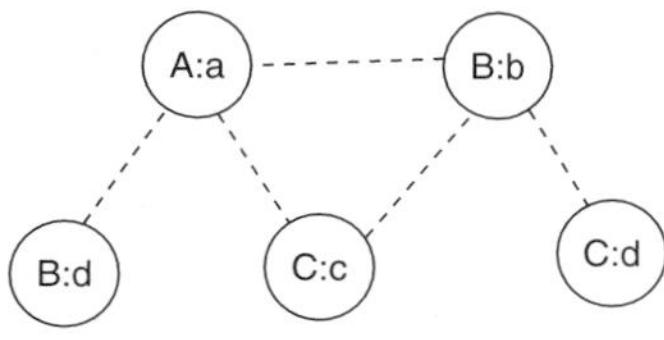

astigmatism: Astigmatism is a refractive error where the light is focused within an optical system, such as in this example.

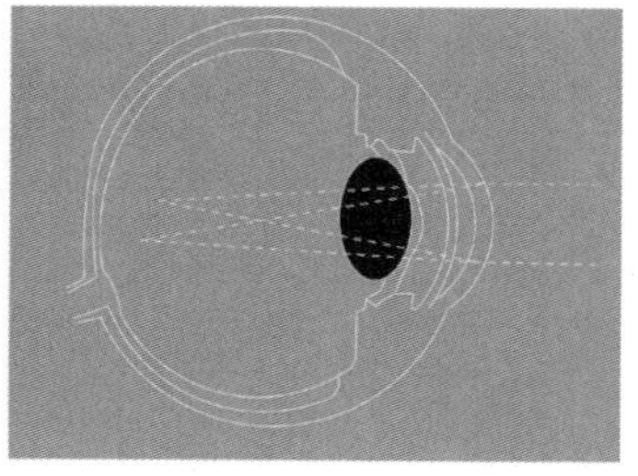

It occurs when a lens has irregular curvature causing light rays to focus at an area, rather than at a point. It may be corrected with a toric lens, which has a greater refractive index on one axis than the others. In human eyes, astigmatism often occurs with nearsightedness and farsightedness.

atlas based segmentation: A segmentation technique used in medical image processing, especially with brain images. Automatic tissue segmentation is achieved using a model of the brain structure and imagery (see atlas registration) compiled with the assistance of human experts. See also image segmentation.

atlas registration: An image registration technique used in medical image processing, especially to register brain images. An atlas is a model (perhaps statistical) of the characteristics of multiple brains, providing examples of normal and pathological structures. This makes it possible to take into account anomalies that single-image registration could not. See also medical image registration.

ATR: See automatic target recognition.

attention: See visual attention.

attenuation: The reduction of a particular phenomenon, for instance, noise attenuation as the reduction of image noise.

attributed graph: A graph useful for representing different properties of an image. Its nodes are attributed pairs of image segments, their color or shape for example. The relations between them, such as relative texture or brightness are encoded as arcs.

augmented reality: Primarily a projection method that adds graphics or sound, etc as an overlay to original image or audio. For example, a firefighter's helmet display could show exit routes registered to his/her view of the building.

autocalibration: The recovery of a camera's calibration using only point (or other feature) correspondences from multiple uncalibrated images and geometric consistency constraints (*e.g.*, that the camera settings are the same for all images in a sequence).

autocorrelation: The extent to which a signal is similar to shifted copies of itself. For an infinitely long 1D signal $f(t) : \mathbb{R} \mapsto \mathbb{R}$, the autocorrelation at a shift Δt is

$$R_f(\Delta t) = \int_{-\infty}^{\infty} f(t) f(t + \Delta t) \mathrm{d}t$$

The autocorrelation function R_f always has a maximum at 0. A *peaked* autocorrelation function decays quickly away from $\Delta t = 0$. The sample autocorrelation function of a finite set of values $f_{1..n}$ is $\{r_f(d) | d = 1, \ldots, n-1\}$ where

$$r_f(d) = \frac{\sum_{i=1}^{n-d} (f_i - \bar{f})(f_{i+d} - \bar{f})}{\sum_{i=1}^{n} (f_i - \bar{f})^2}$$

and $\bar{f} = \frac{1}{n} \sum_{i=1}^{n} f_i$ is the sample mean.

autofocus: Automatic determination and control of image sharpness in an optical or vision system. There are two major variations in this control system: active focusing and passive focusing. Active autofocus is performed using sonar or infrared signal to determine the object distance. Passive autofocus is performed by analyzing the image itself to optimize differences between adjacent pixels in the CCD array.

automatic: Performed by a machine without human intervention. The opposite of "manual".

automatic target recognition (ATR): Sensors and algorithms used for detecting hostile objects in a scene. Sensors are of many different types, sampling in infrared, visible light and using sonar and radar.

autonomous vehicle: A mobile robot controlled by computer, with human input operating only at a very high level, stating the ultimate destination or task for example. Autonomous navigation requires the visual tasks of route detection, self-localization, landmark location and obstacle detection, as well as robotics tasks such as route planning and motor control.

autoregressive model: A model that uses statistical properties of past behavior of some variable to predict future behavior of that variable. A signal x_t at time t satisfies an autoregressive model if $x_t = \sum_{n=1}^{p} \alpha_n x_{t-n} + \omega_t$, where ω_t is noise.

autostereogram: An image similar to a random dot stereogram in which the corresponding features are combined into a single image. Stereo fusion allows the perception of a 3D shape in the 2D image.

average smoothing: See mean smoothing.

AVI: Microsoft format for audio and video files ("audio video interleaved"). Unlike MPEG, it is not a standard, so that compatibility of AVI video files and AVI players is not always guaranteed.

axial representation: A region representation that uses a curve to describe the image region. The axis may be a skeleton derived from the region by a thinning process.

axis of elongation: 1) The line that minimizes the second moment of the data points. If $\{\vec{x}_i\}$ are the data points, and $d(\vec{x}, L)$ is the distance from point $\vec{x}$ to line L, then the axis of elongation A minimizes $\sum_i d(\vec{x}_i, A)^2$. Let $\vec{\mu}$ be the mean of $\{\vec{x}_i\}$. Define the scatter matrix $\mathbf{S} = \sum_i (\vec{x}_i - \vec{\mu})(\vec{x}_i - \vec{\mu})^T$. Then the axis of elongation is the eigenvector of **S** with the largest eigenvalue. See also principal component analysis. The figure below shows this axis of elongation for a set of points. 2) The longer midline of the bounding box with largest length-to-width ratio. A possible axis of elongation is the line in this figure:

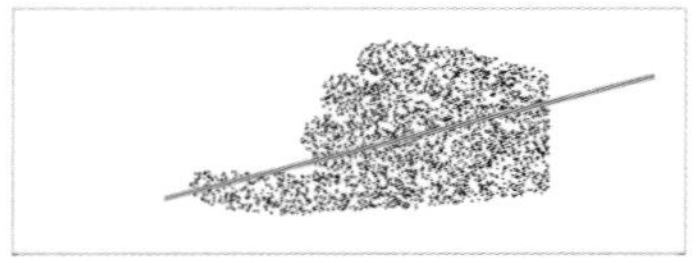

axis of rotation: A line about which a rotation is performed. Equivalently, the line whose points are fixed under the action of a rotation. Given a 3D rotation matrix **R**, the axis is the eigenvector of **R** corresponding to the eigenvalue 1.

axis-angle curve representation: A rotation representation based on the amount of twist θ about the axis of rotation, here a unit vector $\vec{a}$. The quaternion rotation representation is similar.

B

B-rep: See surface boundary representation.

b-spline: A curve approximation spline represented as a combination of basis functions:

$$\vec{c}(t) = \sum_{i=0}^{m} \vec{a}_i B_i(x)$$

where B_i are the basis functions and $\vec{a}_i$ are the control points. B-splines do not necessarily pass through any of the control points; however, if b-splines are calculated for adjacent sets of control points the curve segments will join up and produce a continuous curve.

b-spline fitting: Fitting a b-spline to a set of data points. This is useful for noise reduction or for producing a more compact model of the observed curve.

b-spline snake: A snake made from b-splines.

back projection: 1) A form of display where a translucent screen is illuminated from the side not facing the viewer. 2) The computation of a 3D quantity from its 2D projection. For example, a 2D homogeneous point $\vec{x}$ is the projection of a 3D point $\vec{X}$ by a perspective projection matrix **P**, so $\vec{x} = \mathbf{P}\vec{X}$. The backprojection of $\vec{x}$ is the 3D line $\{\text{null}(\mathbf{P}) + \lambda \mathbf{P}^{+}\vec{x}\}$ where $\mathbf{P}^{+}$ is the pseudoinverse of **P**. 3) Sometimes used interchangeably with triangulation. 4) Technique to compute the attenuation coefficients from intensity profiles covering a total cross section under various angles. It is used in CT and MRI to recover 3D from essentially 2D images. 5) Projection of the estimated 3D position of a shape back into the 2D image from which the shape's pose was estimated.

background: In computer vision, generally used in the context of object recognition. The background is either (1) the area of the scene behind an object or objects of interest or (2) the part of the image whose pixels sample from the background in the scene. As opposed to foreground. See also figure/ground separation.

background labeling: Methods for differentiating objects in the foreground of images or those of interest from those in the background.

background modeling: Segmentation or change detection method where the scene behind the objects of interest is modeled as a fixed or slowly changing background, with possible foreground occlusions. Each pixel is modeled as a distribution which is then used to decide if a given observation belongs to the background or an occluding object.

background normalization: Removal of the background by some image processing technique to estimate the background image and then dividing or subtracting the background from an original image. The technique is useful for when the background is non-uniform. The images below illustrate this where the first shows the input image,

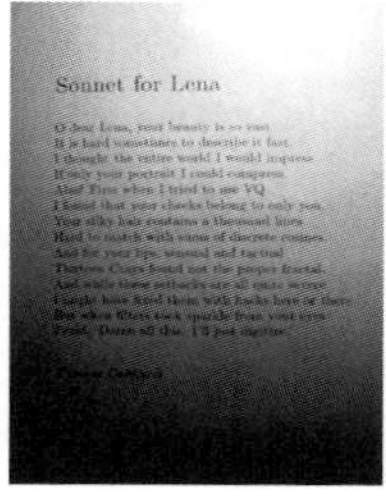

Sonnet for Lena

O dear Lena, your beauty is so vast
It is hard sometimes to describe it fast.
I thought the entire world I would impress
If only your portrait I could compress.
Alas! First when I tried to use VQ
I found that your cheeks belong to only you.
Your silky hair contains a thousand lines
Hard to match with sums of discrete cosines.
And for your lips, sensual and tactual
Thirteen Crays found not the proper fractal.
And while these setbacks are all quite severe
I might have fixed them with hacks here or there
But when filters took sparkle from your eyes
I said, 'Damn all this. I'll just digitize.'

Thomas Colthurst

the second is the background estimate obtained by dilation with ball $(9,9)$ structuring element and the third is the (normalized) division of the input image by the background image.

backlighting: A method of illuminating a scene where the background receives more illumination than the foreground. Commonly this is used to produce silhouettes of opaque objects against a lit background, for easier object detection.

bandpass filter: A signal processing filtering technique that allows signals between two specified frequencies to pass but cuts out signals at all other frequencies.

back-propagation: One of the best-studied neural network training algorithms for supervised learning. The name arises from using the propagation of the discrepancies between the computed and desired responses at the network output back to the network inputs. The discrepancies are one of the inputs into the network weight recomputation process.

back-tracking: A basic technique for graph searching: if a terminal but non-solution node is reached, search does not terminate with failure, but continues with still unexplored children of a previously visited non-terminal node. Classic back-tracking algorithms are breadth-first, depth-first, and A*. See also graph, graph searching, search tree.

bar: A raw primal sketch primitive that represents a dark line segment against a lighter background (or its inverse). Bars are also one of the primitives in Marr's theory of vision. The following is a small dark bar observed inside a receptive field:

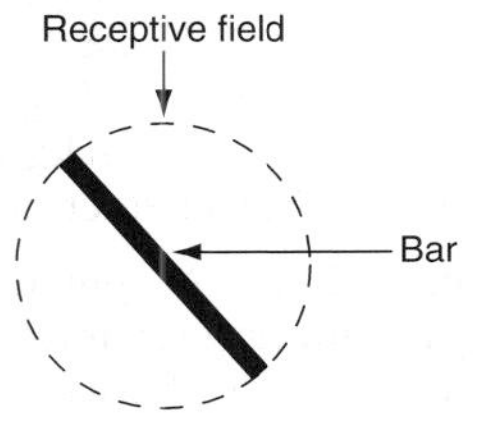

bar detector: 1) Method or algorithm that produces maximum excitation when a bar is in its receptive field. 2) Device used by thirsty undergraduates.

bar-code reading: Methods and algorithms used for the detection, imaging and interpretation of black parallel lines of different widths arranged to give details on products or other objects. Bar codes themselves have many different coding standards and arrangements. An example bar code is:

barycentrum: See center of mass.

barrel distortion: Geometric lens distortion in an optical system that causes the outlines of an object to curve outward, forming a barrel shape. See also pincushion distortion.

bas-relief ambiguity: The ambiguity in reconstructing a 3D object with Lambertian reflectance using shading from an image under orthographic projection. If the true surface is $z(x, y)$, then the family of surfaces $az(x, y) + bx + cy$ generate identical images under these viewing conditions, so any reconstruction, for any values of a, b, c is equally valid. The ambiguity is thus up to a three-parameter family.

baseline: Distance between two cameras used in a binocular stereo system.

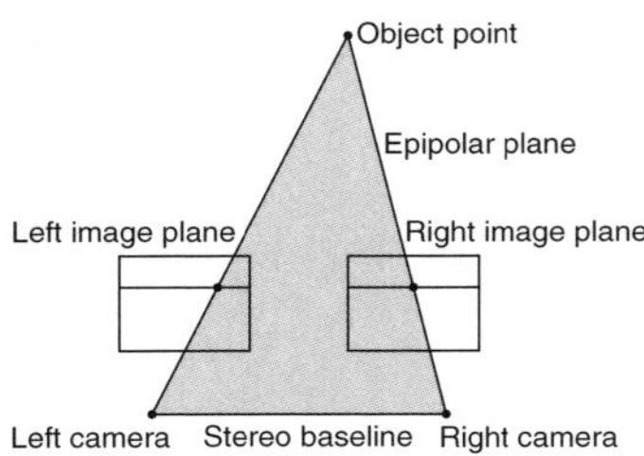

basis function representation: A method of representing a function as a sum of simple (usually orthonormal) ones. For example the Fourier transform represents functions as a weighted sum of sines and cosines.

Bayes' rule: The relationship between the conditional probability of event A given B and the conditional probability of event B given event A. This expressed as

$$P(A|B) = \frac{P(B|A)P(A)}{P(B)}$$

providing that $P(B) \neq 0$.

Bayesian classifier: A mathematical approach to classifying a set of data, by selecting the class most likely to have generated that data. If $\vec{x}$ is the data and c is a class, then the probability of that class is $p(c|\vec{x})$. This probability can be hard to compute so Bayes' rule can then be used here, which says that $p(c|\vec{x}) = \frac{P(\vec{x}|c)p(c)}{p(\vec{x})}$. Then we can compute the probability of the class $p(c|\vec{x})$ in terms of the probability of having observed the given data $\vec{x}$ with, $P(\vec{x}|c)$, and without, $p(\vec{x})$ assuming the class c plus the *a priori* likelihood, $p(c)$, of observing the class. The Bayesian classifier is the most common statistical classifier currently used in computer vision processes.

Bayesian filtering: A probabilistic data fusion technique. It uses a formulation of probabilities to represent the system state and likelihood functions to represent their relationships. In this form, Bayes' rule can be applied and further related probabilities deduced.

Bayesian model: A statistical modeling technique based on two input models:

1. a *likelihood model* $p(y|x, h)$, describing the density of observing y given x and h. Regarded as a function of h, for a fixed y and x, the density is also known as the likelihood of h.
2. a *prior model*, $p(h|D_0)$ which specifies the *a priori* density of h given some known information denoted by D_0 before any new data are taken into account.

The aim of the Bayesian model is to predict the density for outcomes y in test situations x given data $D = D_T, D_0$ with both pre-known and training data.

Bayesian model learning: See probabilistic model learning.

Bayesian network: A belief modeling approach using a graph structure. Nodes are variables and arcs are implied causal dependencies and are given probabilities. These networks are useful for fusing multiple data (possibly of different types) in a uniform and rigorous manner.

BDRF: See bidirectional reflectance distribution function.

beam splitter: An optical system that divides unpolarized light into two orthogonally polarized beams, each at 90° to the other, as in this example:

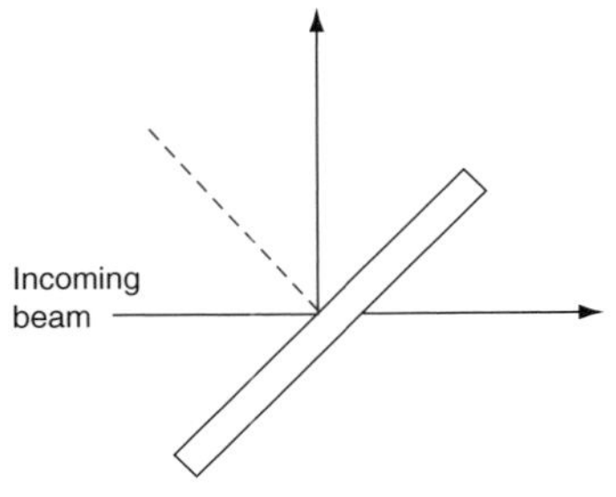

behavior analysis: Model based vision techniques for identifying and tracking behavior in humans. Often used for threat analysis.

behavior learning: Generation of goal-driven behavior models by some learning algorithm, for example reinforcement learning.

Beltrami flow: A noise suppression technique where images are treated as surfaces and the surface area is minimized in such a way as to preserve edges. See also diffusion smoothing.

bending energy: 1) A metaphor borrowed from the mechanics of thin metal plates. If a set of landmarks is distributed on two infinite flat metal plates and the differences in the coordinates between the two sets are vertical displacements of the plate, one Cartesian coordinate at a time, then the bending energy is the energy required to bend the metal plate so that the landmarks are coincident. When applied to images, the sets of landmarks may be sets of features. 2) Denotes the amount of energy that is stored due to an object's shape.

best next view: See next view planning.

Bhattacharyya distance: A measure of the (dis)similarity of two probability distributions. Given two arbitrary distributions $p_i(\mathbf{x})_{i=1,2}$ the Bhattacharyya distance between them is

$$d^2 = -log \int \sqrt{(p_1(\mathbf{x})p_2(\mathbf{x})}.\mathbf{dx}$$

bicubic spline interpolation: A special case of surface interpolation that uses cubic spline functions in two dimensions. This is like bilinear surface interpolation except that the interpolating surface is curved, instead of flat.

bidirectional reflectance distribution function (BRDF): If the energy arriving at a surface patch, denoted $E(\theta_i, \phi_i)$, and the energy radiated in a particular direction is denoted $L(\theta_e, \phi_e)$ in polar coordinates, then BRDF is defined as the ratio of the energy radiated from a patch of a surface in some direction to the amount of energy arriving there. The radiance is determined from the irradiance by

$$L(\theta_e, \phi_e) = f(\theta_i, \phi_i, \theta_e, \phi_e) E(\theta_e, \phi_e)$$

where the function f is the bidirectional reflectance distribution function. This function often only depends on the difference between the incident angle ϕ_i of the ray falling on the surface and the angle ϕ_e of the reflected ray. The geometry is illustrated by:

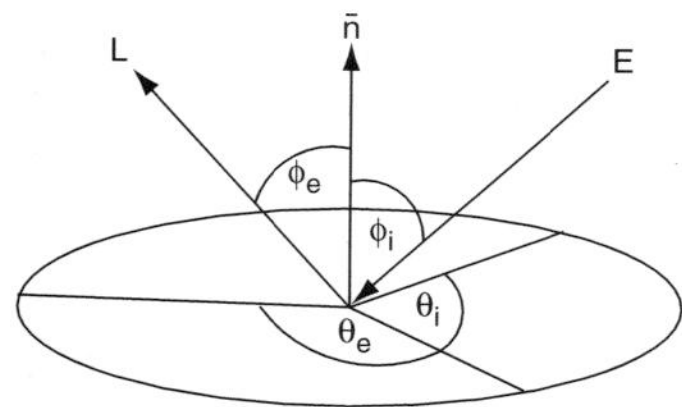

bilateral filtering: A non-iterative alternative to anisotropic filtering where images can be smoothed but edges present in them are preserved.

bilateral smoothing: See bilateral filtering.

bilinear surface interpolation: To determine the value of a function $f(x, y)$ at an arbitrary location (x, y), of which only discrete samples $f_{ij} = \{f(x_i, y_j)\}_{i=1}^{n}{}_{j=1}^{m}$ are available. The samples are arranged on a 2D grid, so the value at point (x, y) is interpolated from the values at the four surrounding points. In the diagram below $f_{\text{bilinear}}(x, y) =$

$$\frac{A+B}{(d_1+\overline{d}_1)(d_2+\overline{d}_2)}$$

where

$$A = d_1 d_2 f_{11} + \overline{d_1} d_2 f_{21}$$

$$B = d_1 \overline{d_2} f_{12} + \overline{d_1 d_2} f_{22}$$

The gray lines offer an easy *aide memoire*: each function value f_{ij} is multiplied by the two closest d values.

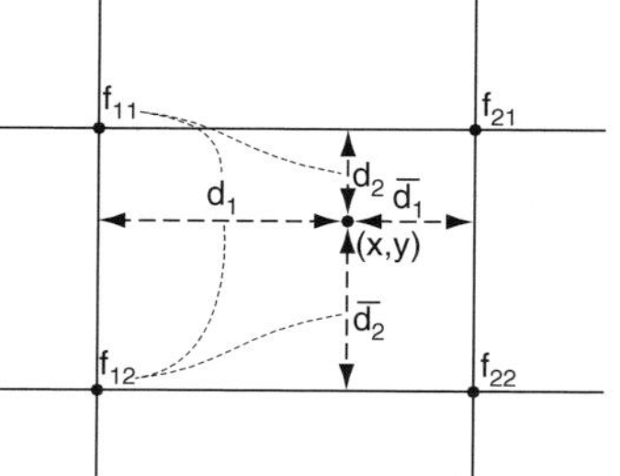

bilinearity: A function of two variables x and y is bilinear in x and y if it is linear in y for fixed x and linear in x for fixed y. For example, if $\vec{x}$ and $\vec{y}$ are vectors and $\mathbf{A}$ is a matrix such that $\vec{x}^\top \mathbf{A} \vec{y}$ is defined, then the function $f(\vec{x}, \vec{y}) = \vec{x}^\top \mathbf{A} \vec{y} + \vec{x} + \vec{y}$ is bilinear in $\vec{x}$ and $\vec{y}$.

bimodal histogram: A histogram with two pronounced peaks, or modes. This is a convenient intensity histogram for determining a binarizing threshold. An example is:

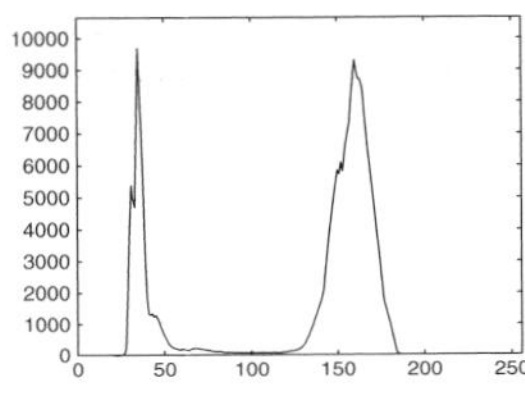

bin-picking: The problem of getting a robot manipulator equipped with vision sensors to pick parts, for instance screws, bolts, components of a given assembly, from a random pile. A classic challenge for hand–eye robotic systems, involving at least segmentation, object recognition in clutter and pose estimation.

binarization: See thresholding.

binary image: An image whose pixels can either be in an 'on' or 'off' state, represented by the integers 1 and 0 respectively. An example is:

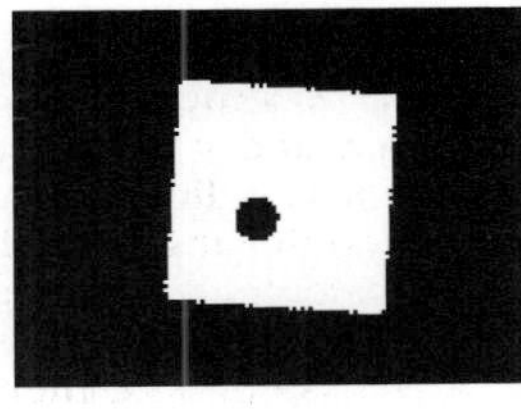

binary mathematical morphology: A group of shape-based operations that can be applied to binary images, based around a few simple mathematical concepts from set theory. Common usages include noise reduction, image enhancement and image segmentation. The two most basic operations are dilation and erosion. These operators take two pieces of data as input: the input binary image and a structuring element (also known as a kernel). Virtually all other mathematical morphology operators can be defined in terms of combinations of erosion and dilation along with set operators such as intersection and union. Some of the more important are opening, closing and skeletonization. Binary morphology is a special case of gray scale mathematical morphology. See also mathematical morphology.

binary moment: Given a binary image $B(i, j)$, there is an infinite family of moments indexed by the integer values p and q. The pqth moment is given by $m_{pq} = \sum_i \sum_j i^p j^q B(i, j)$.

binary noise reduction: A method of removing salt-and-pepper noise from binary images. For example, a point could have its value set to the median value of its eight neighbors.

binary object recognition: Model based techniques and algorithms used to recognize objects from their binary images.

binary operation: An operation that takes two images as inputs, such as image subtraction.

binary region skeleton: See skeleton.

binocular: A system that has two cameras looking at the same scene simultaneously usually from a similar viewpoint. See also stereo vision.

binocular stereo: A method of deriving depth information from a pair of calibrated cameras set at some distance apart and pointing in approximately the same direction. Depth information comes from the parallax between the two images and relies on being able to derive the same feature in both images.

binocular tracking: A method that tracks objects or features in 3D using binocular stereo.

biometrics: The science of discriminating individuals from accurate measurement of their physical features. Example biometric measurements are retinal lines, finger lengths, fingerprints, voice characteristics and facial features.

bipartite matching: Graph matching technique often applied in model based vision to match observations with models or stereo to solve the correspondence problem. Assume a set V of nodes partitioned into two non-intersecting subsets V^1 and V^2. In other words, $V = V^1 \cup V^2$ and $V^1 \cap V^2 = 0$. The only arcs E in the graph lie between the two subsets, *i.e.*, $E \subset \{V^1 \times V^2\} \cup \{V^2 \times V^1\}$. This is the bipartite graph. The bipartite matching problem is to find a maximal matching in the bipartite graph, in other words, a maximal set of nodes from the two subsets connected by arcs such that each node is connected by exactly one arc. One maximal matching in the graph below with sets $V^1 = \{A, B, C\}$ and $V^2 = \{X, Y\}$ pairs (A, Y) and (C, X). The selected arcs are solid, and other arcs are dashed.

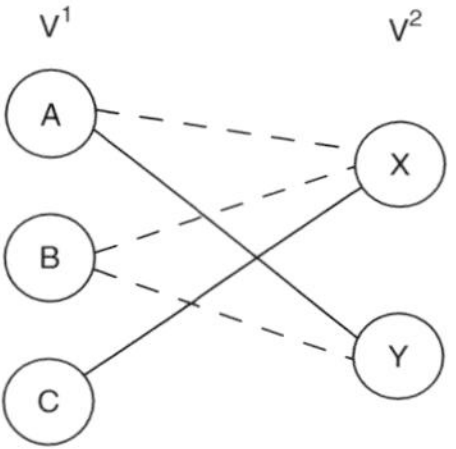

bit map: An image with one bit per pixel.

bit-plane encoding: An image compression technique where the image is broken into bit planes and run length coding is applied to each plane. To get the bit planes of an 8-bit gray scale image, the picture has a boolean AND operator applied with the binary value corresponding to the desired plane. For example, ANDing the image with 00010000 gives the fifth bit plane.

bitangent: See curve bitangent.

bitshift operator: The bitshift operator shifts the binary representation of each pixel to the left or right by a set number of bit positions. Shifting

01010110 right by 2 bits gives 00010101. The bitshift operator is a computationally cheap method of dividing or multiplying an image by a power of 2. A shift of n positions is a multiplication or division by 2^n.

blanking: Clearing a CRT or video device. The vertical blanking interval (VBI) in television transmission is used to carry data other than audio and video.

blending operator: An image processing operator that creates a third image C by a weighted combination of the input images A and B. In other words, $C(i,j) = \alpha A(i,j) + \beta B(i,j)$ for two scalar weights α and β. Usually, $\alpha + \beta = 1$. The results of some process can be illustrated by blending the original and result images. An example of blending that adds a detected boundary to the original image is:

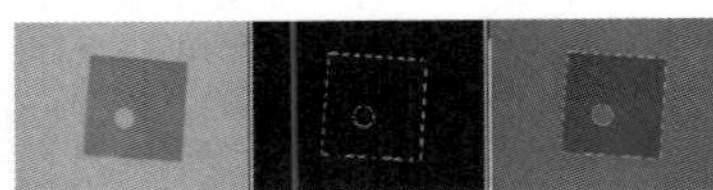

blob analysis: Blob analysis is a group of algorithms used in medical image analysis. There are four steps in the process: derive optimum foreground/background threshold to segment objects from their background; binarize the images by applying a thresholding operation; perform region growing and assign a labels to each discrete group (blob) of connected pixels; extract physical measurements from the blobs.

blob extraction: A part of blob analysis. See connected component labeling.

block coding: A class of signal coding techniques. The input signal is partitioned into fixed-size blocks, and each block is transmitted after translation to a smaller (for compression) or larger (for error-correction) block size.

blocks world: The blocks world is the simplified problem domain in which much early artificial intelligence and computer vision research was done. The essential feature of the blocks world is the restriction of analysis to simplified geometric objects such as polyhedra and the assumption that geometric descriptions such as image edges can be easily recovered from the image. An example blocks world scene is:

blooming: Blooming occurs when too much light enters a digital optical system. The light saturates CCD pixels, causing charge to overspill into surrounding elements giving either vertical or horizontal streaking in the image (depending on the orientation of the CCD).

Blum's medial axis: See medial axis transform

blur: A measure of sharpness in an image. Blurring can arise from the sensor being out of focus, noise in the environment or image capture process, target or sensor motion, as a side effect of an image processing operation, etc. A blurred image is:

border detection: See boundary detection.

border tracing: Given a prelabeled (or segmented) image, the border is the inner layer of each region's connected pixel set. It can be traced using a simple 8-connective or 4-connective stepping procedure in a 3×3 neighborhood.

boundary: A general term for the lower dimensional structure that separates two objects, such as the curve between neighboring surfaces, or surface between neighboring volume.

boundary description: Functional, geometry based or set-theoretic description of a region boundary. For an example, see chain code.

boundary detection: An image processing algorithm that finds and labels the edge pixels between two neighboring image segments after segmentation. The boundary represents physical discontinuities in the scene, for example changes in color, depth, shape or texture.

boundary grouping: An image processing algorithm that attempts to complete a fully connected image-segment boundary from many broken pieces. A boundary might be broken because it is commonplace for sharp transitions in property values to appear in the image as slow transitions, or sometimes disappear due to noise, blurring, digitization artifacts, poor lighting or surface irregularities, etc.

boundary length: The length of the boundary of an object. See also perimeter.

boundary matching: See curve matching.

boundary property: Characteristics of a boundary, such as arc length, curvature, etc.

boundary representation: See boundary description and B-Rep.

boundary segmentation: See curve segmentation.

boundary-region fusion: Region growing segmentation approach where two adjacent regions are merged when their characteristics are close enough to pass some similarity test. The candidate neighborhood for testing similarity can be the pixels lying near the shared region boundary.

bounding box: The smallest rectangular prism that completely encloses either an object or a set of points. The ratio of the length of box sides is often used as a classification metric in model based recognition.

bottom-up: Reasoning that proceeds from the data to the conclusions. In computer vision, describes algorithms that use the data to generate hypotheses at a low level, that are refined as the algorithm proceeds. Compare top-down.

BRDF: See bidirectional reflectance distribution function.

break point detection: See curve segmentation.

breast scan analysis: See mammogram analysis.

Brewster's angle: When light reflects from a dielectric surface it will be polarized perpendicularly to the surface normal. The degree of polarization depends on the incident angle and the refractive indices of the air and reflective medium. The angle of maximum polarization is called Brewster's angle and is given by

$$\theta_B = tan^{-1}\left(\frac{n_1}{n_2}\right)$$

where n_1 and n_2 are the refractive indices of the two materials.

brightness: The quantity of radiation reaching a detector after incidence on a surface. Often measured in lux or ANSI lumens. When translated into an image, the values are scaled to fit the bit patterns available. For example, if an 8-bit byte is used, the maximum value is 255. See also luminance.

brightness adjustment: Increase or decrease in the luminance of an image. To decrease, one can linearly interpolate between the image and a pure black image. To increase, one can linearly extrapolate from a black image and the target. The extrapolation function is

$$v = (1-\alpha) * i_0 + \alpha * i_1$$

where α is the blending factor (often between 0 and 1), v is the output pixel value and i_0 and i_1 are the corresponding image and black pixels. See also gamma correction and contrast enhancement.

Brodatz texture: A well-known set of texture images often used for testing texture-related algorithms.

building detection: A general term for a specific, model-based set of algorithms for finding buildings in data. The range of data used is large, encompassing stereo images, range images, aerial and ground-level photographs.

bundle adjustment: An algorithm used to optimally determine the three dimensional coordinates of points and camera positions from two dimensional image measurements. This is done by minimizing some cost function that includes the model fitting error and the camera variations. The *bundles* are the light rays between detected 3D features and each camera center. It is these bundles that are iteratively adjusted (with respect to both camera centers and feature positions).

burn-in: 1) A phenomenon of early tube-based cameras and monitors where, if the same image was presented for long periods of time it became permanently burnt into the phosphorescent layer. Since the advent of modern monitors (1980s) this no longer happens. 2) The practice of shipping only electronic components that have been tested for long periods, in the hope that any defects will manifest themselves early in the component's life (*e.g.*, 72 hours of typical use). 3) The practice of discarding the first several samples of an MCMC process in the hope that a very low-probability starting point will be converge to a high-probability point before beginning to output samples.

butterfly filter: A linear filter designed to respond to "butterfly" patterns in images. A small butterfly filter convolution kernel is

$$\begin{matrix} 0 & -2 & 0 \\ 1 & 2 & 1 \\ 0 & -2 & 0 \end{matrix}$$

It is often used in conjunction with the Hough transform for finding peaks in the Hough feature space, particularly when searching for lines. The line parameter values of (p, θ) will generally give a butterfly shape with a peak at the approximate correct values.

C

CAD: See computer aided design.

calculus of variations: See variational approach.

calibration object: An object or small scene with easily locatable features used for camera calibration.

camera: 1) The physical device used to acquire images. 2) The mathematical representation of the physical device and its characteristics such as position and calibration. 3) A class of mathematical models of the projection from 3D to 2D, such as affine-, orthographic- or pinhole camera.

camera calibration: Methods for determining the position and orientation of cameras and range sensors in a scene and relating them to scene coordinates. There are essentially four problems in calibration:

1. *Interior orientation*. Determining the internal camera geometry, including its principal point, focal length and lens distortion.
2. *Exterior orientation*. Determining the orientation and position of the camera with respect to some absolute coordinate system.
3. *Absolute orientation*. Determining the transformation between two coordinate systems, the position and orientation of the sensor in the absolute coordinate system from the calibration points.
4. *Relative orientation*. Determining the relative position and orientation between two cameras from projections of calibration points in the scene.

These are classic problems in the field of photogrammetry.

camera coordinates: 1) A viewer-centered representation relative to the camera. The camera coordinate system is positioned and oriented relative to the scene coordinate system and this relationship is determined by camera calibration. 2) An image coordinate system that places the camera's principal point at the origin $(0, 0)$, with unit aspect ratio and zero skew. The focal length in camera coordinates may or may not equal 1. If image coordinates are such that the 3×4 projection matrix is of the form

$$\begin{bmatrix} f & 0 & 0 \\ 0 & f & 0 \\ 0 & 0 & 1 \end{bmatrix} \left[\mathbf{R} \mid \vec{t}\right]$$

then the image and camera coordinate systems are identical.

camera geometry: The physical geometry of a camera system. See also camera model.

camera model: A mathematical model of the projection from 3D (real world) space to the camera image plane. For example see pinhole camera model.

camera motion compensation: See sensor motion compensation.

camera motion estimation: See sensor motion estimation.

camera position estimation: Estimation of the optical position of the camera relative to the scene or observed structure. This generally consists of six degrees of freedom (three for rotation, three for translation). It is often a component of camera calibration. Camera position is sometimes called the extrinsic parameters of the camera. Multiple camera positions may be estimated simultaneously with the reconstruction of 3D scene structure in structure- and- motion algorithms.

Canny edge detector: The first of the modern edge detectors. It took account of the trade-off between sensitivity of edge detection *versus* the accuracy of edge localization. The edge detector consists of four stages: 1) Gaussian smoothing to reduce noise and remove small details, 2) gradient magnitude and direction calculation, 3) non-maximal suppression of smaller gradients by larger ones to focus edge localization and 4) gradient magnitude thresholding and linking that uses hysteresis so as to start linking at strong edge positions, but then also track weaker edges. An example of the edge detection results is:

canonical configuration: A stereo camera configuration in which the optical axes of the cameras are parallel, the baselines are parallel to the image planes and the horizontal axes of the image planes are parallel. This results in epipolar lines that are parallel to the horizontal axes, hence simplifying the search for correspondences.

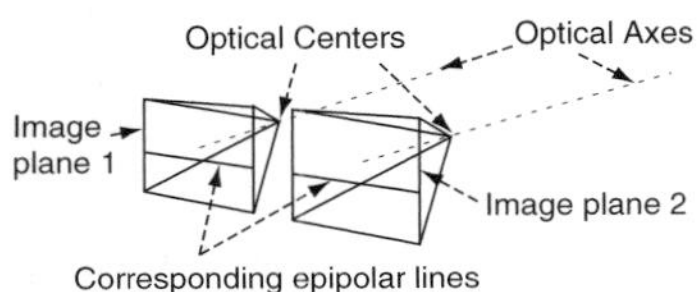

cardiac image analysis: Techniques involving the development of 3D vision algorithms for tracking the motion of the heart from NMR and echocardiographic images.

Cartesian coordinates: A position description system where an n-dimensional point, $\vec{P}$, is described by exactly n coordinates with respect to n linearly independent and often orthonormal vectors, known as axes.

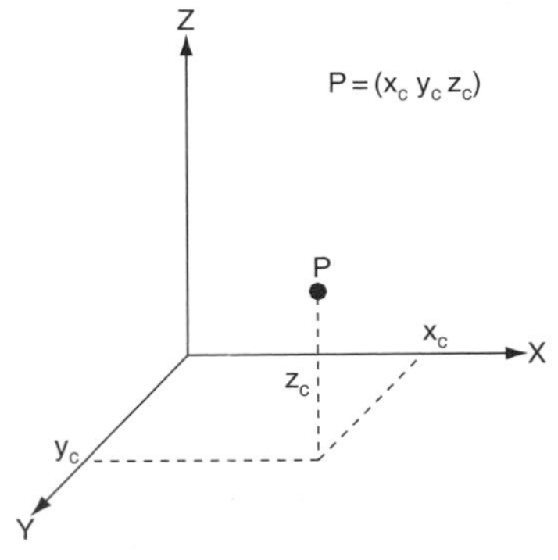

cartography: The study of maps and map-building. Automated cartography is the development of algorithms that reduce the manual effort in map building.

cascaded Hough transform: An application of several successive Hough transforms, with the output of one transform used as input to the next.

cascading Gaussians: A term referring to the fact that the convolution of a Gaussian with itself is another Gaussian.

CAT: See X-ray CAT.

catadioptric optics: The general approach of using mirrors in combination with conventional imaging systems to get wide viewing angles (*e.g.*, 180°). It is desirable that a catadioptric system has a single viewpoint because it permits the generation of geometrically correct perspective images from the captured images.

categorization: The subdivision of a set of elements into clearly distinct groups, or categories, defined by specific properties. Also the assignment of an element to a category or recognition of its category.

category: A group or class used in a classification system. For example, in mean and Gaussian curvature shape classification, the local shape of a surface is classified into four main categories: planar, ellipsoidal, hyperbolic, and cylindrical. Another example is the classification of observed grazing animals into one of {sheep, cow, horse}. See also categorization.

CBIR: See content based image retrieval.

CCD: Charge-Coupled Device. A solid state device that can record the number of photons falling on it.

A 2D matrix of CCD elements are used, together with a lens system, in digital cameras where each pixel value in the final images corresponds to the output one or more of the elements.

CCIR camera: Camera fulfilling color conversion and pixel formation criteria laid out by the *Comité Consultatif International des Radio*.

cell microscopic analysis: Automated image processing procedures for finding and analyzing different cell types from images taken by a microscope vision system. Common examples are the analysis of pre-cancerous cells and blood cell analysis.

cellular array: A massively parallel computing architecture, composed of a high number of processing elements. Particularly useful in machine vision applications when a simple 1:N mapping is possible between image pixels and processing elements. See also systolic array and SIMD.

center line: See medial line.

center of curvature: The center of the circle of curvature (or osculating circle) at a point $\vec{P}$ of a plane curve at which the curvature is nonzero. The circle of curvature is tangent to the curve at $\vec{P}$, has the same curvature as the curve at $\vec{P}$, and lies towards the concave (inner) side of the curve. This figure shows the circle and center of curvature, $\vec{C}$, of a curve at point $\vec{P}$:

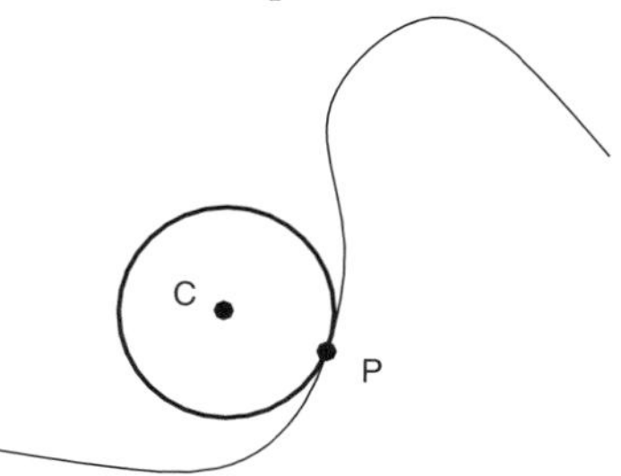

center of mass: The point within an object at which the force of gravity appears to act. If the object can be described by a multi-dimensional point set $\{\vec{x}_i\}$ containing N points, the center of mass is $\frac{1}{N}\sum_{i=0}^{N}\vec{x}_i f(\vec{x}_i)$, where $f(\vec{x}_i)$ is the value of the image (*e.g.*, binary or gray scale) at point $\vec{x}_i$.

center of projection: The origin of the camera reference frame in the pinhole camera model. In such a camera, the projection of a point in space is determined by the line passing through the point itself and the center of projection. See:

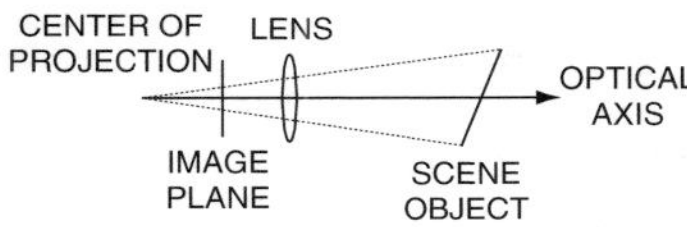

center-surround operator: An operator that is particularly sensitive to spot-like image features that have higher (or lower) pixel values in the center than the surrounding areas. A simple convolution mask that can be used as an orientation independent spot detector is:

$$\begin{matrix} -\frac{1}{8} & -\frac{1}{8} & -\frac{1}{8} \\ -\frac{1}{8} & 1 & -\frac{1}{8} \\ -\frac{1}{8} & -\frac{1}{8} & -\frac{1}{8} \end{matrix}$$

central moments: A family of image moments that are invariant to translation because the center of mass has been subtracted during the calculation. If $f(c, r)$ is the input image pixel value (binary or gray scale) at row r and column c then the pq^{th} central moment is $\sum_c \sum_r (c - \hat{c})^p (r - \hat{r})^q f(c, r)$ where $(\hat{c}, \hat{r})$ is the center of mass of the image.

central projection: It is defined by projection of an image on the surface of a sphere onto a tangential plane by rays from the center of the sphere. A great circle is the intersection of a plane with the sphere. The image of the great circle under central projection will be a line. Also known as the gnomonic projection.

centroid: See center of mass.

certainty representation: Any of a set of techniques for encoding the belief in a hypothesis, conclusion, calculation, etc. Example representation methods are probability and fuzzy logic.

chain code: An efficient method for contour coding where an arbitrary curve is represented by a sequence of small vectors of unit length in a limited set of possible directions. Depending on whether the 4 connected or the 8 connected grid is employed, the chain code is defined as the digits from 0 to 3 or 0 to 7, assigned to the 4 or 8 neighboring grid points in a counter-clockwise sense. For example, the string 222233000011 describes the small curve shown below using a 4 connected coding scheme, starting from the upper right pixel

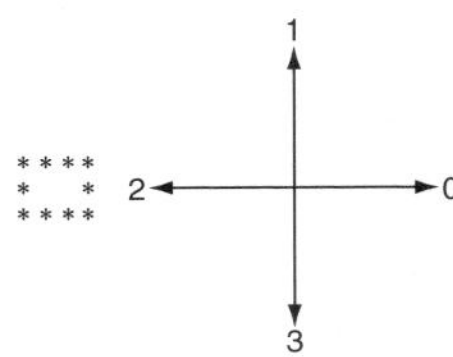

chamfer matching: A matching technique based on the comparison of contours, and based on the concept of chamfer distance assessing the similarity of two sets of points. This can be used for matching edge images using the distance transform. See also Hausdorff distance. To find the parameters (for example, translation and scale below) that register a library image and a test image, the binary edge map of the test image is compared to the distance transform. Edges are detected on image 1, and the distance transform of the edge pixels is computed. The edges from image 2 are then matched. (See plate section for a color version of these figures.)

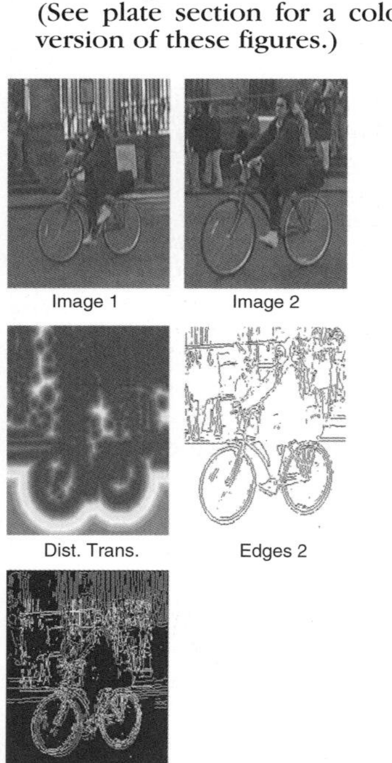

chamfering: See distance transform.

change detection: See motion detection.

character recognition: See optical character recognition.

character verification: A process used to confirm that printed or displayed characters are within some tolerance that guarantees that they are readable by humans. It is used in applications such as labeling.

characteristic view: An approach to object representation in which an object is encoded by a set of views of the object. The views are chosen so that small changes in viewpoint do not cause large changes in appearance (*e.g.*, a singularity event). Real objects have an unrealistic number of singularities, so practical approaches to creating characteristic views require approximations, such as only using views on a tessellated viewsphere, or only representing the viewpoints that are reasonable stable over large ranges on the viewsphere. See also aspect graph and appearance based recognition.

chess board distance metric: See Manhattan metric.

chi-squared distribution: The chi-squared (χ^2) probability distribution describes the distribution of squared lengths of vectors drawn from a normal distribution. Specifically let the cumulative distribution function of the χ^2 distribution with d degrees of freedom be denoted $\chi^2(d, u)$. Then the probability that a point $\vec{x}$

drawn from a d-dimensional Gaussian distribution will have squared norm $|\vec{x}|^2$ less than a value τ is given by $\chi^2(d, \tau)$. Empirical and theoretical plots of the χ^2 probability density function with five degrees of freedom are here:

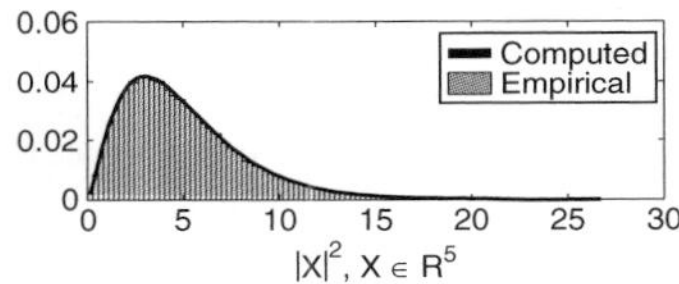

chi-squared test: A statistical test of the hypothesis that a set of sampled values has been drawn from a given distribution. See also chi-squared distribution.

chip sensor: A CCD or other semiconductor based light sensitive imaging device.

chord distribution: A 2D shape description technique based on all chords in the shape (that is all pairwise segments between points on the boundary). Histograms of their lengths and orientations are computed. The values in the length histogram are invariant to rotations and scale linearly with the size of object. The orientation histogram values are invariant to scale and shifts.

chroma: The color portion of a video signal that includes hue and saturation, requiring luminance to make it visible. It is also referred to as chrominance.

chromatic aberration: A focusing problem where light of different wavelengths (color) is refracted by different amounts and consequently images at different places. As blue light is refracted more than red light, objects may be imaged with color fringes at places where there are strong changes in lightness.

chromaticity diagram: A 2D slice of a 3D color space. The CIE 1931 chromaticity diagram is the slice through the xyz color space of the CIE where $x+y+z=1$. This slice is shown below. (See plate section for a colour version of this figure.) The color gamut of standard 0–1 RGB values in this model is the bright triangle in the center of the horseshoe-like shape. Points outside the triangle have had their saturations truncated. See also CIE chromaticity coordinates.

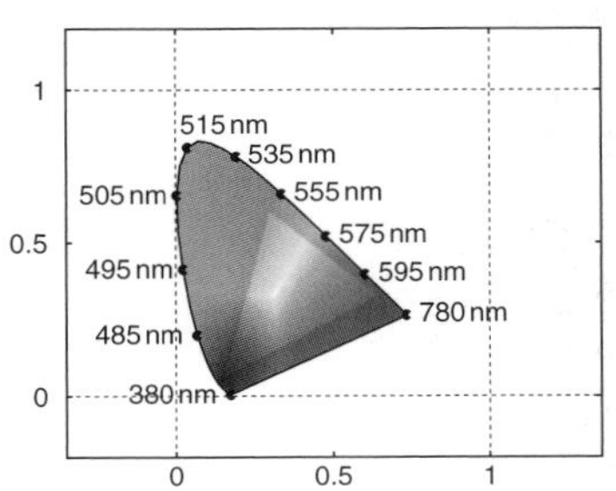

chrominance: 1) The part of a video signal that carries color. 2) One or both of the color axes in a 3D color space that distinguishes intensity and color. See also chroma.

chromosome analysis: Vision technique used for the diagnosis of some genetic disorders from microscope images. This usually includes sorting the chromosomes into the 23 pairs and displaying them in a standard chart.

CID: *Charge Injection Device*. A type of semiconductor imaging device with a matrix of light-sensitive cells. Every pixel in a CID array can be individually addressed via electrical indexing of row and column electrodes. It is unlike a CCD because it transfers collected charge out of the pixel during readout, thus erasing the image.

CIE chromaticity coordinates: Coordinates in the CIE color space with reference to three ideal standard colors X, Y and Z. Any visible color can be expressed as a weighted sum of these three ideal colors, for example, for a color $p = w_1X + w_2Y + w_3Z$. The normalized values are given by

$$x = \frac{w_1}{w_1 + w_2 + w_3}$$

$$y = \frac{w_2}{w_1 + w_2 + w_3}$$

$$z = \frac{w_3}{w_1 + w_2 + w_3}$$

since $x + y + z = 1$, we only need to know two of these values, say (x, y). These are the chromaticity coordinates.

CIE L*A*B* model: A color representation model based on that proposed by the Commission Internationale d'Eclairage (CIE) as an international standard for color measurement. It is designed to be device-independent and perceptually uniform (*i.e.*, the separation between two points in this space corresponds to the perceptual difference between the colors). L*A*B* color consists of a luminance, L*, and two chromatic components: A* component, from green to red; B* component, from blue to yellow. See also CIE L*U*V* model.

CIE L*U*V* model: A color representation system where colors are represented by luminance (L*) and two chrominance components(U*V*). A given change in value in any component corresponds approximately to the same perceptual difference. See also CIE L*A*B* model.

circle: A curve consisting of all points on a plane lying a fixed radius r from the *center* point C. The arc defining the entire circle is known as the *circumference* and is of length

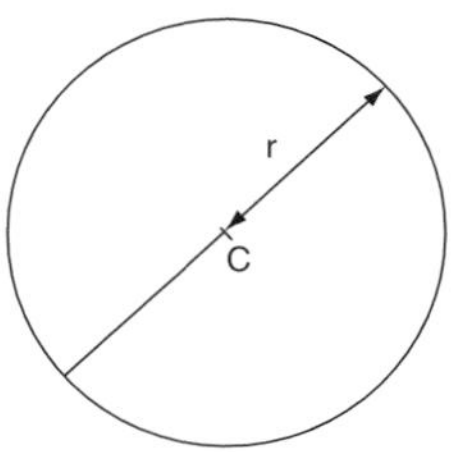

$2\pi r$. The area contained inside the curve is given by $A = \pi r^2$. A circle centered at the point (h, k) has equation $(x-h)^2 + (y-k)^2 = r^2$. The circle is a special case of the ellipse.

circle detection: A class of algorithms, for example the Hough transform, that locate the centers and radii of circles in digital images. In general images, scene circles usually appear as ellipses, as in this example:

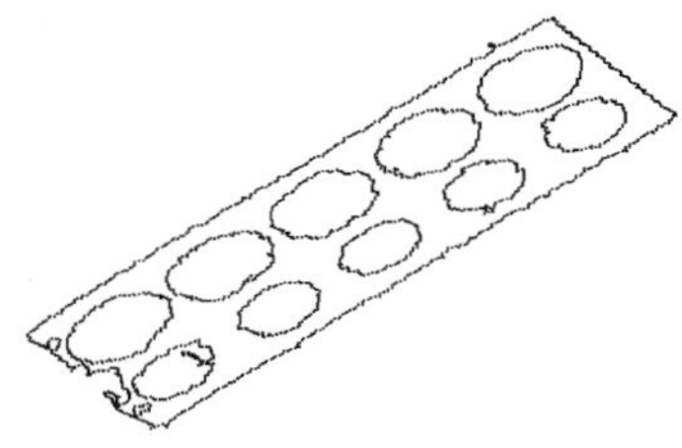

circle fitting: Techniques for deriving circle parameters from either 2D or 3D observations. As with all fitting problems, one can either search the parameter space using a good metric (using, for example, a Hough transform), or can solve a well-posed least-squares problem.

circular convolution: The circular convolution (c_k) of two vectors $\{x_i\}$ and $\{y_i\}$ that are of length n is defined as $c_k = \sum_{i=0}^{n-1} x_i y_j$ where $0 \leq k < n$ and $j = (i-k) \bmod n$.

circularity: One measure C of the degree to which a 2D shape is similar to a circle is given by

$$C = 4\pi \left(\frac{A}{P^2} \right)$$

where C varies from 0 (noncircular) to 1 (perfectly circular). A is the object area and P is the object perimeter.

city block distance: See Manhattan metric.

classification: A general term for the assignment of a label (or class) to structures (*e.g.*, pixels, regions, lines, etc.). Example classification problems include: a) labelling pixels as road, vegetation or sky, b) deciding whether cells are cancerous based on cell shapes or c) the person with the observed face is an allowed system user.

classifier: An algorithm assigning a class among several possible to an input pattern or data. See also classification, unsupervised classification, clustering, supervisedclassification and rule-based classification.

clipping: Removal or non-rendering of objects that do not coincide with the display area.

clique: A clique of a graph G is a fully connected subgraph of G. In a fully connected graph, every vertex is a neighbor of all others. The graph below has a clique with five nodes. (There are other cliques in the graph with fewer nodes, *e.g.*, ABac with four nodes, etc.).

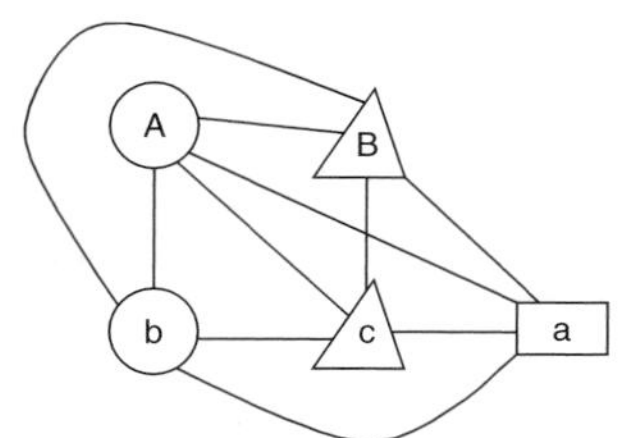

close operator: The application of two binary morphology operators, dilation followed by erosion, which has the effect of filling small holes in an image. This figure shows the result of closing with a mask 22 pixels in diameter:

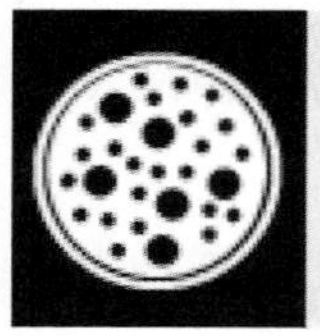

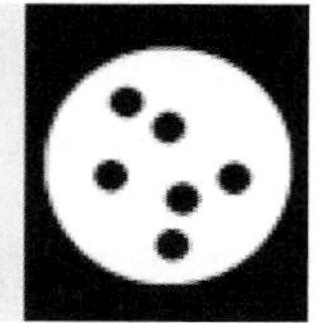

clustering: 1) Grouping together images regions or pixels into larger, homogeneous regions sharing some property. 2) Identifying the subsets of a set of data points $\{\vec{x}_i\}$ based on some property such as proximity.

clutter: A generic term for unmodeled or uninteresting elements in an image. For example, a face detector generally has a model for faces, and not for other objects, which are regarded as clutter. The background of an image is often expected to include "clutter". Loosely speaking, clutter is more structured than "noise".

CMOS: Complementary metal-oxide semiconductor. A technology used in making image sensors and other computer chips.

CMY: See CMYK.

CMYB: See CMYK.

CMYK: Cyan, magenta, yellow and black color model. It is a subtractive model where colors are absorbed by a medium, for example pigments in paints. Where the RGB color model adds hues to black to generate a particular color, the CMYK model subtracts from white. Red, green and blue are secondary colors in this model. (See plate section for a colour version of this figure.)

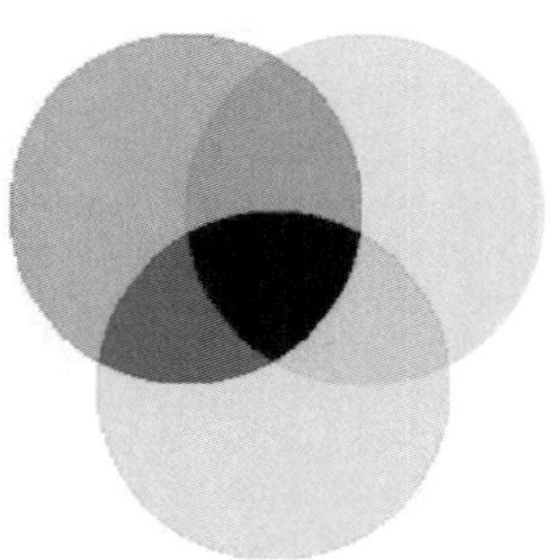

coarse-to-fine processing: Multi-scale algorithm application that begins by processing at a large or coarse level and then, iteratively, to a small or fine level. Importantly, results from each level must be propagated to ensure a good final result. It is used for computing, for example, optical flow.

coaxial illumination: Front lighting with the illumination path running along the imaging optical axis. Advantages of this technique are no visible shadows or direct specularities from the camera's viewpoint.

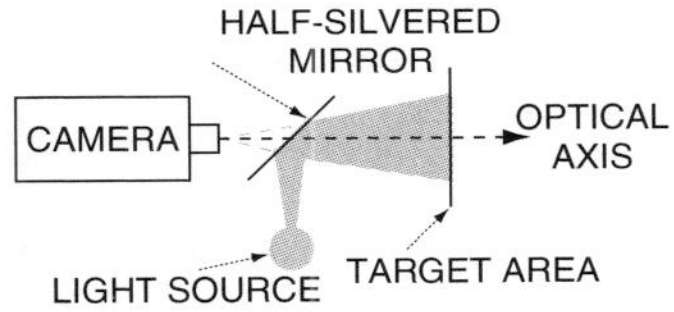

cognitive vision: A part of computer vision focusing techniques for recognition and categorization of objects, structures and events, learning and knowledge representation, control and visual attention.

coherence detection: Stereo vision technique where maximal patch correlations are searched for across two images to generate features. It relies on having a good correlation measure and a suitably chosen patch size.

coherent fiber optics: Many fiber optic elements bound into a single cable component with the individual fiber spatial positions aligned, so that it can be used to transmit images.

coherent light: Light, for example generated by a laser, in which the emitted light waves have the same wavelength and are in phase. Such light waves can remain focused over long distances.

coincidental alignment: When two structures seem to be related, but in fact the structures are independent or the alignment is just a consequence of being in some special viewpoint. Examples are random edges being collinear or surfaces coplanar, or object corners being nearby. See also non-accidentalness.

collimate: To align the optics of a vision system, especially those in a telescopic system.

collimated lighting: Collimated lighting (*e.g.*, directional backlighting) is a special form of structured light. A collimator produces light in which all the rays are parallel.

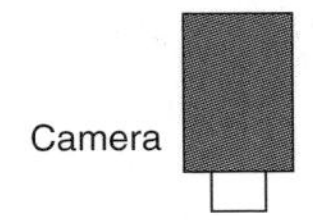

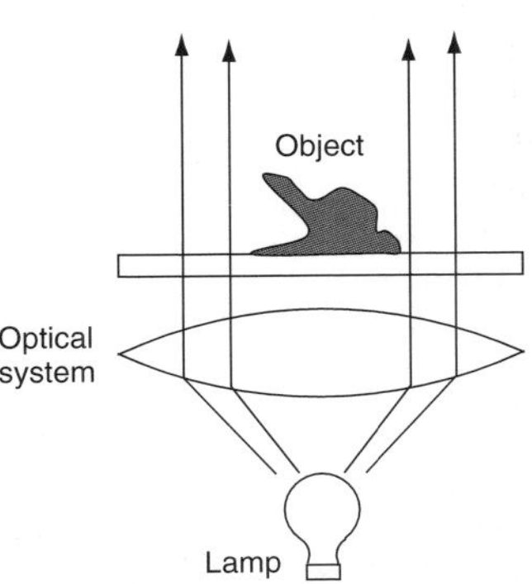

It is used to produce well defined shadows that can be cast directly onto either a sensor or an object.

collinearity: The property of lying along the same straight line.

collineation: See projective transformation.

color: Color is both a physical and psychological phenomenon. Physically, color refers to the nature of an object texture that allows it to reflect or absorb particular parts of the light incident on it. (See also reflectance.) The psychological aspect is characterized by the visual sensation experienced when light of a particular frequency or wavelength is incident on the retina. The key paradox here concerns why light of slightly different wavelengths should be be so perceptually different (*e.g.*, red *versus* blue).

color based database indexing: See color based image retrieval.

color based image retrieval: An example of the more general image database indexing process, where one of the main indices into the image database comes from either color samples, the color distribution from a sample image, or by a set of text color terms (*e.g.*, "red"), etc.

color clustering: See color image segmentation.

color constancy: The ability of a vision system to assign a color description to an object that is independent of the lighting environment. This will allow the system to recognize objects under many different lighting conditions. The human vision system does this automatically, but most machine vision systems cannot. For example, humans observing a red object in a cluttered scene under a blue light will still see the object as red. A machine vision system might see it as a very dark blue.

color co-occurrence matrix: A matrix (actually a histogram) whose elements represent the sum of color values existing, in a given image in a sequence, at a certain pixel position relative to another color existing at a different position in the image. See also co-occurrence matrix.

color correction: 1) Adjustment of colors to achieve color constancy. 2) Any change to the colors of an image. See also gamma correction.

color differential invariant: A type of differential invariant based on color information, such as $\frac{\nabla R \cdot \nabla G}{||\nabla R|| \, ||\nabla G||}$ that has the same value invariant to translation, rotation and variations in uniform illumination.

color doppler: A method for noninvasively imaging blood flow through the heart or other body parts by displaying flow data on the two dimensional echocardiographic image. Blood flow in different directions will be displayed in different colors.

color edge detection: The process of edge detection in color images. A simple approach is combine (*e.g.*, by addition) the edge strengths of the individual RGB color planes.

color efficiency: A tradeoff that is made with lighting systems, where conflicting design constraints require energy efficient production of light while simultaneously producing sufficiently broad spectrum illumination that the the colors look natural. An obvious example of a skewed tradeoff is with low pressure sodium street lighting. This is energy efficient but has poor color appearance.

color gamut: The subset of all possible colors that a particular display device (CRT, LCD, printer) can display. Because of physical difference in how various devices produce colors, each scanner, display, and printer has a different gamut, or range of colors, that it can represent. The RGB color gamut can only display approximately 70% of the colors that can be perceived. The CMYK color gamut is much smaller, reproducing about 20% of perceivable colors. The color gamut achieved with premixed inks (like the Pantone Matching System) is also smaller than the RGB gamut.

color halftoning: See dithering.

color histogram matching: Used in color image indexing where the similarity measure is the distance between color histograms of two images, *e.g.*, by using the Kullback–Leibler divergence or Bhattacharyya distance.

color image: An image where each element (pixel) is a tuple of values from a set of color bases.

color image restoration: See image restoration.

color image segmentation: Segmenting a color image into homogeneous regions based on some similarity criteria. The boundaries around typical regions are shown here (see plate section for a colour version of this figure):

color indexing: Using color information, *e.g.*, color histograms, for image database indexing. A key issue is varying illumination. It is possible to use ratios of colors from neighboring locations to obtain illumination invariance.

color matching: Due to the phenomenon of trichromacy, any color stimulus can be matched by a mixture of the three primary stimuli. Color matching is expressed as :

$$C = R\mathbf{R} + G\mathbf{G} + B\mathbf{B}$$

where a color stimulus C is matched by R units of primary stimulus **R** mixed with G units of primary stimulus **G** and B units of primary stimulus **B**.

color mixture model: A mixture model based on distributions in some color representation system that specifies both the color groups in a model as well as their relationships to each other. The conditional probability of a observed pixel $\vec{x}_i$ belonging to an object Ow is modeled as a mixture with K components.

color models: See color representation system.

color moment: A color image description based on moments of each color channel's histogram, *e.g.*, the mean, variance and skewness of the histograms.

color normalization:Techniques for normalizing the distribution of color values in a color image, so that the image description is invariant to illumination. One simple method for producing invariance to lightness is to use vectors of unit length for color entries, rather than coordinates in the color representation system.

color quantization: The process of reducing the number of colors in an image by selecting a subset of colors, then representing the original image using only them. This has the side-effect of allowing image compression with fewer bits. A color image encoded with progressively fewer numbers of colors is shown here (see plate section for a colour version of these figures):

16,777,216 colors

256 colors

16 colors

4 colors

color re-mapping: An image transformation where each original color is replaced by another color from a colormap. If the image has indexed colors, this can be a very fast operation and can provide special graphical effects for very low processing overhead. (See

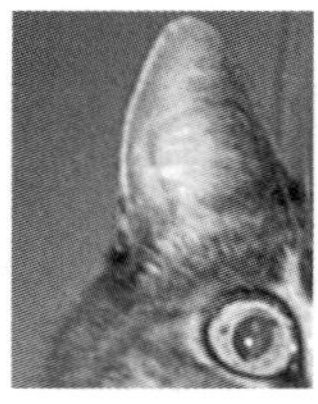

Original

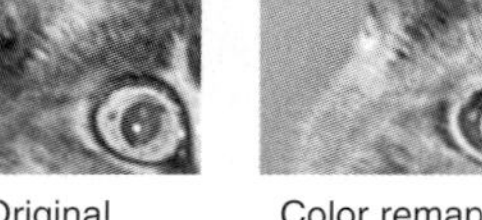

Color remapped

plate section for a colour version of these figures.)

color representation system: A 2D or 3D space used to represent a set of absolute color coordinates. RGB and CIE are examples of such spaces.

color spaces: See color representation system.

color temperature: A scalar measure of colour. 1) The colour temperature of a given colour C is the temperature in kelvins at which a heated black body would emit light that is dominated by colour C. It is relevant to computer vision in that the illumination color changes the appearance of the observed objects. The color temperature of incandescent lights is about 3200 kelvins and sunlight is about 5500 kelvins. 2) Photographic color temperature is the ratio of blue to red intensity.

color texture: Variations (texture) in the appearance of a surface (or region illumination, etc.) arising because of spatial variations in either the color, reflectance or lightness of a surface.

colorimetry: The measurement of color intensity relative to some standard.

combinatorial explosion: When used correctly, this term refers to how the computational requirements of an algorithm increases very quickly relative to the increase in the number of elements to be processed, as a consequence of having to consider all combinations of elements. For example, consider matching M model features to D data features with $D \geq M$, each data feature can be used at most once and all model features must be matched. Then the number of possible matchings that need to be considered is $D \times (D-1) \times (D-2) \cdots \times (D-M+1)$. Here, if M increases by only one, approximately D times as much matching effort is needed. Combinatorial explosion is also loosely used for other non-combination algorithms whose effort grows rapidly with even small increases in input data sizes.

compactness: A scale, translation and rotation invariant descriptor based on the ratio $\frac{perimeter^2}{area}$.

compass edge detector: A class of edge detectors based on combining the response of separate edge operators applied at several orientations. The edge response at a pixel is commonly the maximum of the responses over the several orientations.

composite filter: Hardware or software image processing method based on a mixture of components such as noise reduction, feature detection, grouping, etc.

composite video: A television video transmission method created as a backward-compatible solution for the transition from black-and-white to color television. The black-and-white TV sets ignore the color component while color TV sets separate out the color information and display it with the black-and-white intensity.

compression: See image compression.

computational theory: An approach to computer vision algorithm description promoted by Marr. A process can be described at three levels, implementation (*e.g.*, as a program), algorithm (*e.g.*, as a sequence of activities) and computational theory. This third level is characterized by the assumptions behind the process, the mathematical relationship between the input and output process and the description of the properties of the input data (*e.g.*, assumptions of statistical distributions). The claimed advantage of this approach is that the computational theory level makes explicit the essentials of the process, that can then be compared to the essentials of other processes solving the same problem. By this method, the implementation details that can confuse comparisons can be ignored.

computational vision: See computer vision.

computer aided design: 1) A general term for object design processes where a computer assists the designer, *e.g.*, in the specification and layout of components. For example, most current mechanical parts are designed by a computer aided design (CAD) process. 2) A term used for distinguishing objects designed with the assistance of a computer.

computer vision: A broad term for the processing of image data. Every professional will have a different definition that distinguishes computer vision from machine vision, image processing or pattern recognition. The boundary is not clear, but the main issues that lead to this term being used are more emphasis on 1) underlying theories of optics, light and surfaces, 2) underlying statistical, property and shape models, 3) theory-based algorithms, as contrasted to commercially exploitable algorithms and 4) issues related to what humans broadly relate to "understanding" as contrasted with "automation".

computed axial tomography: Also known as CAT. An X-ray procedure used in conjunction with vision techniques to build a 3D volumetric image from multiple X-ray images taken from different viewpoints. The procedure can be used to produce a series of cross sections of a selected part of the human body, that can be used for medical diagnosis.

concave mirror: The type of mirror used for imaging, in which a concave surface is used to reflect light to a focus.

The reflecting surface usually is rotationally symmetric about the optical or principal axis and mirror surface can be part of a sphere, paraboloid, ellipsoid, hyperboloid or other surfaces. It is also known as a converging mirror because it brings light to a focus. In the case of the spherical mirror, half way between the vertex and the sphere center, C, is the mirror focal point, F, as shown here:

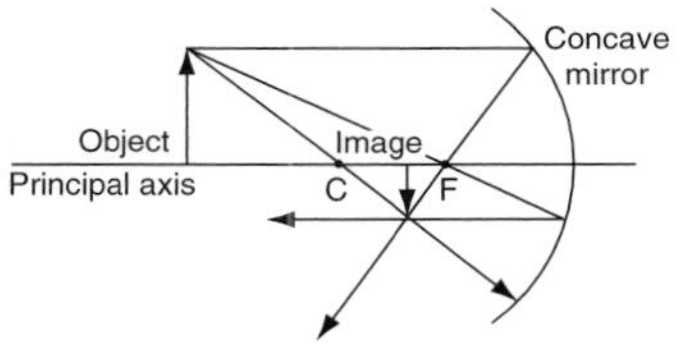

concave residue: The set difference between a shape and its convex hull. For a convex shape, the concave residue is empty. Some shapes (in black) and their concave residues (in gray) are shown here:

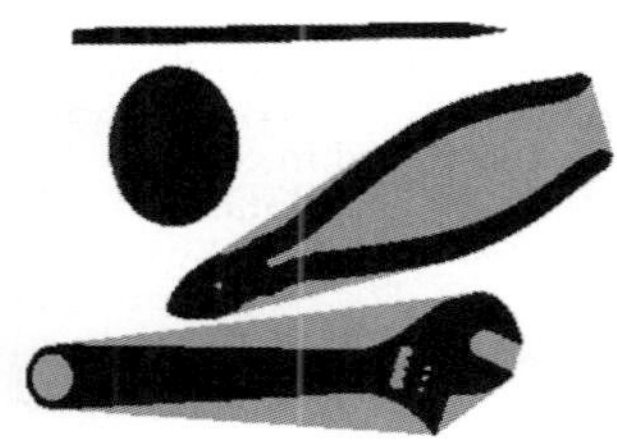

concavity: Loosely, a depression, dent, hollow or hole in a shape or surface. More precisely, a connected component of a shape's concave residue.

concavity tree: An hierarchical description of an object in the form of a tree. The concavity tree of a shape has the convex hull of its shape as the parent node and the concavity trees of its concavities as the child nodes. These are subtracted from the parent shape to give the original object. The concavity tree of a convex shape is the shape itself. The concavity tree of the gray shape below is shown:

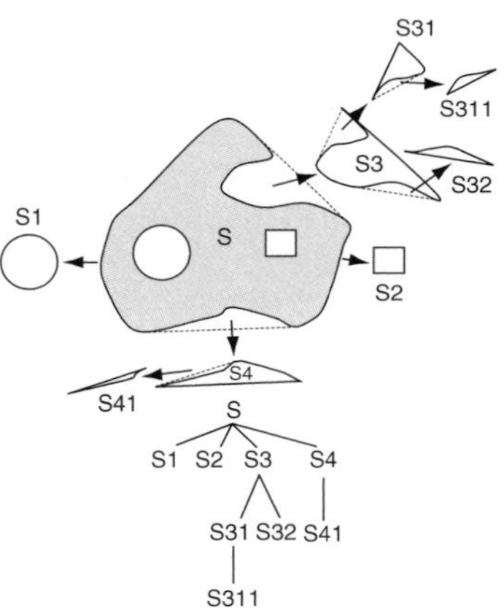

concurrence matrix: See co-occurrence matrix.

condensation tracking: Conditional density propagation tracking. The particle filter technique applied by Blake and Isard to edge tracking. A framework for object tracking with multiple simultaneous hypotheses that switches between multiple continuous autoregressive process motion models according to a discrete transition matrix. Using importance sampling it is possible to keep only the N strongest hypotheses.

condenser lens: An optical device used to collect light over a wide angle and produce a collimated output beam.

conditional dilation: A binary image operation that is a combination of the dilation operator and a logical AND operation with a mask, that only allows dilation into pixels that belong to the mask. This process can be described by the formula: dilate$(X, J) \cap M$, where X is the original image, M is the mask and J is the structuring element.

conditional distribution: A distribution of one variable given the values of one or more other variables.

conditional replenishment: A method for coding of video signals, where only the portion of a video image that has changed since the previous frame is transmitted. Effective for sequences with largely stationary backgrounds, but more complex sequences require more sophisticated algorithms that perform motion compensation.

conformal mapping: A function from the complex plane to itself, $f: \mathbb{C} \mapsto \mathbb{C}$, that preserves local angles. For example, the complex function $y = \sin(z) = -\frac{1}{2}i(e^{iz} - e^{-iz})$ is conformal.

conic: Curves arising from the intersection of a cone with a plane (also called conic sections). This is a family of curves including the circle, ellipse, parabola and hyperbola. The general form for a conic in 2D is $ax^2 + bxy + cy^2 + dx + ey + f = 0$. Some example conics are:

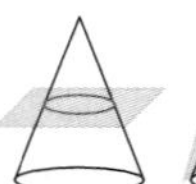

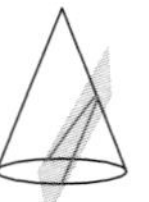
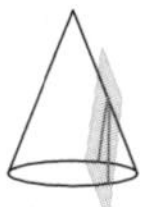

circle ellipse parabola hyperbola

conic fitting: The fitting of a geometric model of a conic section $ax^2+bxy+cy^2+dx+ey+f=0$ to a set of data points $\{(x_i, y_i)\}$. Special cases include fitting circles and ellipses.

conic invariant: An invariant of a conic section. If the conic is in canonical form

$$ax^2+bxy+cy^2+dx+ey+f=0$$

with $a^2 + b^2 + c^2 + d^2 + e^2 + f^2 = 1$, then the two invariants to rotation and translation are functions of the eigenvalues of the leading quadratic form matrix $\mathbf{A} = \begin{bmatrix} a & b \\ b & c \end{bmatrix}$. For example, the trace and determinant are invariants that are convenient to compute. For an ellipse, the eigenvalues are functions of the radii. The only invariant to affine transformation is the *class* of the conic (hyperbola, ellipse, parabola, etc.). The invariant to projective transformation is the set of signs of the eigenvalues of the 3×3 matrix representing the conic in homogeneous coordinates.

conical mirror: A mirror in the shape of (possibly part of) a cone. It is particularly useful for robot navigation since a camera placed facing the apex of the cone aligning the cone's axis and the optical axis and oriented towards its base can have a full 360° view. Conical mirrors were used in antiquity to produce cipher images known as anamorphoses.

conjugate direction: Optimization scheme where a set of independent directions are identified on the search space. A pair of vectors $\vec{u}$ and $\vec{v}$ are conjugate with respect to matrix **A** if $\vec{u}^\top \mathbf{A} \vec{v} = 0$. A conjugate direction optimization method is one in which a series of optimization directions are devised that are conjugate with respect to the normal matrix but do not require the normal matrix in order for them to be determined.

conjugate gradient: A basic technique of numerical optimization in which the minimum of a numerical target function is found by iteratively descending along non-interfering (conjugate) directions. The conjugate gradient method does not require second derivatives and can find the optima of an N dimensional quadric form in N iterations. By comparison, a Newton method requires one iteration and gradient descent can require an arbitrarily large number of iterations.

connected component labeling: 1) A standard graph problem. Given a graph consisting of nodes and arcs, the problem is to identify nodes forming a connected set. A node is in a set if it has an arc connecting it to another node in the set. 2) Connected component labeling is used in binary and gray scale image processing to join together neighboring pixels into regions. There are several efficient sequential algorithms for this procedure. In this image, the pixels in each connected component have a different gray shade:

connectivity: See pixel connectivity.

conservative smoothing: A noise filtering technique whose name derives from the fact that it employs a fast filtering algorithm that sacrifices noise suppression power to preserve the image detail. A simple form of conservative smoothing replaces a pixel that is larger (smaller) than its 8 connected neighbors by the largest (smallest) value amongst those neighbors. This process works well with impulse noise but is not as effective with Gaussian noise.

constrained least squares: It is sometimes useful to minimize $||\mathbf{A}\vec{x} - \vec{b}||_2$ over some subset of possible solutions $\vec{x}$ that are predetermined. For example, one may already know the function values at certain points on the parameterized curve. This leads to an equality constrained version of the least squares problem, stated as: minimize $||\mathbf{A}\vec{x} - \vec{b}||_2$ subject to $\mathbf{B}\vec{x} = \vec{c}$. There are several approaches to the solution of this problem such as *QR* factorization and the SVD. As an example, this regression technique can be useful in least squares surface fitting where the plane described by $\vec{x}$ is constrained to be perpendicular to some other plane.

constrained matching: A generic term for recognition approaches where two objects are compared under a constraint on either or both. One example of this would be a search for moving vehicles under 20 feet in length.

constrained optimization: Optimization of a function f subject to constraints on the parameters of the function. The general problem is to find the x that minimizes (or maximizes) $f(x)$ subject to $g(x) = 0$ and $h(x) >= 0$, where the functions f, g, h may all take vector-valued arguments, and g and h may also be vector-valued, encoding multiple constraints to be satisfied. Optimization subject to equality constraints is achieved by the method of Lagrange multipliers. Optimization of a quadratic form subject to equality constraints results in a generalized eigensystem. Optimization of a general f subject to general g and h may be achieved by iterative methods, most notably sequential quadratic programming.

constraint satisfaction: An approach to problem solving that consists of three components: 1) a list of what "variables" need values, 2) a set of allowable values for each "variable" and 3) a set of relationships that must hold between the values for each "variable" (*i.e.*, the constraints). For example, in computer vision, this approach has been used for different structure labelling (*e.g.*, line labelling, region labelling) and geometric model recovery tasks (*e.g.*, reverse engineering of 3D parts or buildings from range data).

constructive solid geometry (CSG): A method for defining 3D shapes in terms of a mathematically defined set of primitive shapes. Boolean set theoretic operations of intersection, union and difference are used to combine shapes to make more complex shapes. For example:

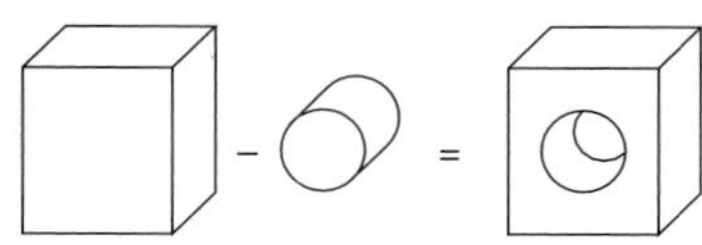

content-based image retrieval: Image database searching methods that produce matches based on the contents of the images in the database, as contrasted with using text descriptors to do the indexing. For example, one can use descriptors based on color moments to select images with similar invariants.

context: In vision, the elements, information, or knowledge occurring together with or accompanying some data, contributing to the data's full meaning. For example, in a video sequence one can speak of spatial context of a pixel, indicating the intensities at surrounding location in a given frame (image), or of temporal context, indicating the intensities at that pixel location (same coordinates) but in previous and following frames. Information deprived of appropriate context can be ambiguous: for instance, differential optical flow methods can only estimate the normal flow; the full flow can be estimated considering the spatial context of each pixel. At the level of scene understanding, knowing that the image data comes from a theater performance provides context information that can help distinguish between a real fight and a stage act.

contextual image classification: Algorithms that take into account the source or setting of images in their search for features and relationships in the image. Often this context is composed of region identifiers, color, topology and spatial relationships as well as task-specific knowledge.

contextual method: Algorithms that take into account the spatial arrangement of found features in their search for new ones.

continuous convolution: The convolution of two continuous signals. In 2D image processing terms the convolution of two images f and h is:

$$
\begin{aligned}
g(x,y) &= f(x,y) \otimes h(x,y) \\
&= \int_{-\infty}^{\infty}\int_{-\infty}^{\infty} f(\tau_u,\tau_v) \\
&\qquad \times h(x-\tau_u, y-\tau_v)\,\mathrm{d}\tau_u \mathrm{d}\tau_v
\end{aligned}
$$

continuous Fourier transform: See Fourier transform.

continuous learning: A general term describing how a system continually updates its model of a process based on current data. For example, updating a background model (for change detection) as the illumination changes during the day.

contour analysis: Analysis of outlines of image regions.

contour following: See contour linking.

contour grouping: See contour linking.

contour length: The length of a contour in appropriate units of measurements. For instance, the length of an image contour in pixels. See also arc length.

contour linking: Edge detection or boundary detection processes typically identify pixels on the boundary of a region. Connecting these pixels to form a curve is the goal of contour linking.

contour matching: See curve matching.

contour partitioning: See curve segmentation.

contour representation: See boundary representation.

contour tracing: See contour linking.

contour tracking: See contour linking.

contours: See object contour.

contrast: 1) The difference in brightness values between two structures, such as regions or pixels. 2) A texture measure. In a gray scale image, contrast, C, is defined as

$$C = \sum_i \sum_j (i-j)^2 P[i,j]$$

where P is the gray-level co-occurrence matrix.

contrast enhancement: Contrast enhancement (also known as contrast stretching) expands the distribution of intensity values in an image so that a larger range of sensitivity in the output device can be used. This can make subtle changes in an image more obvious by increasing the displayed contrast between image brightness levels. Histogram equalization is one method of contrast enhancement. An example of contrast enhancement is here (see plate section for a colour version of these figures):

Input image

After contrast enhancement

contrast stretching: See contrast enhancement.

control strategy: The guidelines behind the sequence of processes performed by an automatic image analysis or scene understanding system. For instance, control can be top-down (searching for image data that verifies an expected target) or bottom-up (progressively acting on image data or results to derive hypotheses). The control strategy may allow selection of alternative hypotheses, processes or parameter values, etc.

convex hull: Given a set of points, S, the convex hull is the smallest convex set that contains S. a 2D example is shown here:

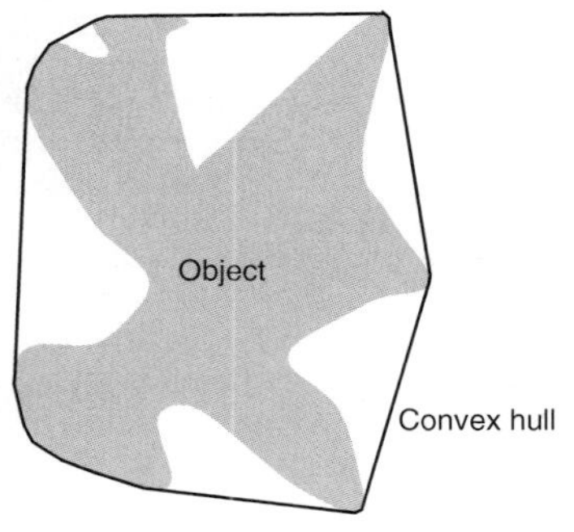

convexity ratio: Also known as solidity. A measure that characterizes deviations from convexity. The ratio for shape X is defined as $\frac{area(X)}{area(C_X)}$, where C_X is the convex hull of X. A convex figure has convexity factor 1, while all other figures have convexity less than 1.

convolution operator: A widely used general image and signal processing operator that computes the weighted sum $y(j) = \sum_i w(i)x(j-i)$ where $w(i)$ are the weights, $x(i)$ is the input signal and $y(j)$ is the result. Similarly, convolutions of image data take the form $y(r, c) = \sum_{i,j} w(i, j)x(r-i, c-j)$. Similar forms using integrals exist for continuous signals and images. By the appropriate choice of the weight values, convolution can compute low pass/smoothing, high pass/differentiation filtering or template matching/matched filtering, as well as many other linear functions. The right image below is the result of convolving (and then inverting) the left image with a | +1 | −1 | mask:

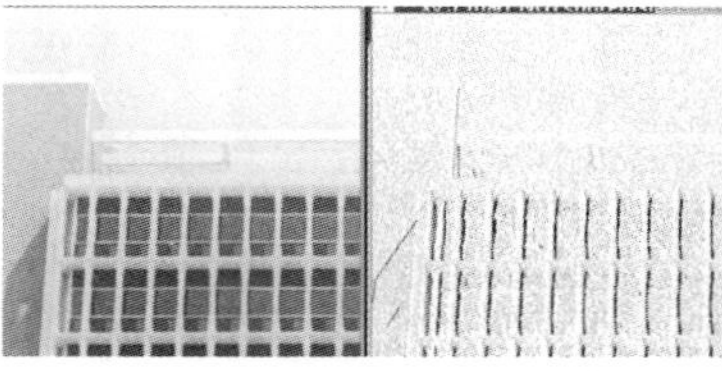

co-occurrence matrix: A representation commonly used in texture analysis algorithms. It records the likelihood (usually empirical) of two features or properties being at a given position relative to each other. For example, if the center of the matrix $\mathbf{M}$ is position (a, b) then the likelihood that the given property is observed at an offset (i, j) from the current pixel is given by matrix value $\mathbf{M}(a+i, b+j)$.

cooperative algorithm: An algorithm that solves a problem by a series of local interactions between adjacent structures, rather than some global process that has access to all data. The value at a structure changes iteratively in response to changing values at the adjacent structures, such as pixels, lines, regions, etc. The expectation is that the process will converge to a good solution. The algorithms are well suited for massive local parallelism (*e.g.*, SIMD), and are sometimes proposed as models for human image processing. An early algorithm to solve the stereo correspondence problem used cooperative processing between elements representing the disparity at a given picture element.

coordinate system: A spanning set of linearly independent vectors defining a vector space. One example is the set generally referred to as the X, Y and Z axes. There are, of course, an infinite number of sets of three linearly independent vectors describing 3D space. The right-handed version of this is shown in the figure.

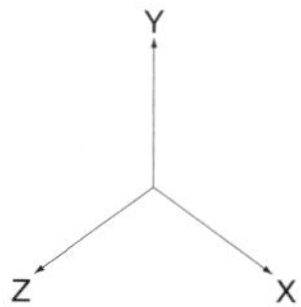

coordinate system transformation: A geometric transformation that maps points, vectors or other structures from one coordinate system to another. It is also used to express the relationship between two coordinate systems. Typical transformations include translation and rotation. See also Euclidean transformation.

coplanarity: The property of lying in the same plane. For example, three vectors $\vec{a}$, $\vec{b}$ and $\vec{c}$ are coplanar if their scalar triple product $(\vec{a} \times \vec{b}) \cdot \vec{c} = 0$ is zero.

coplanarity invariant: A projective invariant that allows one to determine when five corresponding points observed in two (or more) views are coplanar in the 3D space. The five points allow the construction of a set of four collinear points whose cross ratio value can be computed. If the five points are coplanar, then the cross ratio value must be the same in the two views. Here, point A is selected and the lines AB, AC, AD and AE are used to define an invariant cross ratio for any line L that intersects them:

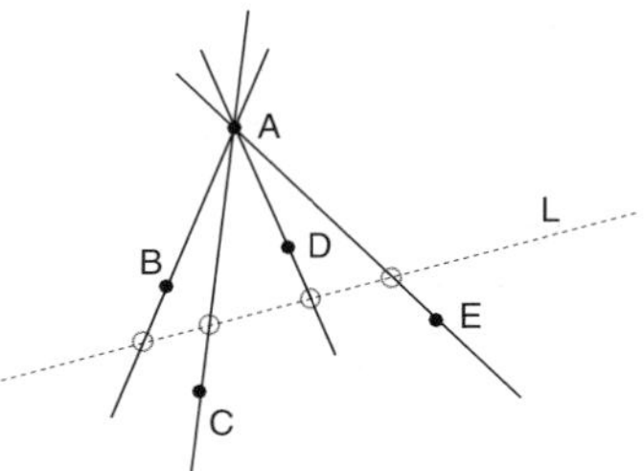

core line: See medial line.

corner detection: See curve segmentation.

corner feature detectors: See interest point feature detectors and curve segmentation.

coronary angiography: A class of image processing techniques (usually based on X-ray data) for visualizing and inspecting the blood vessels surrounding the heart (coronaries). See also angiography.

correlation: See cross correlation.

correlation based optical flow estimation: Optical flow estimated by correlating local image texture at each point in two or more images and noting their relative movement.

correlation based stereo: Dense stereo reconstruction (*i.e.*, at every pixel) computed by cross correlating local image neighborhoods in the two images to find corresponding points, from which depth can be computed by stereo triangulation.

correspondence constraint: See stereo correspondence constraint.

correspondence problem: See stereo correspondence problem.

cosine diffuser: Optical correction mechanism for correcting spatial responsivity to light. Since off-angle light is treated with the same response as normal light, a cosine transfer is used to decrease the relative responsivity to it.

cosine transform: Representation of an signal in terms of a basis of cosine functions. For an even 1D function $f(x)$, the cosine transform is

$$F(u) = 2\int_0^\infty f(x)\cos(2\pi ux)\mathrm{d}x.$$

For a sampled signal $f_{0..(n-1)}$, the discrete cosine transform is the vector $b_{0..(n-1)}$ where, for $k \geq 1$:

$$b_0 = \sqrt{\frac{1}{n}}\sum_{i=0}^{n-1} f_i$$

$$b_k = \sqrt{\frac{2}{n}}\sum_{i=0}^{n-1} f_i \cos\left(\frac{\pi}{2n}(2i+1)k\right)$$

For a 2D signal $f(x, y)$ the cosine transform $F(u, v)$ is

$$4\int_0^\infty \int_0^\infty f(x, y)\cos(2\pi ux)$$
$$\cos(2\pi vy)\mathrm{d}x\mathrm{d}y$$

cost function: The function or metric quantifying the cost of a certain action, move or configuration, that is to be minimized over a given parameter space. A key concept of optimization. See also Newton's optimization method and functional optimization.

covariance: The covariance, denoted σ^2, of a random variable X is the expected value of the square of the deviation of the variable from the mean. If μ is the mean, then $\sigma^2 = E[(X-\mu)^2]$. For a d-dimensional data set represented as a set of n column vectors $\vec{x}_{1..n}$, the sample mean is $\vec{\mu} = \frac{1}{n}\sum_{i=1}^{n}\vec{x}_i$, and the sample covariance is the $d \times d$ matrix $\Sigma = \frac{1}{n-1}\sum_{i=1}^{n}(\vec{x}_i - \vec{\mu})(\vec{x}_i - \vec{\mu})^\top$.

covariance propagation: A method of statistical error analysis, in which the covariance of a derived variable can be estimated from the covariances of the variables from which it is derived. For example, assume that independent variables $\vec{x}$ and $\vec{y}$ are sampled from multi-variate normal distributions with associated covariance matrices $\mathbf{C}_x$ and $\mathbf{C}_y$. Then, the covariance of the derived variable $\vec{z} = a\vec{x} + b\vec{y}$ is $\mathbf{C}_z = a^2\mathbf{C}_x + b^2\mathbf{C}_y$.

crack code: A contour description method that codes not the pixels themselves but the *cracks* between them. This is done as a four-directional scheme as shown below. It can be viewed as a chain code with four directions rather than eight.

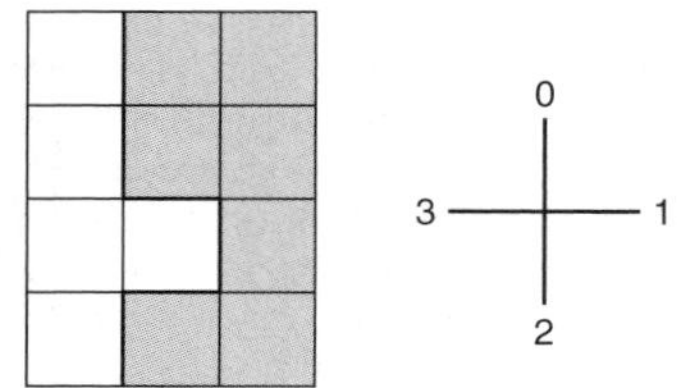

Crack code = {2, 2, 1, 2, 3, 2}

crack edge: A type of edge used in line labeling research to represent where two aligned blocks meet. Here, neither a step edge nor fold edge is seen:

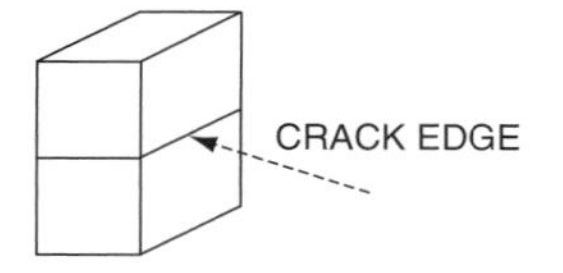

crack following: Edge tracking on the dual lattice or "cracks" between pixels based on the continuous segments of line from a crack code.

Crimmins smoothing operator: An iterative algorithm for speckle (salt-and-pepper noise) reduction. It uses a nonlinear noise reduction technique that compares the intensity of each image pixel with its eight neighbors and either increments or decrements the value to try and make it more representative of its surroundings. The algorithm raises the intensity of pixels that are darker relative to their neighbors and lowers pixels that are relatively brighter. More iterations produce more reduction in noise but at the cost of increased blurring of detail.

critical motion: In the problem of self-calibration of a moving camera, there are certain motions for which calibration algorithms fail to give unique solutions. Sequences for which self-calibration is not possible are known as critical motion sequences.

cross correlation: Standard method of estimating the degree to which two series are correlated. Given two series $\{x_i\}$ and $\{y_i\}$, where $i = 0, 1, 2, .., (N-1)$ the cross correlation, r_d, at a delay d is defined as

$$\frac{\sum_i (x_i - m_x) \cdot (y_{i-d} - m_y)}{\sqrt{\sum_i ((x_i - m_x)^2} \sqrt{\sum_i (y_{i-d} - m_y)^2}}$$

where m_x and m_y are the means of the corresponding sequences.

cross correlation matching: Matching based on the cross correlation of two sets. The closer the correlation is to 1, the better the match is. For example, in correlation based stereo, for each pixel in the first image, the corresponding pixel in the second image is the one with the highest correlation score, where the sets being matched are the local neighborhoods of each pixel.

cross ratio: The simplest projective invariant. It generates a scalar from four points of any 1D projective space (*e.g.*, a projective line). The cross ratio for the four points ABCD below is:

$$\frac{(r+s)(s+t)}{s(r+s+t)}$$

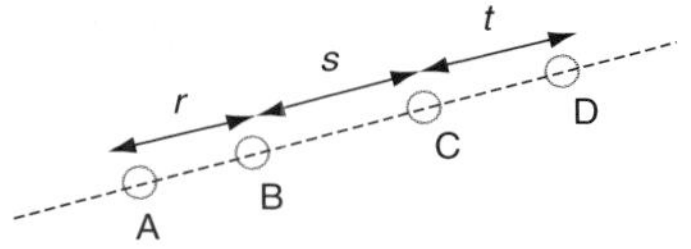

cross section function: Part of the generalized cylinder representation that gives a volumetric based representation of an object. The representation defines the volume by a curved axis, a cross section and a cross section function at each point on that axis. The cross section function defines how the size or shape of the cross section varies as a function of its position along the axis. See also generalized cone. This example shows how the size of the square cross section varies along a straight line to create a truncated pyramid:

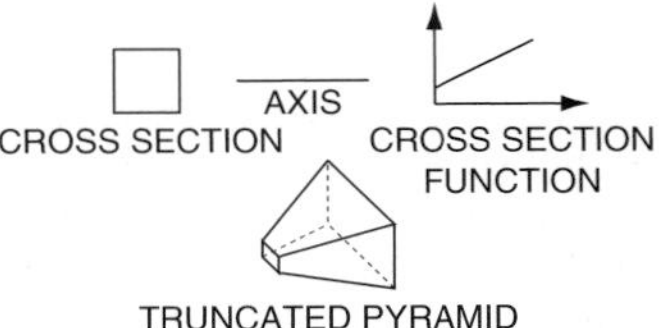

cross-validation: A test of how well a model generalizes to other data (*i.e.*, using samples other than those that were used to create the model). This approach can be used to determine when to stop training/learning, before over-generalization occurs. See also leave-one-out test.

crossing number: The crossing number of a graph is the minimum number of arc intersections in any drawing of that graph. A planar graph has crossing number zero. This graph has a crossing number of one:

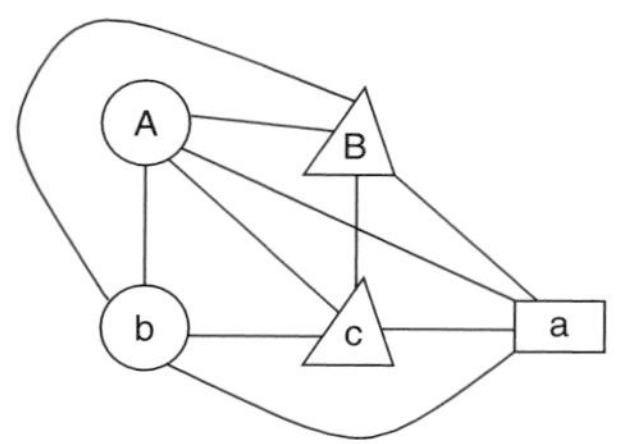

CSG: See constructive solid geometry

CT: See X-ray CAT.

cumulative histogram: A histogram where the bin contains not only the count of all instances having that value but also the count of all bins having a lower index value. This is the discrete equivalent of the cumulative probability distribution. The right figure is the cumulative histogram corresponding to the normal histogram on the left:

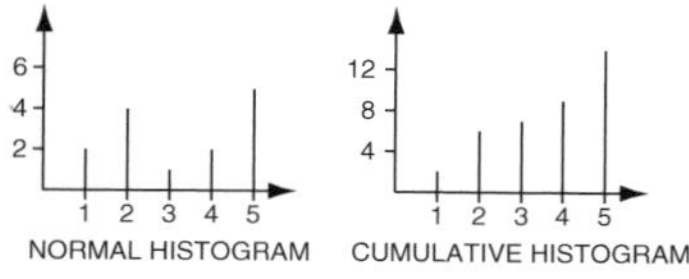

currency verification: Algorithms for checking that printed money and coinage are genuine. A specialist field involving optical character recognition.

curse of dimensionality: The exponential growth of possibilities as a function of dimensionality. This might manifest as several effects as the dimensionality increases: 1) the increased amount of computational effort required, 2) the exponentially increasing amount of data required to populate the data space in order that training works and 3) how all data points tend to become equidistant from each other, thus causing problems for clustering and machine learning algorithms.

cursive script recognition: Methods of optical character recognition whereby handwritten cursive (also called joined-up) characters are automatically classified.

curvature: Usually meant to refer to the change in shape of a curve or surface. Mathematically, the curvature κ of a curve is the length of the second derivative $|\frac{\partial^2 \vec{x}(s)}{\partial s^2}|$ of the curve $\vec{x}(s)$ parameterized as a function of arc length s. A related definition holds for surfaces, only here there are two distinct principal curvatures at each point on a sufficiently smooth surface.

curvature primal sketch: A multi-scale representation of the significant changes in curvature along a planar curve.

curvature scale space: A multi-scale representation of the curvature zero-crossing points of a planar contour as it evolves during smoothing. It is found by parameterizing the contour using arc length, which is then convolved with a Gaussian filter of increasing standard deviation. Curvature zero-crossing points are then recovered and mapped to the scale-space image with the horizontal axis representing the arc length parameter on the original contour and the vertical axis representing the standard deviation of the Gaussian filter.

curvature sign patch classification: A method of local surface classification based on its mean and Gaussian curvature signs, or principal curvature sign class. See also mean and Gaussian curvature shape classification.

curve: A set of connected points in 2D or 3D, where each point has at most two neighbors. The curve could be defined by a set of connected points, by an implicit function (*e.g.*, $y + x^2 = 0$), by an explicit form (*e.g.*, $(t, -t^2)$ for all t), or by the intersection of two surfaces (*e.g.*, by intersecting the planes $X = 0$ and $Y = 0$), etc.

curve binormal: The vector perpendicular to both the tangent and normal vectors to a curve at any given point:

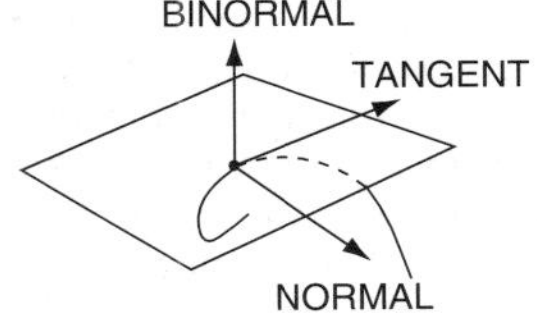

curve bitangent: A line tangent to a curve or surface at two different points, as illustrated here:

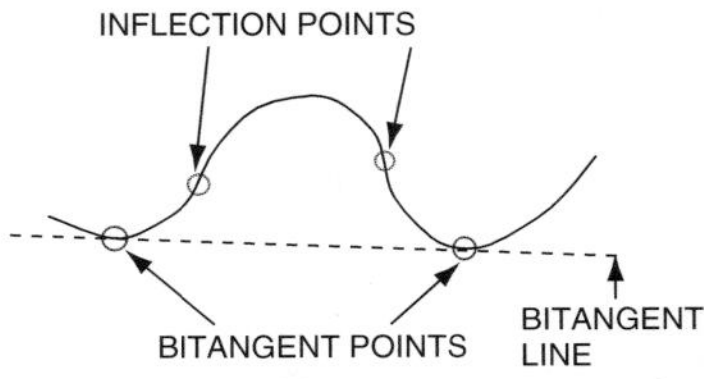

curve evolution: A curve abstraction method whereby a curve can be iteratively simplified, as in this example:

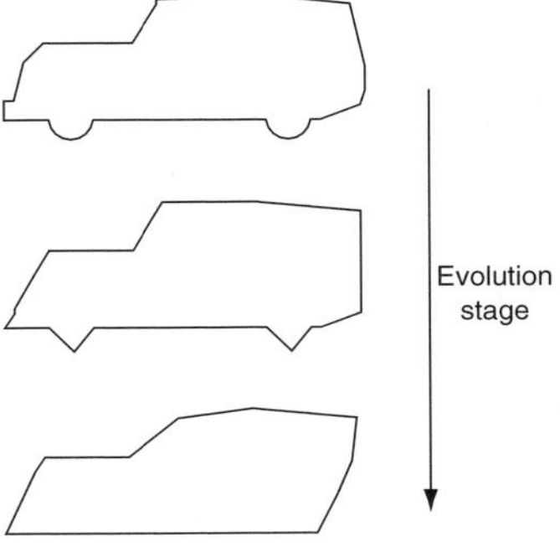

For example, a relevance measure is assigned to every vertex in the curve. The least important can be removed at each iteration by directly connecting its neighbors. This elimination is repeated until the desired stage of abstraction is reached. Another method of curve evolution is to progressively smooth the curve with Gaussian weighting of increasing standard deviation.

curve fitting: Methods for finding the parameters of a best-fit curve through a set of 2D (or 3D) data points. This is often posed as a minimization of the least- squares error between some hypothesized curve and the data points. If the curve, $y(x)$, can be thought of as the sum of a set of m arbitrary basis functions, X_k and written

$$y(x) = \sum_{k=1}^{k=m} a_k X_k(x)$$

then the unknown parameters are the weights a_k. The curve fitting process can then be considered as the minimization of some log-likelihood function giving the best fit to N points whose Gaussian error has standard deviation σ_i. This function may be defined as

$$\chi^2 = \sum_{i=1}^{i=N} \left[\frac{y_i - y(x_i)}{\sigma_i} \right]^2$$

The weights that minimize this can be found from the design matrix D

$$D_{i,j} = \frac{X_j(x_i)}{\sigma_i}$$

by finding the solution to the linear equation

$$\mathbf{Da} = \mathbf{r}$$

where the vector $r_i = \frac{y_i}{\sigma_i}$.

curve inflection: A point on a curve where the curvature is zero as it changes sign from positive to negative, as in the two examples below:

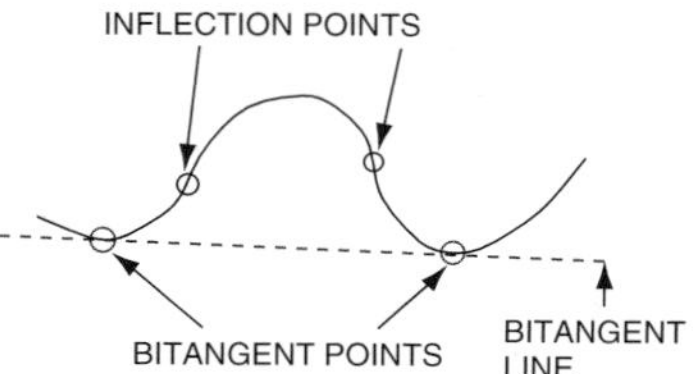

curve invariant: Measures taken over a curve that remain invariant under certain transformations, *e.g.*, arc length and curvature are invariant under Euclidean transformations.

curve invariant point: A point on a curve that has a geometric property that is invariant to changes in projective transformation. Thus, the point can be identified and used for correspondence in multiple views of the same scene. Two well known planar curve invariant points are curvature inflection points and bitangent points, as shown here:

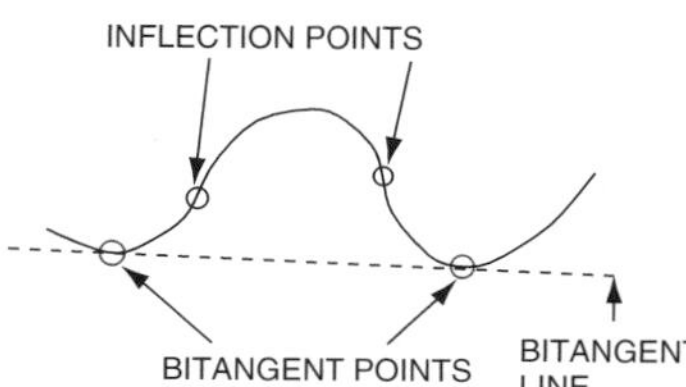

curve matching: The comparison of data sets to previously modeled curves or other curve data sets. If a modeled curve closely corresponds to a data set then an interpretation of similarity can be made. Curve matching differs from curve fitting in that curve fitting involves minimizing the parameters of theoretical models rather than actual examples.

curve normal: The vector perpendicular to the tangent vector to a curve at any given point and that also lies in the plane that locally contains the curve at that point:

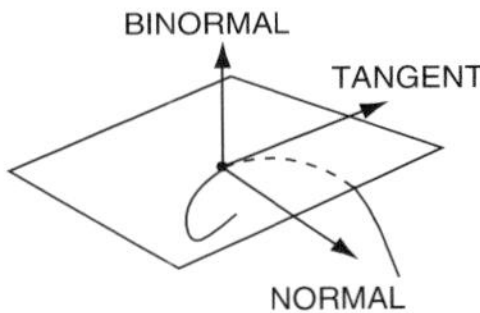

curve representation system: Methods of representing or modeling curves parametrically. Examples include: b-splines, crack codes, cross section functions, Fourier descriptors, intrinsic equations, polycurves, polygonal approximations, radius vector functions, snakes, splines, etc.

curve saliency: A voting method for the detection of curves in a 2D or 3D image. Each pixel is convolved with a curve mask to build a saliency map. This map will hold high values for locations in space where likely candidates for curves exist.

curve segmentation: Methods of identifying and splitting curves into different primitive types. The location of changes between one primitive type and another is particularly important. For example, a good curve segmentation algorithm should detect the four lines that make up a square. Methods include: corner detection, Lowe's method and recursive splitting.

curve smoothing: Methods for rounding polygon approximations or vertex-based approximations of surface boundaries. Examples include Bezier curves in 2D and NURBS in 3D. See also curve evolution. An example of a polygonal data curve smoothed by a Bezier curve is:

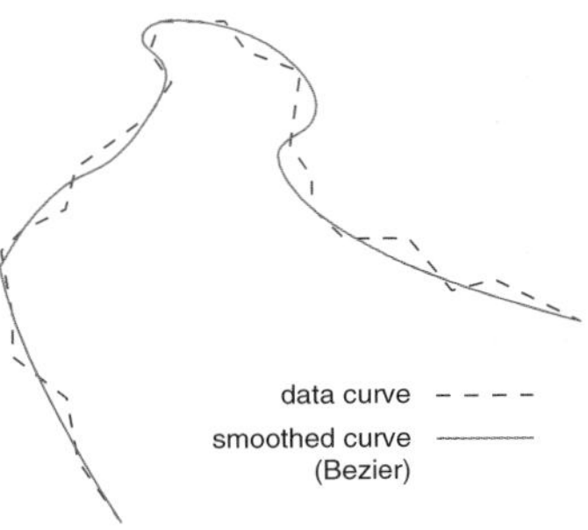

curve tangent vector: The vector that is instantaneously parallel to a curve at any given point:

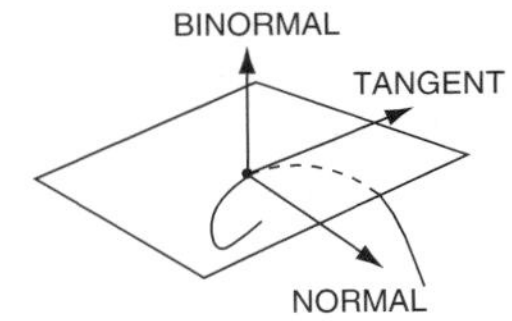

cut detection: The identification of the frames in film or video where the camera viewpoint suddenly changes, either to a new viewpoint within the current scene or to a new scene.

cyclopean view: A term used in stereo image analysis, based on the mythical one-eyed Cyclops. When stereo reconstruction of a scene occurs based on two cameras, one has to consider what coordinate system to use to base the reconstructed 3D coordinates, or what viewpoint to use when presenting the reconstruction. The cyclopean viewpoint is located at the midpoint of the baseline between the two cameras.

cylinder extraction: Methods of identifying the cylinders and the constituent data points from 2.5D and 3D images that are samples from 3D cylinders.

cylinder patch extraction: Given a range image or a set of 3D data points, cylinder patch extraction finds (usually connected) sets of points that lie on the surface of a cylinder, and usually also the equation of that cylinder. This process is useful for detecting and modelling pipework in range images of industrial scenes.

cylindrical mosaic: A photomosaicing approach where individual 2D images are projected onto a cylinder. This is possible only when the camera rotates about a single axis or the camera center of projection remains approximately fixed with respect to the distance to the nearest scene points.

cylindrical surface region: A region of a surface that is locally cylindrical. A region in which all points have zero Gaussian curvature, and nonzero mean curvature.

D

darkfield illumination: A specialized illumination technique that uses oblique illumination to enhance contrast in subjects that are not imaged well under normal illumination conditions.

data fusion: See sensor fusion.

data integration: See sensor fusion.

data parallelism: Reference to the parallel structuring of either the input to programs, the organization of programs themselves or the programming language used. Data parallelism is a useful model for much image processing because the same operation can be applied independently and in parallel at all pixels in the image.

data reduction: A general term for processes that 1) reduce the number of data points, *e.g.*, by subsampling or by using cluster centers of mass as representative points or by decimation, or 2) reduce the number of dimensions in each data point, *e.g.*, by projection or principal component analysis (PCA).

data structure: A fundamental concept in programming: a collection of computer data organized in a precise structure, for instance a tree (see for instance quadtree), a queue, or a stack. Data structures are accompanied by sets of procedures, or libraries, implementing various types of data manipulation, for instance storage and indexing.

DCT: See discrete cosine transform.

deblur: To remove the effect of a known blurring function on an image. If an observed image I is the convolution of an unknown image I' and a known blurring kernel B, so that $I = I' * B$, then deblurring is the process of computing I' given I and B.

Dictionary of Computer Vision and Image Processing R.B. Fisher, K. Dawson-Howe, A. Fitzgibbon, C. Robertson and E. Trucco

See deconvolution, image restoration, Wiener filtering.

decentering distortion (lens): Lens decentering is a common cause of tangential distortion. It arises when the lens elements are not perfectly aligned and creates an asymmetric component to the distortion.

decimation: 1) In digital signal processing, a filter that keeps one sample out of every N, where N is a fixed number. See also subsampling. 2) "Mesh" decimation: merging of similar adjacent surface patches or mesh vertices in order to reduce the size of a model. Often used as a processing step when deriving a surface model from a range image.

decision tree: Tools for helping to choose between several courses of action. They are an effective structure within which an agent can search options and investigate the possible outcomes. They also help to balance the risks and rewards associated with each possible course of action.

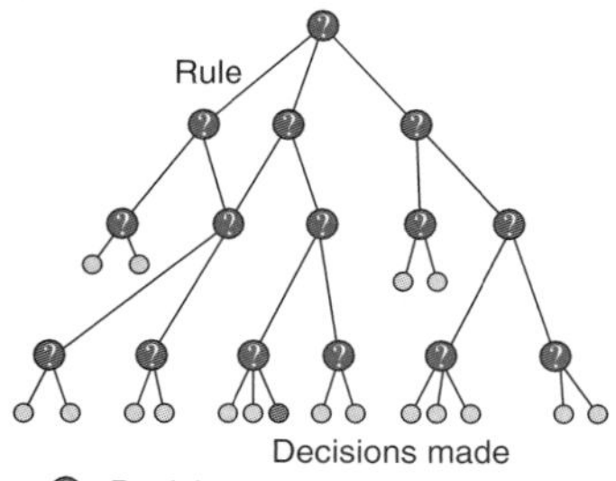

decoding: Converting a signal that has been encoded back into its original form (lossless coding) or into a form close to the original (lossy coding). See also image compression.

decomposable filters: A complex filter that can be applied as a number of simpler filters applied one after the other. For example the 2D Laplacian of Gaussian filter can be decomposed into four simpler filters.

deconvolution: The inverse process of convolution. Deconvolution is used to remove certain signals (for example blurring) from images by inverse filtering (see deblur). For a convolution producing image $h = f * g + \eta$ given f and g, the image and convolution mask, η is the noise and $*$ is the convolution, deconvolution attempts to estimate f. Deconvolution is often an ill-posed problem and may not have a unique solution. See also image restoration.

defocus: Blurring of an image, either accidental or deliberate, by incorrect focus or viewpoint parameters use or estimation. See also shape from focus, shape from defocus.

defocus blur: Deformation of an image due to the predictable behavior of optics when incorrectly adjusted. The blurring is the result of light rays that, after entering the optical system, misconverge on the imaging plane. If the camera parameters are known in advance, the blurring can be partially corrected.

deformable model: Object descriptors that model a specific class of deformable objects (*e.g.*, eyes, hands) where the shapes vary according to the values of the parameters. If the general, but not specific, characteristics of an object type are known then a deformable model can be constructed and used as a matching template for new data. The degree of deformation needed to match the shape can be used as matching score. See also modal deformable model, geometric deformable model.

deformable shape: See deformable model.

deformable superquadric: A type of superquadric volumetric model that can be deformed by bending, twisting, etc. in order to fit to the data being modeled.

deformable template model: See deformable model.

deformation energy: The metric that must be minimized when determining an active shape model. Comprised of terms for both internal energy (or force) arising from the model shape deformation and external energy (or force) arising from the discrepancy between the model shape and the data.

degradation: A loss of quality suffered by an image, the content of which gets corrupted by unwanted processes. For instance, MPEG compression–decompression can alter some intensities, so that the image is degraded. (See also JPEG image compression), image noise.

degree of freedom: A free variable in a given function. For instance, rotations in 3D space depend on three angles, so that a rotation matrix has nine entries but only three degrees of freedom.

Delaunay triangulation: The Delaunay graph of the point set can be constructed from its Voronoi diagram by connecting the points in adjacent polygons. The connections form the Delaunay triangulation. The triangulation has the property that the circumcircle of every triangle contains no other points. The approach can be used to construct a polyhedral surface approximation from a set of 3D sample points. The solid lines connecting the points below are the Delaunay triangulation and the dashed lines are the boundaries of the Voronoi diagram.

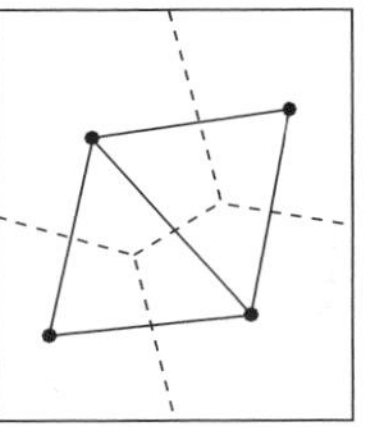

demon: A program that runs in the background, for instance performing checks or guaranteeing the correct functioning of a module of a complex system.

demosaicing: The process of converting a single color per pixel image (as captured by most digital cameras) into a three color per pixel image.

Dempster–Shafer: A belief modeling approach for testing a hypothesis that allows information, in the form of beliefs, to be combined into a plausibility measure for that hypothesis.

dense reconstruction: A class of techniques estimating depth at each pixel of an input image or sequence, thus generating a dense sampling of the 3D surfaces imaged. This can be achieved, for instance, by range sensing, or stereo vision.

dense stereo matching: A class of methods establishing the correspondence (see stereo correspondence problem) between all pixels in a stereo pair of images. The generated disparity map can then be used for depth estimation.

densitometry: A class of techniques that estimate the density of a material from images, for instance bone density in the medical domain (bone densitometry).

depth: Distance of scene points from either the camera center or the camera imaging plane. In a range image, the intensity value in the image is a measure of depth.

depth estimation: The process of estimating the distance between a sensor (*e.g.*, a stereo pair) and a part of the scene being imaged. Stereo vision and range sensing are two well-known ways to estimate depth.

depth from defocus: The depth from defocus method uses the direct relationships among the depth, camera parameters and the amount of blurring in images to derive the depths from parameters that can be directly measured.

depth from focus: A method to determine distance to one point by taking many images in better and better focus. This is also called autofocus or software focus.

depth image: See range image.

depth image edge detector: See range image edge detector.

depth map: See range image.

depth of field: The distance between the nearest and the farthest point in focus for a given camera:

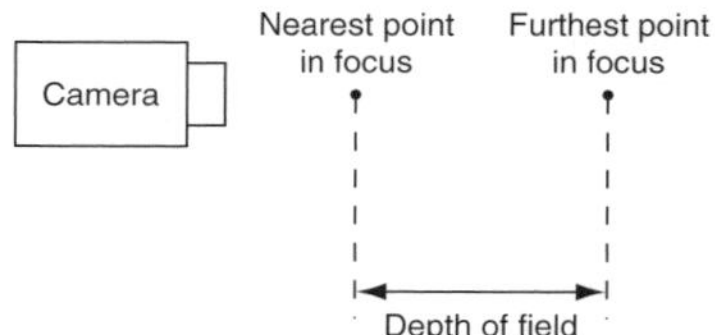

depth perception: The ability to perceive distances from visual stimuli, for instance motion or stereo vision.

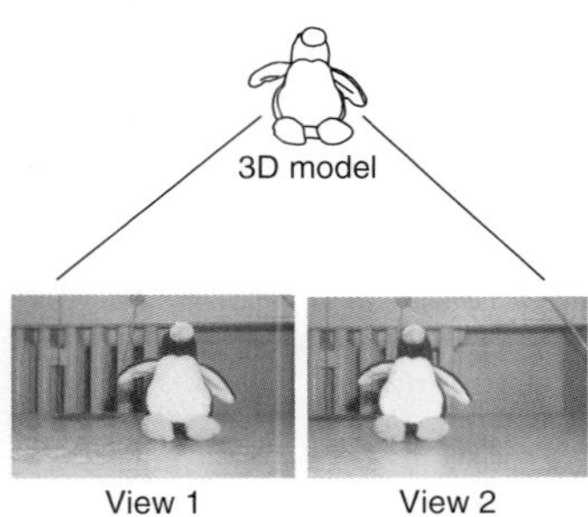

depth sensor: See range sensor.

Deriche edge detector: Convolution filter for edge finding similar to the Canny edge detector. Deriche uses a different optimal operator where the filter is assumed to have infinite extent. The resulting convolution filter is sharper than the derivative of the Gaussian that Canny uses

$$f(x) = Axe^{-\frac{|x|}{\sigma}}$$

See also edge detection.

derivative based search: Numerical optimization methods assuming that the gradient can be estimated. An example is the quasi-Newton approach, that attempts to generate an estimate of the inverse Hessian matrix. This is then used to determine the next iteration point.

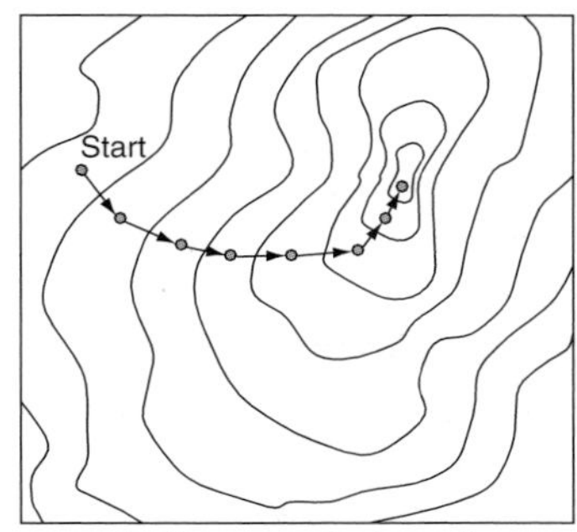

Conjugate gradient search

DFT: See discrete Fourier transform.

diagram analysis: Syntactic analysis of images of line drawings, possibly with text in a report or other document. This field is closely related to the analysis of visual languages.

dichroic filter: A dichroic filter selectively transmits light of a given wavelength.

dichromatic model: The dichromatic model states that the light reflected from a surface is the sum of two components, body and interface reflectance. Body reflectance follows Lambert's law. Interface reflectance models highlights. The model has been applied to several computer vision tasks including color constancy, shape recovery and color image segmentation. See also color.

difference image: An image computed as pixelwise difference of two other images, that is, each pixel in the difference image is the difference between the pixels at the same location in the two input images. For example, in the figure below the right image is the difference of the left and middle images (after adding 128 for display purposes).

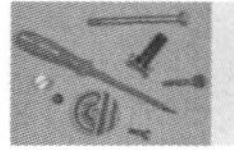  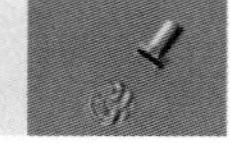

diffeomorphism: A differentiable one-to-one map between manifolds. The map has a differentiable inverse.

difference-of-Gaussians operator: A convolution operator used to locate edges in a gray-scale image using an approximation to the Laplacian of Gaussian operator. In 2D the convolution mask is:

$$c_1 e^{\left(-\frac{(x^2+y^2)}{\sigma_1^2}\right)} - c_2 e^{\left(-\frac{(x^2+y^2)}{\sigma_2^2}\right)}$$

where the constants c_1 and c_2 control the height of the individual Gaussians and σ_1, σ_2 are the standard deviations.

differential geometry: A field of mathematics studying the local derivative-based properties of curves and surfaces, for instance tangent plane and curvature.

differential invariant: Image descriptors that are invariant under geometric transformations as well as illumination changes. Invariant descriptors are generally classified as global invariants (corresponding to object primitives) and local invariants (typically based on derivatives of the image function). The image function is always assumed to be continuous and differentiable.

differential pulse code modulation: A technique for converting an analogue signal to binary by sampling it, expressing the value of the sampled data modulation in binary and then reducing the bit rate by taking account of the fact that consecutive samples do not change much.

differentiation filtering: See gradient filter.

diffraction: The bending of light rays at the edge of an object or through a transparent medium. The amount by which a ray is bent is dependent on wavelength.

diffraction grating: An array of diffracting elements that has the effect of producing periodic alterations in a wave's phase, amplitude or both. The simplest arrangement is an array of slits (see moiré interferometry).

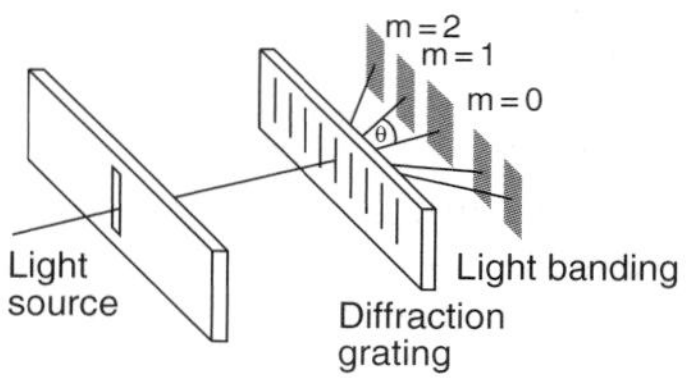

diffuse illumination: Light energy that comes from a multitude of directions, hence not causing significant shading or shadow effects. The opposite of diffuse illumination is directed illumination.

diffuse reflection: Scattering of light by a surface in many directions. Ideal Lambertian diffusion results in the same energy being reflected in every direction regardless of the direction of the incoming light energy.

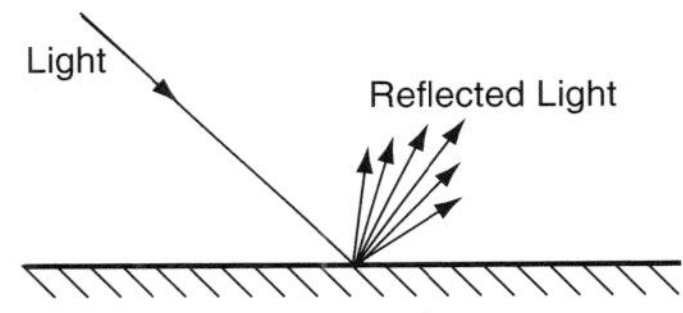

diffusion smoothing: A technique achieving Gaussian smoothing as the solution of a diffusion equation with the image to be filtered as the initial boundary condition. The advantage is that, unlike repeated averaging, diffusion smoothing allows the construction of a continuous scale space.

digital camera: A camera in which the image sensing surface is made up of individual semiconductor sampling elements (typically one per pixel of the image), and quantized versions of the sensed values are recorded when an image is captured.

digital elevation map: A sampled and quantized map where every point represents a height above a reference ground plane (*i.e.*, the elevation).

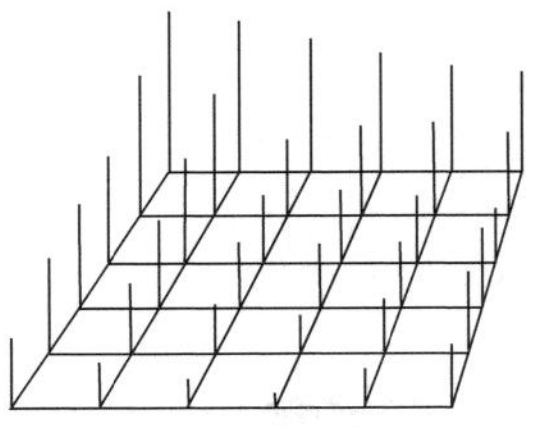

digital geometry: Geometry (points, lines, angles, surfaces, etc.) in a sampled and quantized domain.

digital image: Any sampled and quantized image.

41	43	45	51	56	49	45	40
56	48	65	85	55	52	44	46
59	77	99	81	127	83	46	56
52	116	44	54	55	186	163	163
51	129	46	48	71	164	86	97
50	85	192	140	167	99	51	44
57	63	91	126	102	56	54	49
146	169	213	246	243	139	180	163
41	44	54	56	47	45	36	54

digital image processing: Image processing restricted to the domain of digital images.

digital signal processor: A class of co-processors designed to execute processing operations on digitized signals efficiently. A common characteristic is the provision of a fast multiply and accumulate function, *e.g.*, $a \leftarrow a + b \times c$.

digital subtraction angiography: A basic technique used in medical image processing to detect, visualize and inspect blood vessels, based on the subtraction of a background image from the target image, usually where the blood vessels are made more visible by using an X-ray contrast medium. See also medical image registration.

digital terrain map: See digital elevation map.

digital topology: Topology (*i.e.*, how things are connected/ arranged) in a digital domain (*e.g.*, in a digital image). See also connectivity.

digital watermarking: The process of embedding a signature/watermark into digital data. In the domain of digital images this is most normally done for copyright protection. The digital watermark may be invisible or visible (as shown).

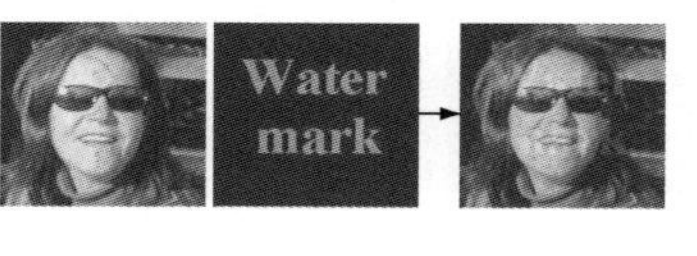

digitization: The process of making a sampled digital version of some analog signal (such as an image).

dihedral edge: The edge made by two planar surfaces. A "fold" in a surface:

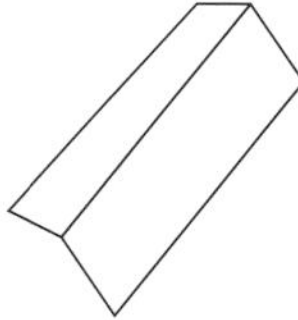

dilate operator: The operation of expanding a binary or gray-scale object with respect to the background. This has the effect of filling in any small holes in the object(s) and joining any object regions that are close together. Most frequently described as a morphological transformation, and is the dual of the erode operator.

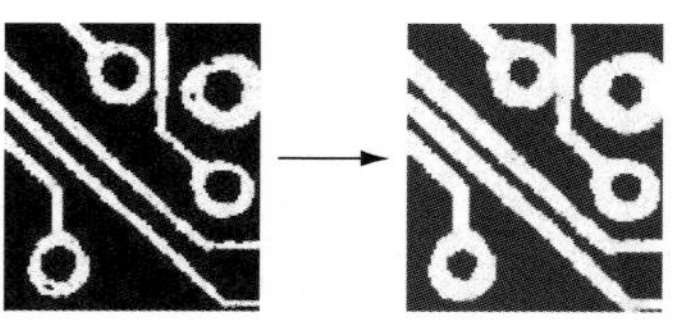

dimensionality: The number of dimensions that need to be considered. For example 3D object location is often considered as a seven dimensional problem (three dimensions for position, three for orientation and one for the object scale).

direct least square fitting: Direct fitting of a model to some data by a method that has a closed form or globally convergent solution.

directed illumination: Light energy that comes from a particular direction hence causing relatively sharp shadows. The opposite of this form of illumination is diffuse illumination.

directional derivative: A derivative taken in a specific direction, for instance, the component of the gradient along one coordinate axis. The images on the right are the vertical and horizontal directional derivatives of the image on the left.

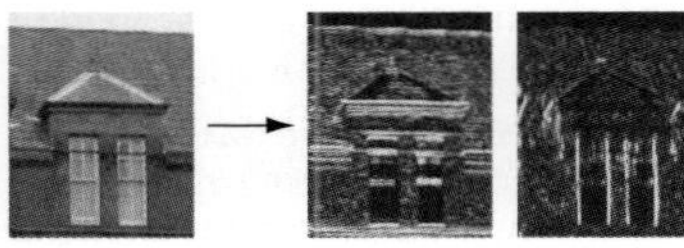

discontinuity detection: See edge detection.

discontinuity preserving regularization: A method for preserving edges (discontinuities) from being blurred as a result of some regularization operation (such as the recovery of a dense disparity map from a sparse set of disparities computed at matching feature points).

discontinuous event tracking: Tracking of events (such as a moving person) through a sequence of images. The discontinuous nature of the tracking is caused by the distance that a person (or hand, arm, etc.) can travel between frames and also be the possibility of occlusion (or self-occlusion).

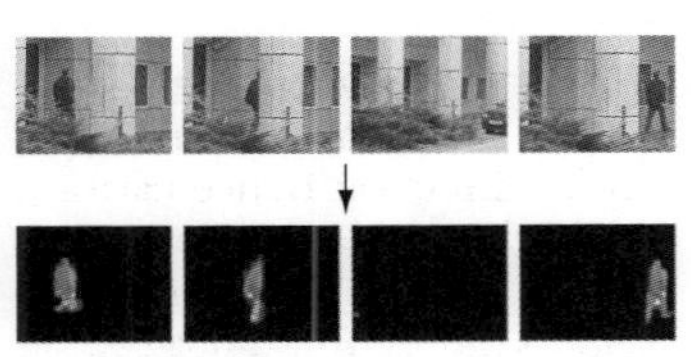

discrete cosine transform (DCT): A transformation that converts digital images into the frequency domain in terms of the coefficients of discrete cosine functions. Used, for example, within JPEG image compression.

discrete Fourier transform (DFT): A version of the Fourier transform for sampled data.

discrete relaxation: A technique for labeling objects in which the possible type of each object is iteratively constrained based on relationships with other objects in the scene. The aim is to obtain a globally consistent interpretation (if possible) from locally consistent relationships.

discrimination function: A binary function separating data into two classes. See classifier.

disparity: The image distance shifted between corresponding points in stereo image pairs.

Left image features

Right image features

Disparity

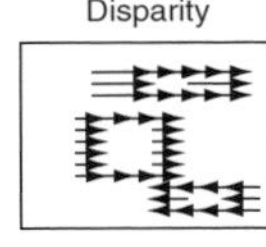

disparity gradient: The gradient of a disparity map for a stereo pair, that estimates the surface slope at each image point. See also binocular stereo.

disparity gradient limit: The maximum allowed disparity gradient in a potential stereo feature match.

disparity limit: The maximum allowed disparity in a potential stereo feature match. The notion of a disparity limit is supported by evidence from the human visual system.

dispersion: Scattering of light by the medium through which it is traveling.

distance function: See distance metric.

distance map: See range image.

distance metric: A measure of how far apart two things are in terms of physical distance or similarity. A metric can be other functions besides the standard Euclidean distance, such as the algebraic or Mahalanobis distances. A true metric must satisfy: 1) $d(x, y) + d(y, z) \geq d(x, z)$, 2) $d(x, y) = d(y, x)$, 3) $d(x, x) = 0$ and 4) $d(x, y) = 0$ implies $x = y$, but computer vision processes often use functions that do not satisfy all of these criteria.

distance transform: An image processing operation normally applied to binary images in which every object point is transformed into a value representing the distance from the point to the nearest object boundary. This operation is also referred to as chamfering (see chamfer matching).

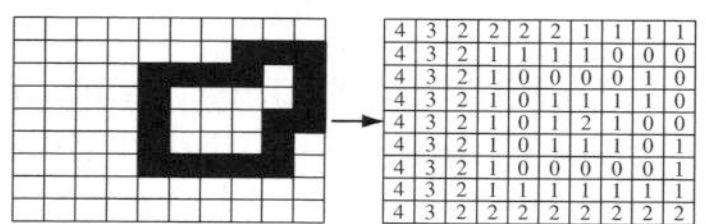

distortion coefficient: A coefficient in a given image distortion model, for instance k_1, k_2 in the distortion polynomial. See also pincushion distortion, barrel distortion.

distortion polynomial: A polynomial model of radial lens distortion. A common example is $x = x_d(1 + k_1 r^2 + k_2 r^4)$, $y = y_d(1 + k_1 r^2 + k_2 r^4)$. Here, x, y are the undistorted image coordinates, x_d, y_d are the distorted image coordinates, $r^2 = x_d^2 + y_d^2$, and k_1, k_2 are the distortion coefficient. Usually k_2 is significantly smaller than k_1, and can be set to 0 in cases where high accuracy is not required.

distortion suppression: Correction of image distortions (such as non-linearities introduced by a lens). See geometric distortion and geometric transformation.

dithering: A technique simulating the appearance of different shades or colors by varying the pattern of black and white (or different color) dots. This is a common task for inkjet printers.

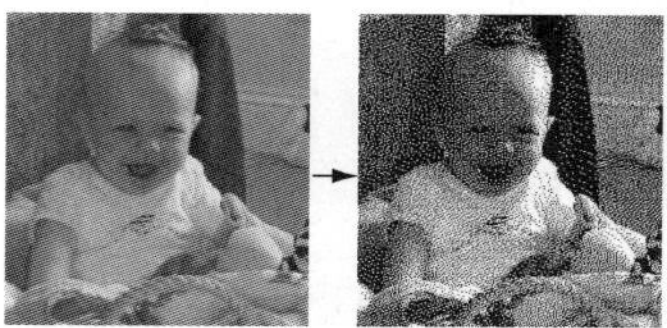

divide and conquer: A technique for solving problems efficiently by subdividing the problem into smaller subproblems, and then recursively solving these subproblems in the expectation that the smaller problems will be easier to solve. An example is an algorithms for deriving a polygonal approximation of a contour in which a straight line estimate is recursively split in the middle (into two segments with the midpoint put exactly on the contour) until the distance between the polygonal representation and the actual contour is below some tolerance.

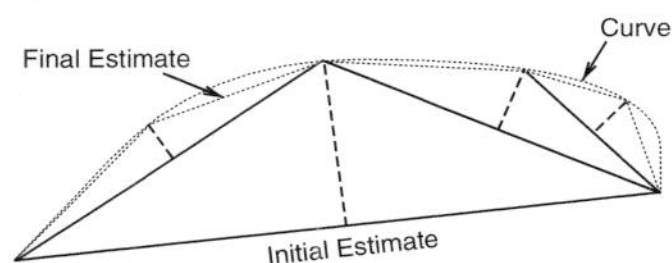

divisive clustering: Clustering/cluster analysis in which all items are initially considered as a single set (cluster) and subsequently divided into component subsets (clusters).

DIVX: An MPEG 4 based video compression technology aiming to achieve sufficiently high compression to enable transfer of digital video contents over the Internet, while maintaining high visual quality.

document analysis: A general term describing operations that attempt to derive information from documents (including for example character recognition and document mosaicing).

document mosaicing: Image mosaicing of documents.

document retrieval: Identification of a document in a database of scanned documents based on some criteria.

DoG: See difference of Gaussians.

dominant plane: A degenerate case encountered in uncalibrated structure and motion recovery where most or all of the tracked image features are co-planar in the scene.

Doppler: A physics phenomenon whereby an instrument receiving acoustic or electromagnetic waves from a source in relative motion measures an increasing frequency if the source is approaching, and decreasing if receding. The acoustic Doppler effect is employed in sonar sensors to estimate target velocity as well as position.

downhill simplex: A method for finding a local minimum using a simplex (a geometrical figure specified by $N+1$ vertices) to bound the optimal position in an N-dimensional space. See also optimization.

DSP: See digital signal processor.

dual of the image of the absolute conic (DIAC): If ω is the matrix representing the image of the absolute conic, then ω^{-1} represents its dual (DIAC). Calibration constraints are sometimes more readily expressed in terms of the DIAC than the IAC.

duality: The property of two concepts or theories having similar properties that can be applied to the one or to the other. For instance, several relations linking points in a projective space are formally the same as those linking lines in a projective space; such relations are dual.

dynamic appearance model: A model describing the changing appearance of an object/scene over time.

dynamic programming: An approach to numerical optimization in which an optimal solution is searched by keeping several competing partial paths throughout and pruning alternative paths that reach the same point with a suboptimal value.

dynamic range: The ratio of the brightest and darkest values in an image. Most digital images have a dynamic range of around 100:1 but humans can perceive detail in dark regions when the range is even 10,000:1. To allow for this we can create high dynamic range images.

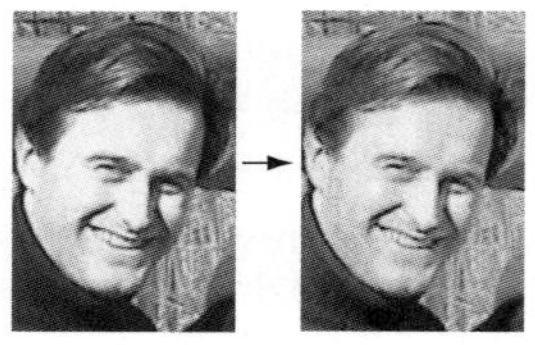

dynamic scene: A scene in which some objects move, in contrast to the common assumption in shape from motion that the scene is rigid and only the camera is moving.

dynamic stereo: Stereo vision for a moving observer. This allows shape from motion techniques to be used in addition to the stereo techniques.

dynamic time warping: A technique for matching a sequence of observations (usually one per time sample) to a model sequence of features, where the hope is for a one-to-one match of observations to features. But, because of variations in rate at which observations are produced, some features may get skipped or others matched to more than one observation. The usual goal is to minimize the amount of skipping or multiple samples matched (time warping). Efficient algorithms to solve this problem exist based on the linear ordering of the sequences. See also hidden Markov models (HMM).

E

early vision: A general term referring to the initial stages of computer vision (*i.e.*, image capture and image processing). Also known as low level vision.

earth mover's distance: A metric for comparing two distributions by evaluating the minimum cost of transforming one distribution into the other (*e.g.*, can be applied to color histogram matching).

Distribution 1 Distribution 2 Transformation

eccentricity: A shape representation that measures how non-circular a shape is. One way of computing this is to take the ratio of the maximum chord length of the shape to the maximum chord length of any orthogonal chord.

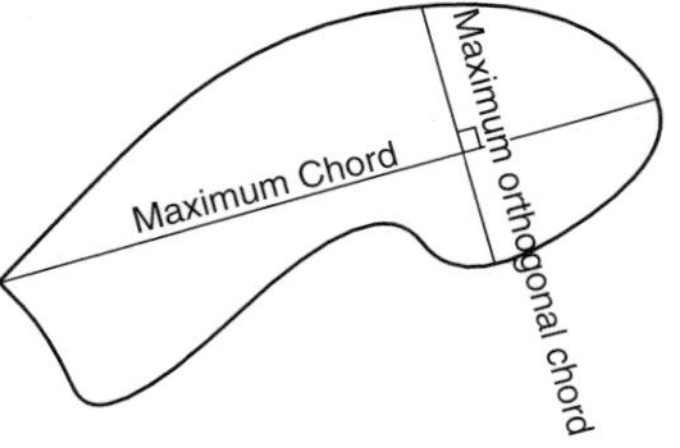

echocardiography: Cardiac ultrasonography (echocardiography) is a non-invasive technique for imaging the heart and surrounding structures. Generally used to evaluate cardiac chamber size, wall thickness, wall motion, valve configuration and motion and the proximal great vessels.

edge: A sharp variation of the intensity function. Represented by its position, the magnitude of the intensity gradient, and the direction of the maximum intensity variation.

edge based segmentation: Segmentation of an image based on the edges detected.

edge based stereo: A type of feature based stereo where the features used are edges.

edge detection: An image processing operation that computes edge vectors (gradient and orientation) for every point in an image. The first stage of edge based segmentation.

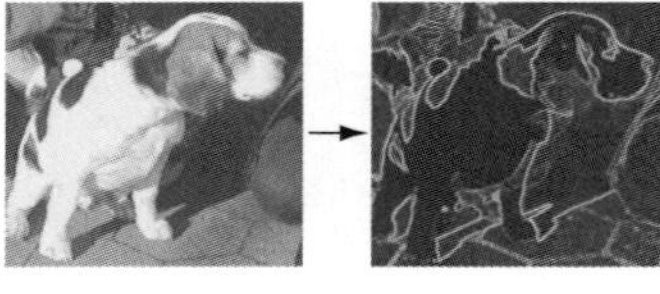

edge direction: The direction perpendicular to the normal to an edge, that is, the direction along the edge, parallel to the lines of constant intensity. Alternatively, the normal direction to the edge, *i.e.*, the direction of maximum intensity change (gradient). See also edge detection, edge point.

edge enhancement: An image enhancement operation that makes the gradient of edges steeper. This can be achieved, for example, by adding some multiple of a Laplacian convolved version of the image $L(i,j)$ to the image $g(i,j)$. $f(i,j) = g(i,j) + \lambda L(i,j)$ where $f(i,j)$ is the enhanced image and λ is some constant.

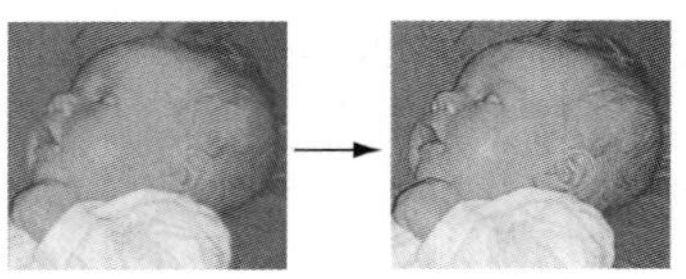

edge finding: See edge detection.

edge following: See edge tracking.

edge gradient image: See edge image.

edge grouping: See edge tracking.

edge image: An image where every pixel represents an edge or the edge magnitude.

edge linking: See edge tracking.

edge magnitude: A measure of the contrast at an edge, typically the magnitude of the intensity gradient at the edge point. See also edge detection, edge point.

edge matching: See curve matching.

edge motion: The motion of edges through a sequence of images. See also shape from motion and the aperture problem.

edge orientation: See edge direction.

edge point: 1) A location in an image where some quantity (*e.g.*, intensity) changes rapidly. 2) A location where the gradient is greater than some threshold.

edge preserving smoothing: A smoothing filter that is designed to preserve the edges in the image while reducing image noise. For example see median filter.

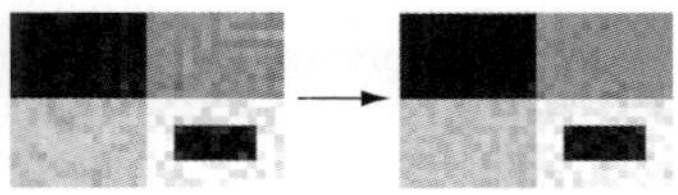

edge sharpening: See edge enhancement.

edge tracking: 1) The grouping of edges into chains of significant edges. The second stage of edge based segmentation. Also known as edge following, edge grouping and edge linking. 2) Tracking how the edge moves in a video sequence.

edge type labeling: Classification of edge points or edges into a limited number of types (*e.g.*, fold edge, shadow edge, occluding edge, etc.).

EGI: See extended Gaussian image.

egomotion: The motion of the observer with respect to the observed scene.

egomotion estimation: Determination of the motion of a camera. Generally based on image features corresponding to static objects in the scene. See also structure and motion. A typical image pair where the camera position is to be estimated is:

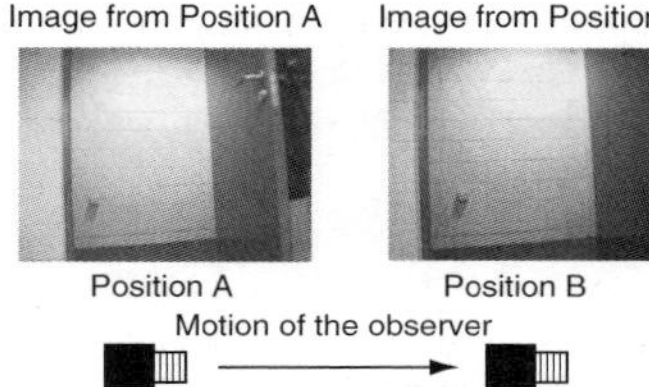

eigenface: An eigenvector determined from a matrix A in which the columns of A are images of faces. These vectors can be used for face recognition.

eigenspace based recognition: Recognition based on an eigenspace representation.

eigenspace representation: See principal component representation.

eigenvalue: A scalar λ that for a matrix A satisfies $Ax = \lambda x$ where x is a nonzero vector (eigenvector).

eigenvector: A non-zero vector x that for a matrix A satisfies $Ax = \lambda x$ where λ is a scalar (the eigenvalue).

eigenvector projection: Projection onto the PCA basis vectors.

electromagnetic spectrum: The entire range of frequencies of electromagnetic waves including X-rays, ultraviolet, visible light, infrared, microwave and radio waves.

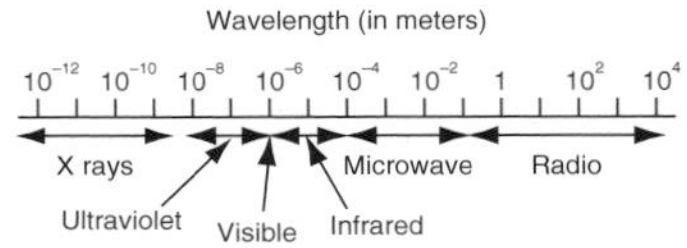

ellipse fitting: Fitting of an ellipse model to the boundary of some shape, data points, etc.

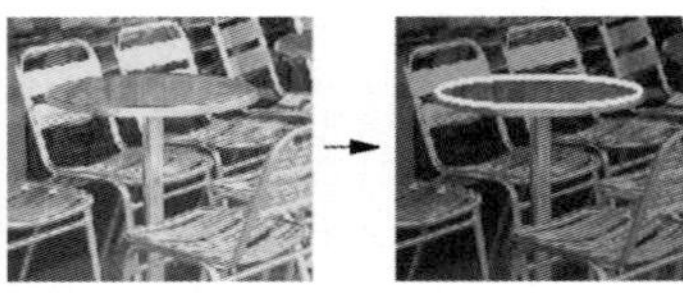

ellipsoid: A 3D volume in which all plane cross sections are ellipses or circles. An ellipsoid is the set of points (x, y, z) satisfying $\frac{x^2}{a^2} + \frac{y^2}{b^2} + \frac{z^2}{c^2} = 1$. Ellipsoids are used in computer vision as a basic shape primitive and can be combined with other primitives in order to describe a complex shape.

elliptic snake: An active contour model of an ellipse whose parameters are estimated through energy minimization from an initial position.

elongatedness: A shape representation that measures how long a shape is with respect to its width (*i.e.*, the ratio of the length of the bounding box to its width), as illustrated below. See also eccentricity.

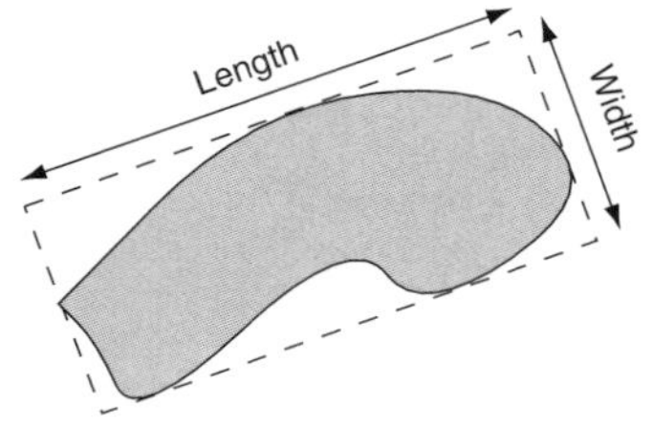

EM: See expectation maximization.

empirical evaluation: Evaluation of computer vision algorithms in order to characterize their performance by comparing the results of several algorithms on standardized test problems. Careful evaluation is a difficult research problem in its own right.

encoding: Converting a digital signal, represented as a set of values, from one form to another, often to compress the signal. In *lossy* encoding, information is lost in the process and the decoding algorithm cannot recover it. See also MPEG and JPEG image compression.

endoscope: An instrument for visually examining the interior of various bodily organs. See also fiberscope.

energy minimization: The problem of determining the absolute minimum of a multivariate function representing (by a potential energy-like penalty) the distance of a potential solution from the optimal solution. It is a specialization of the optimization problem. Two popular minimization algorithms in computer vision are the Levenberg–Marquardt and Newton optimization methods.

entropy: 1. Colloquially, the amount of disorder in a system. 2. A measure of the information content of a random variable X. Given that X has a set of possible values or outcomes $\mathbb{X}$, with probabilities $\{P(x), x \in \mathbb{X}\}$, the *entropy* $H(X)$ of X is defined as

$$\sum_{x \in \mathbb{X}} [-P(x) \log P(x)]$$

with the understanding that $0 \log 0 := 0$. For a multivariate distribution, the *joint entropy* $H(X, Y)$ of X, Y is

$$\sum_{(x,y) \in \mathbb{X} \times \mathbb{Y}} [-P(x, y) \log P(x, y)]$$

For a set of values represented as a histogram, the entropy of the set may be defined as the entropy of the probability distribution function represented by the histogram.

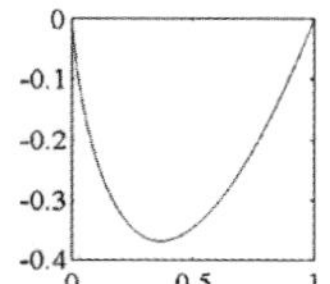

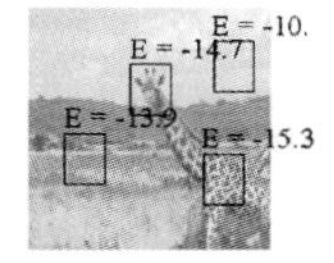

Left: $p \log p$ as a function of p. Probabilities near 0 and 1 signal high entropy, probabilities between are less entropic. Right: The entropy of the gray scale histograms in some windows on an image.

epipolar constraint: A geometric constraint reducing the dimensionality of the stereo correspondence problem. For any point in one image, the possible matching points in the other image are constrained to lie on a line known as the epipolar line. This constraint may be described mathematically using the fundamental matrix. See also epipolar geometry.

epipolar correspondence matching: Stereo matching using the epipolar constraint.

epipolar geometry: The geometric relationship between two perspective cameras.

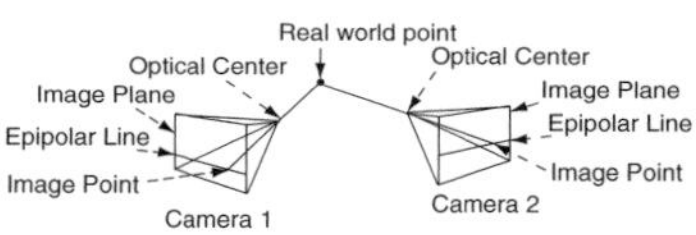

epipolar line: The intersection of the epipolar plane with the image plane. See also epipolar constraint.

epipolar plane: The plane defined by any real world scene point together with the optical centers of two cameras.

epipolar plane image (EPI): An image that shows how a particular line from a camera changes as the camera position is changed such that the image line remains on the same epipolar plane. Each line in the EPI is a copy of the relevant line from the camera at a different time. Features that are distant from the camera will remain in the same position in each line, and features that are close to the camera will move from line to line (the closer the feature the further it will move).

Image 1 Image 8

EPI from 8 images for highlighted line:

epipolar plane image analysis: An approach to determining shape from motion in which epipolar plane images (EPIs) are analyzed. The slope of lines in an EPI is proportional to the distance of the object from the camera, where vertical lines corresponding to features at infinity

epipolar plane motion: See epipolar plane image analysis.

epipolar rectification: The image rectification of stereo images so that the epipolar lines are aligned with the image rows (or columns).

epipolar transfer: The transfer of corresponding epipolar lines in a stereo pair of images, defined by a homography. See also stereo and stereo vision.

epipole: The point through which all epipolar lines from a camera appear to pass. See also epipolar geometry.

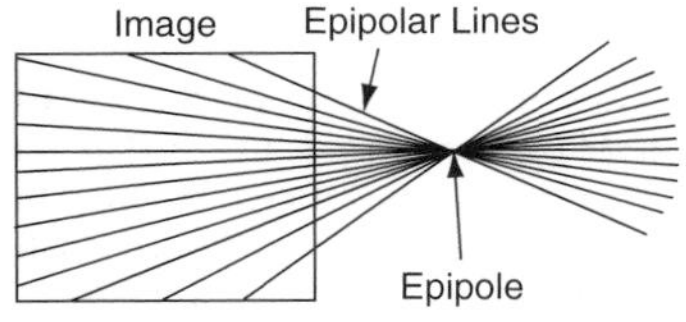

epipole location: The operation of locating the epipoles.

equalization: See histogram equalization.

erode operator: The operation of reducing a binary or gray scale object with respect to the background. This has the effect of removing any isolated object regions and separating any object regions that are only connected by a thin section. Most frequently described as a morphological transformation and is the dual of the dilate operator.

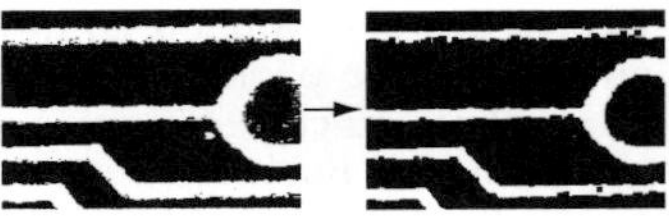

error propagation: 1) The propagation of errors resulting from one computation to the next computation. 2) The estimation of the error (*e.g.*, variance) of a process based on the estimates of the error in the input data and intermediate computations.

essential matrix: In binocular stereo, a matrix E expressing a bilinear constraint between corresponding image points u, u' in camera coordinates: $u'Eu = 0$. This constraint is the basis for several reconstruction algorithms. E is a function of the translation and rotation of the camera in the world reference frame. See also the fundamental matrix.

Euclidean distance: The geometric distance between two points (x_1, y_1) and (x_2, y_2), *i.e.*, $\sqrt{(x_1 - x_2)^2 + (y_1 - y_2)^2}$. For n-dimensional vectors $\vec{x}_1$ and $\vec{x}_2$, the distance is $(\sum_{i=1}^{n}(x_{1,i} - x_{2,i})^2)^{\frac{1}{2}}$.

Euclidean reconstruction: 3D reconstruction of a scene using a Euclidean frame of reference, as opposed to an affine reconstruction or projective reconstruction. The most complete reconstruction achievable. For example, using stereo vision.

Euclidean space: A representation of the space of all n-tuples (where n is the dimensionality). For example the three dimensional Euclidean space (X, Y, Z) is typically used to describe the real world. Also known as Cartesian space (see Cartesian coordinates).

Euclidean transformation: A transformation that operates in Euclidean space (*i.e.*, maintaining the Euclidean spatial arrangements). Examples include rotation and translation. Often applied to homogeneous coordinates.

Euler angle: The Euler angles (α, β, γ) are a particular set of angles describing rotations in three dimensional space.

Euler–Lagrange: The Euler–Lagrange equations are the basic equations in the calculus of variations, a branch of calculus concerned with maxima and minima of definite integrals. They occur, for instance, in Lagrangian mechanics and have been used in computer vision for a variety of optimizations, including for surface interpolation. See also variational approach and variational problem.

Euler number: The number of contiguous parts (regions) less the number of holes. Also known as the genus.

even field: The first of the two fields in an interlaced video signal.

even function: A function where $f(x) = f(-x)$ for all x.

event analysis: See event understanding.

event detection: Analysis of a sequence of images to detect activities in the scene.

event understanding: Recognition of an event (such as a person walking) in a sequence of images. Based on the data provided by event detection.

exhaustive matching: Matching where all possibilities are considered. As an alternative see hypothesize and verify.

expectation maximization (EM): A method of finding a maximum likelihood estimate of some parameters based on a sample data set. This method works well even when there are missing values.

expectation value: The mean value of a function (*i.e.*, the average expected value). If $p(x)$ is the probability density function of a random variable x, the expectation of x is $\bar{x} = \int p(x)x\,\mathrm{d}x$.

expert system: A system that uses available knowledge and heuristics to solve problems. See also knowledge based vision.

exponential smoothing: A method for predicting a data value (P_{t+1}) based on the previous observed value (D_t) and the previous prediction (P_t). $P_{t+1} = \alpha D_t + (1-\alpha)P_t$ where α is a weighting value between 0 and 1.

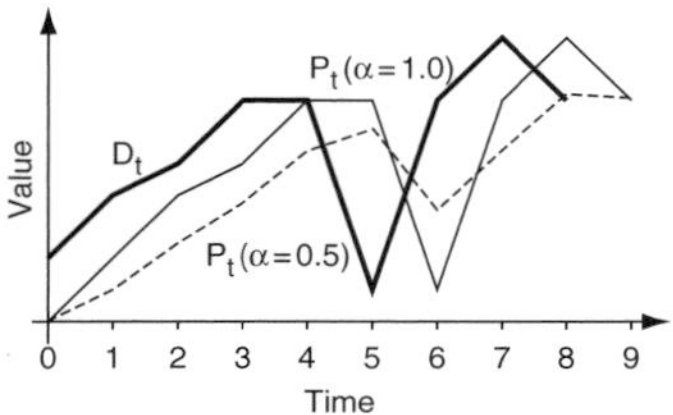

exponential transformation: See pixel exponential operator.

expression understanding: See facial expression analysis.

extended Gaussian image (EGI): Use of a Gaussian sphere for histogramming surface normals. Each surface normal is considered from the center of the sphere and the value associated with the surface patch with which it intersects is incremented.

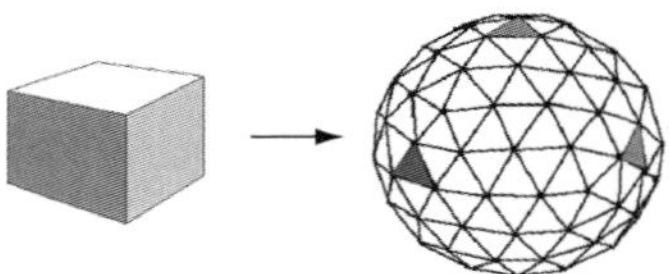

extended light source: A light source that has a significant size relative to the scene, *i.e.*, is not approximated well by a point light source. In other words this type of light source has a diameter and hence can produce fuzzy shadows. Contrast with: point light sources.

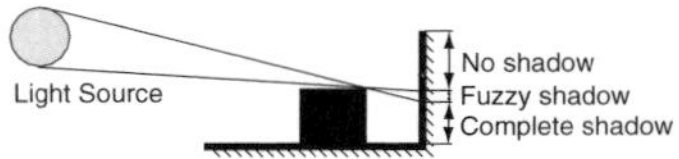

exterior orientation: The position of a camera in a global coordinate system. That which is determined by an absolute orientation calculation.

external energy (or force): A measure of fit between the image data and an active shape model that is part of the model's deformation energy. This measure is used to deform the model to the image data.

extremal point: Points that lie on the boundary of the smallest convex region enclosing a set of points (*i.e.*, that lie on the convex hull).

extrinsic parameters: See exterior orientation.

eye location: The task of finding eyes in images of faces. Approaches include blink detection, face feature detection, etc.

eye tracking: Tracking the position of the eyes in a face image sequence. Also, tracking the gaze direction.

F

face analysis: A general term covering the analysis of face images and models. Often used to refer to facial expression analysis.

face authentication: Verification that (the image of) a face corresponds to a particular individual. This differs from the face recognition in that here only the model of a single person is considered.

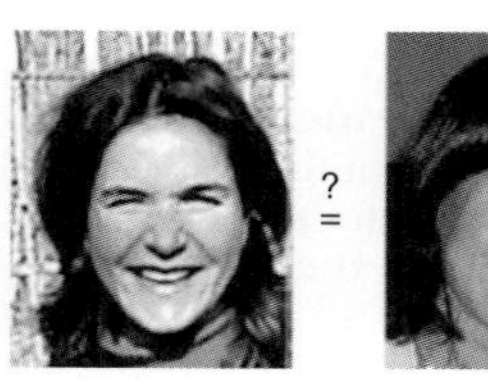

face detection: Identification of faces within an image or series of images. This often involved a combination of human motion analysis and skin color analysis.

face feature detection: The location of features (such as eyes, nose, mouth) from a human face. Normally performed after face detection although it can be used as part of face detection.

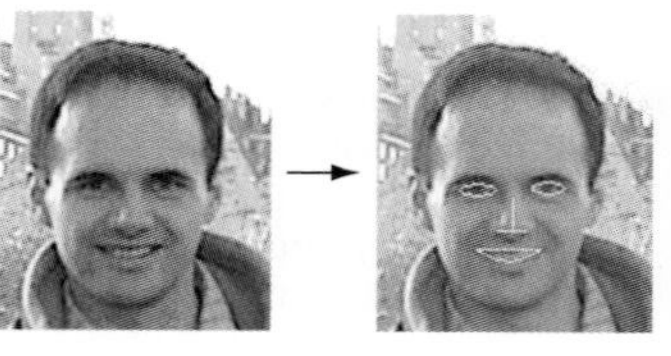

face identification: See face recognition.

face indexing: Indexing from a database of known faces as a precursor to face recognition.

face modeling: Representing a face using some type of model typically derived from an image (or images). These models are

Dictionary of Computer Vision and Image Processing R.B. Fisher, K. Dawson-Howe, A. Fitzgibbon, C. Robertson and E. Trucco

used in face authentication, face recognition, etc.

face recognition: The task of recognizing a face from an image as an instance of a person recorded in a database of faces.

face tracking: Tracking of a face in a sequence of images. Often used as part of a human–computer interface.

face verification: See face authentication.

facet model based extraction: The extraction of a model based on facets (small simple surfaces; *e.g.*, see planar facet model) from range data. See also planar patch extraction.

facial animation: The way in which facial expressions change. See also face feature detection.

facial expression analysis: Study or identification of the facial expression(s) of a person from an image or sequence of images.

factorization: See motion factorization.

false alarm: See false positive.

false negative: A binary classifier $c(x)$ returns + or – for examples x. A *false negative* occurs when the classifier returns –for an example that is in reality +.

false positive: A binary classifier $c(x)$ returns + or – for examples x. A *false positive* occurs when the classifier returns + for an example that is in reality –.

fast Fourier transform (FFT): A version of the Fourier transform for discrete samples that is significantly more efficient (order $N\log_2 N$) than the standard discrete Fourier transform (which is order N^2) on data sets with N points.

fast marching method: A type of level set method in which the search can move in only one direction (hence making it faster).

feature: 1) A distinctive part of something (*e.g.*, the nose and eyes are distinctive features of the face), or an attribute derived from an object/shape (*e.g.*, circularity). See also image feature. 2) A numerical property (possibly combined with others to form a feature vector) and generally used in a classifier.

feature based optical flow estimation: Calculation of optical flow in a sequence of images from image features.

feature based stereo: A solution to the stereo correspondence problem in which image features are compared from the two images. The main alternative approach is correlation based stereo.

feature based tracking: Tracking the motion of image features through a sequence.

feature contrast: The difference between two features. This can be measured in many domains (*e.g.*, intensity, orientation, etc.).

feature detection: Identification of given features in an image (or model). For example see corner detection.

feature extraction: See feature detection.

feature location: See feature detection.

feature matching: Matching of image features in several images of the same object (for instance, feature based stereo), or of features from an unknown object with features from known objects (feature based recognition).

feature orientation: The orientation of an image feature with respect to the image frame of reference.

feature point: The image location at which a particular feature is found.

feature point correspondence: Matching feature points in two or more images. The assumption is that the feature points are the image of the same scene point. Having the correspondence allows the estimation of the depth from binocular stereo, fundamental matrix, homography or trifocal tensor in the case of 3D scene structure recovery or of the 3D target motion in the case of target tracking.

feature point tracking: Tracking of individual image features in a sequence of images.

feature selection: Selection of suitable features (properties) for a specific task, for example, classification. Typically features should be independent, detectable, discriminatory and reliable.

feature similarity: How much two features resemble each other. Measures of feature similarity are required for feature based stereo, feature based tracking, feature matching, etc.

feature space: The dimensions of a feature space are the feature (property) values of a given problem. An object or shape is mapped to feature space by computing the values of the set of features defining the space, typically for recognition and classification.

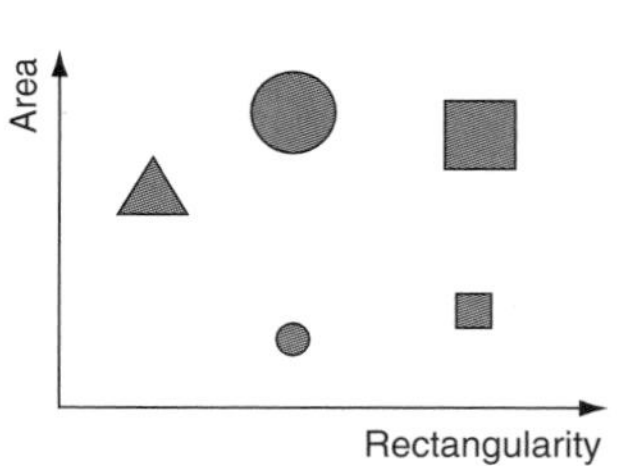

In the example above, different shapes are mapped to a 2D feature space defined by area and rectangularity.

feature stabilization: A technique for stabilizing the position of an image feature in an image sequence so that it remains in a particular position on a display (allowing/causing the rest of the image to move relative to that feature).

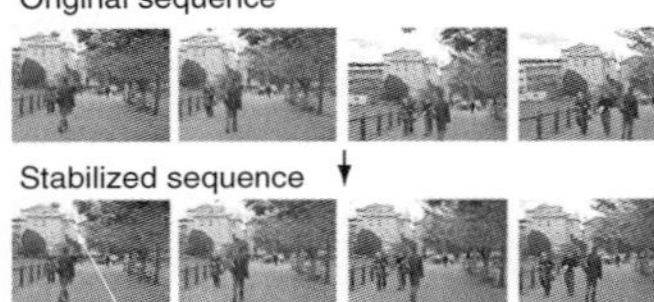

feature tracking: See feature based tracking.

feature vector: A vector formed by the values of a number of image features (properties), typically all associated with the same object or image.

feedback: The use of outputs from a system to control the system's actions.

Feret's diameter: The distance between two parallel lines at the extremities of some shape that are tangential to the boundary of the shape. Maximum, minimum and mean values of Feret's diameter are often used (where every possible pair of parallel tangent lines is considered).

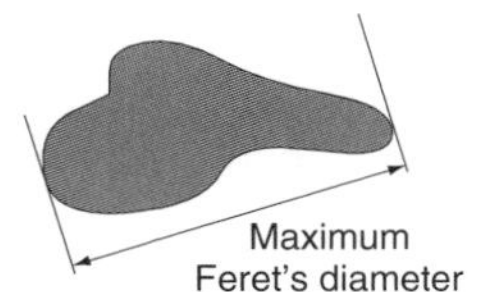

FERET: A standard database of face images with a defined experimental protocol for the testing and comparison of face recognition algorithms.

FFT: See fast Fourier transform.

fiber optics: A medium for transmitting light that consists of very thin glass or plastic fibers. It can be used to provide much higher bandwidth for signals encoded as patterns of light pulses. Alternately, it can be used to transmit images directly through rigidly connected bundles of fibers, so as to see around corners, past obstacles, etc.

fiberscope: A flexible fiber optic instrument allowing parts of an object to be viewed that would normally be inaccessible. Most often used in medical examinations.

fiducial point: A reference point for a given algorithm, *e.g.*, a fixed, known, easily detectable pattern for a calibration algorithm.

figure–ground separation: The segmentation of the area

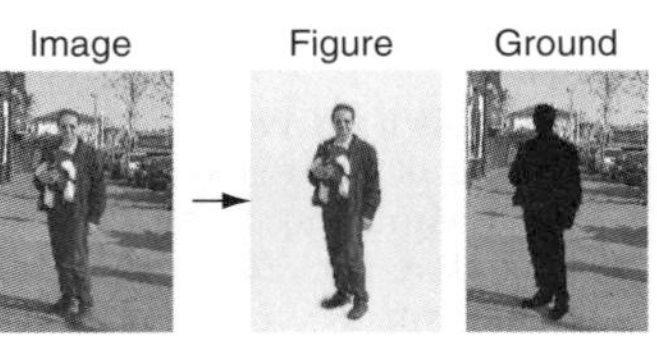

of the image representing the object of interest (the figure) from the remainder of the image (the background).

figure of merit: Any scalar that is used to characterize the performance of an algorithm.

filter: In general, any algorithm that transforms a signal into another. For instance, bandpass filters remove/reduce the parts of an input signal outside a given frequency interval; gradient filters allow only image gradients to pass through; smoothing filters attenuate high frequencies.

filter ringing: A type of distortion caused by the application of a steep recursive filter. Normally this term applies to electronic filters in which certain components (*e.g.*, capacitors and inductors) can store energy and later release it, but there are also digital equivalents to this effect.

filtering: Application of a filter.

fingerprint database indexing: Indexing into a database of fingerprints using a number of features derived from the fingerprints. This allows a smaller number of fingerprints to be considered when attempting fingerprint identification within the database.

fingerprint identification: Identification of an individual through comparison of an unknown fingerprint (or fingerprints) with previously known fingerprints.

fingerprint indexing: See fingerprint database indexing.

finite element model: A class of numerical methods for solving differential problems. Another relevant class is finite difference methods.

finite impulse response filter (FIR): A filter that produces an output value (y_n) based on the current and past input values (x_i). $y_n = \sum_{i=0}^{p} a_i x_{n-i}$ where a_i are weights. See also infinite impulse response filters.

FIR: See finite impulse response filter.

Firewire (IEEE 1394): A serial digital bus system supporting 400 Mbits per second. Power, control and data signals are carried in a single cable. The bus system makes it possible to address up to 64 cameras from a single interface card and multiple computers can acquire images from the same camera simultaneously.

first derivative filter: See gradient filter.

first fundamental form: See surface curvature.

Fisher linear discriminant (FLD): A classification method that maps high dimensional data into a single dimension in such a way as to maximize class separability.

fisheye lens: See wide angle lens.

flat field: 1) An object of uniform color, used for photometric calibration of optical systems. 2) A camera system is *flat field correct* if the gray scale output at each pixel is the same for a given light input.

flexible template: A model of a shape in which the relative position of points is not fixed (*e.g.*, defined in probabilistic form). This approach allows for variations in the appearance of the shape.

FLIR: Forward Looking Infrared. An infrared system mounted on a vehicle looking ahead along the direction of travel.

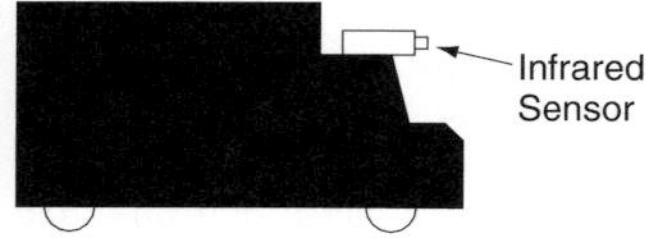

flow field: See optical flow field.

flow histogram: A histogram of the optical flow in an image sequence. This can be used, for example, to provide a qualitative description of the motion of the observer.

flow vector field: Optical flow is described by a vector (magnitude and orientation) for each image point. Hence a flow vector field is the same as an optical flow field.

fluorescence: The emission of visible light by a substance caused by the absorption of some other (possibly invisible) electromagnetic wavelength. This property is sometimes used in industrial machine vision.

fMRI: Functional Magnetic Resonance Imaging, or fMRI, is a technique for identifying which parts of the brain are activated by different types of physical stimulation, *e.g.*, visual or acoustic stimuli. A MRI scanner is set up to register the increased blood flow to the activated areas of the brain on Functional MRI scans. See also nuclear magnetic resonance.

FOA: See focus of attention.

FOC: See focus of contraction.

focal length: 1) The distance between the camera lens and the focal plane. 2) The distance from a lens at which an object viewed at infinity would be in focus.

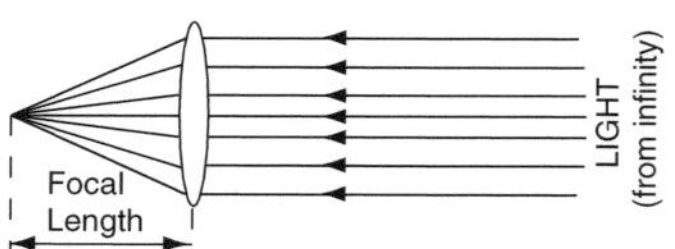

focal point: The point on the optical axis of a lens where light rays from an object at infinity (also placed on the optical axis) converge.

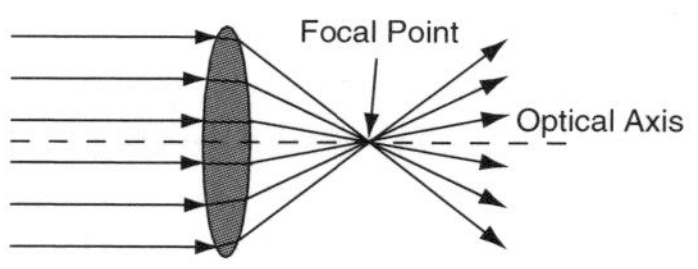

focal plane: The plane on which an image is focused by a lens system. Generally this consists of an array of photosensitive elements. See also image plane.

focal surface: A term most frequently used when a concave mirror is used to focus an image (*e.g.*, in a reflector telescope). The focal surface in this case is the surface of the mirror.

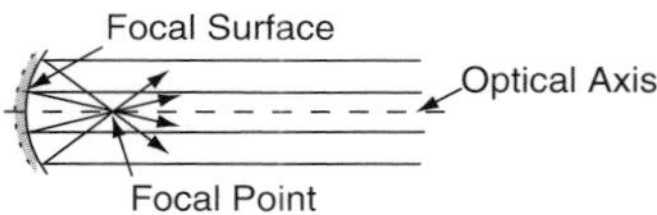

focus: To focus a camera is to arrange for the focal points of various image features to converge on the focal plane. An image is considered to be in focus if the main subject of interest is in focus. Note that focus (or lack of focus) can be used to derive useful information (*e.g.*, see depth from focus).

In focus Out of focus

focus control: The control of the focus of a lens system usually by moving the lens along the optical axis or by adjusting the focal length. See also autofocus.

focus following: A technique for slowly changing the focus of a camera as an object of interest moves. See also depth from focus.

focus invariant imaging: Imaging systems that are designed to be invariant to focus. Such systems have large depths of field.

focus of attention (FOA): The feature or object or area to which the attention of a visual system is directed.

focus of contraction (FOC): The point of convergence of the optical flow vectors for a translating camera. The component of the translation along the optical axis must be nonzero. Compare focus of expansion.

focus of expansion (FOE): The point from which all optical flow vectors appear to emanate in a static scene where the observer is moving. For example if a camera system was moving directly forwards along the optical axis then the optical flow vectors would all emanate from the principal point (usually near the center of the image).

FOE: See focus of expansion.

fold edge: A surface orientation discontinuity. An edge where two locally planar surfaces meet. The figure below shows a fold edge.

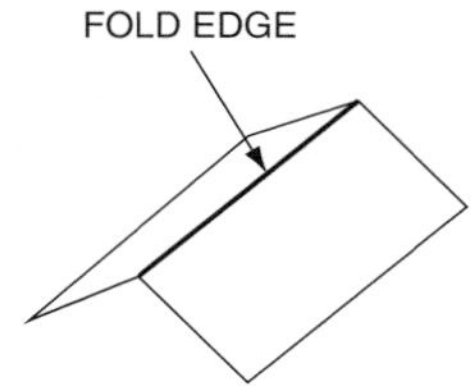

foreground: In computer vision, generally used in the context of object recognition. The area of the scene or image in which the object of interest lies. See figure–ground separation.

foreshortening: A typical perspective effect whereby distant objects appear smaller than closer ones.

form factor: The physical size or arrangement of an object. This term is frequently used with reference to computer boards.

Förstner operator: A feature detector used for corner detection as well as other edge features.

forward looking radar: A radar system mounted on a vehicle looking ahead along the direction of travel. See also side looking radar.

Fourier–Bessel transform: See Hankel transform.

Fourier domain convolution: Convolution in the Fourier domain involves simply multiplication of the Fourier transformed image by the Fourier transformed filter. For very large filters this operation is much more efficient than convolution in the original domain.

Fourier domain inspection: Identification of defects based on features in the Fourier transform of an image.

Fourier image processing: Image processing in the Fourier domain (*i.e.*, processing images that have been transformed using the Fourier transform).

Fourier matched filter object recognition: Object recognition in which correlation is determined using a matched filter that is the conjugate of the Fourier transform of the object being located.

Fourier shape descriptor: A boundary representation of a shape in terms of the coefficients of a Fourier transformation.

Fourier slice theorem: A slice at an angle θ of a 2D Fourier transform of an object is equal to a 1D Fourier transform of a parallel projection of the object taken at the same angle. See also slice based reconstruction.

Fourier space: The frequency domain space in which an image (or other signal) is represented after application of the Fourier transform.

Fourier space smoothing: Application of a smoothing filter (*e.g.*, to remove high-frequency noise) in a Fourier transformed image.

Fourier transform: A transformation that allows a signal to be considered in the frequency domain as a sum of sine and cosine waves or equivalently as a sum of exponentials. For a two dimensional image $F(u, v) = \int_{-\infty}^{\infty}\int_{-\infty}^{\infty} f(x, y)e^{-2\pi i(xu+yv)}\,dx\,dy$. See also fast Fourier transform, discrete Fourier transform and inverse Fourier transform.

fovea: The high-resolution central region of the human retina. The analogous region in an artificial sensor that emulates the retinal arrangement of photoreceptor, for example a log-polar sensor.

foveal image: An image in which the sampled pattern is inspired by the arrangement of the human fovea, *i.e.*, sampling is most dense in the image center and gets progressively sparser towards the periphery of the image.

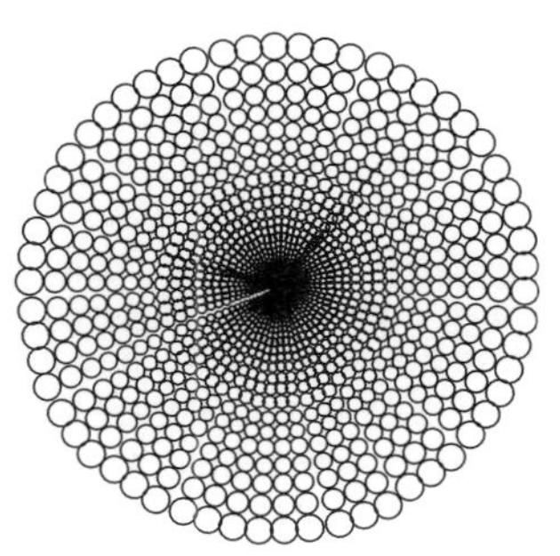

foveation: 1) The process of creating a foveal image. 2) Directing the camera optical axis to a given direction.

fractal image compression: An image compression method based on exploiting self-similarity at different scales.

fractal measure/dimension: A measure of the roughness of a shape. Consider a curve whose length (L_1 and L_2) is measured at two scales (S_1 and S_2). If the curve is rough the length will grow as the scale is increased. The fractal dimension is $D = \frac{log(L_1 - L_2)}{log(S_2 - S_1)}$.

fractal representation: A representation based on self-similarity. For example a fractal representation of an image could be based on similarity of blocks of pixels.

fractal surface: A surface model that is defined progressively using fractals (*i.e.*, the surface displays self-similarity at different scales).

fractal texture: A texture representation based on self-similarity between scales.

frame: 1) A complete standard television video image consisting of both the even and odd video fields. 2) A knowledge representation technique suitable for recording a related set of facts, rules of inference, preconditions, etc.

frame buffer: A device that stores a video frame for access, display and processing by a computer. For example such devices are used to store the frame from which a video display is refreshed. See also frame store.

frame grabber: See frame store.

frame of reference: A coordinate system defined with respect to some object, the camera or with respect to the real world.

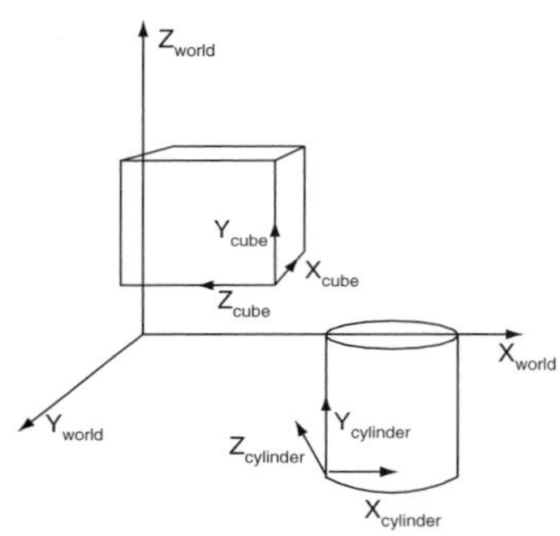

frame store: An electronic device for recording a frame from an imaging system. Typically such devices are used as interfaces between CCIR cameras and computers.

freeform surface: A surface that does not follow any particular mathematical form; for example, the folds of a piece of fabric, as shown below.

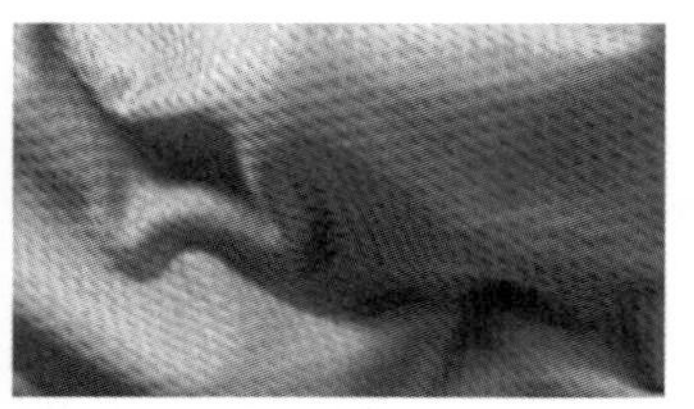

Freeman code: A type of chain code in which a contour is represented by coordinates for the first point followed by a series of direction codes (typically 0 through 7). In the following figure we show the Freeman codes relative to the center point on the left and an example of the codes derived from a chain of points on the right.

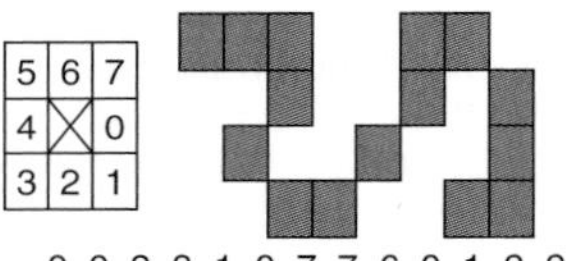

Frenet frame: A triplet of mutually orthogonal unit vectors (the normal, the tangent and the binormal/bitangent) describing a point on a curve.

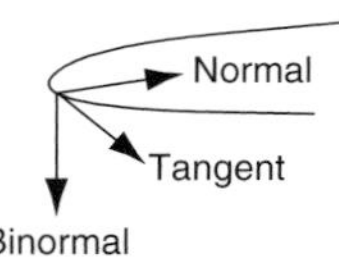

frequency domain filter: A filter defined by its action in the Fourier space. See high pass filter and low pass filter.

frequency spectrum: The range of (electromagnetic) frequencies.

front lighting: A general term covering methods of lighting a scene where the lights are on the same side of the object as the camera. As an alternative consider backlighting. For example,

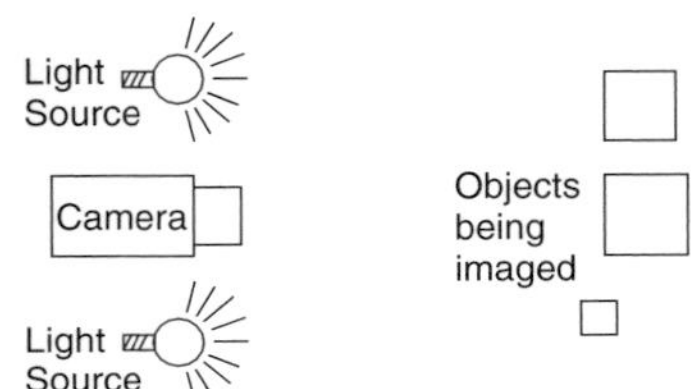

frontal: Frontal presentation of a planar surface is one in which the plane is parallel to the image plane.

full primal sketch: A representation described as part of Marr's theory of vision, that is made up of the raw primal sketch primitives together with grouping information. The sketch contains described image structures that could correspond with scene structures (*e.g.*, image regions with scene surfaces).

function based model: An object representation based on the object functionality (*e.g.*, an object's purpose or the way in which an object moves and interacts with other objects) rather than its geometric properties.

function based recognition: Object recognition based on object functionality rather than geometric properties. See also function based model.

functional optimization: An analytical technique for optimizing (maximizing or minimizing) complex functions of continuous variables.

functional representation: See function based model.

fundamental form: A metric that useful in determining local properties of surfaces. See also first fundamental form and second fundamental form.

fundamental matrix: A bilinear relationship between corresponding points (u, u') in binocular stereo images. The fundamental matrix, $\mathbf{F}$, incorporates the two sets of camera parameters $(\mathbf{K}, \mathbf{K}')$ and the relative position $(\vec{t})$ and orientation $(\mathbf{R})$ of the cameras. Matching points $\vec{u}$ from one image and $\vec{u}'$ from the other image satisfy $\vec{u}^T \mathbf{F} \vec{u}' = 0$ where $\mathbf{S}(\vec{t})$ is the skew symmetric matrix of $\vec{t}$ and $\mathbf{F} = (\mathbf{K}^{-1})^T \mathbf{S}(\vec{t}) \mathbf{R}^{-1} (\mathbf{K}')^{-1}$. See also the essential matrix.

fusion: Integration of data from multiple sources into a single representation.

fuzzy logic: A form of logic that allows a range of possibilities between true and false (*i.e.*, a degree of truth).

fuzzy morphology: A type of mathematical morphology that is based on fuzzy logic rather than the more conventional Boolean logic.

fuzzy set: A grouping of data (into a set) where each item in the set has an associated grade/likelihood of membership in the set.

fuzzy reasoning: See fuzzy logic.

G

Gabor filter: A filter formed by multiplying a complex oscillation by an elliptical Gaussian distribution (specified by two standard deviations and an orientation). This creates filters that are local, selective for orientation, have different scales and are tuned for intensity patterns (*e.g.*, edges, bars and other patterns observed to trigger responses in the simple cells of the mammalian visual cortex) according to the frequency chosen for the complex oscillation. The filter can be applied in the frequency domain as well as the spatial domain.

Gabor transform: A transformation that allows a 1D or 2D signal (such as an image) to be represented as a weighted sum of Gabor functions.

Gabor wavelets: A type of wavelet formed by a sinusoidal function that is restricted by a Gaussian envelope function.

gaging: Measuring or testing. A standard requirement of industrial machine vision systems.

gait analysis: Analysis of the way in which human subjects move. Frequently used for biometric or medical purposes.

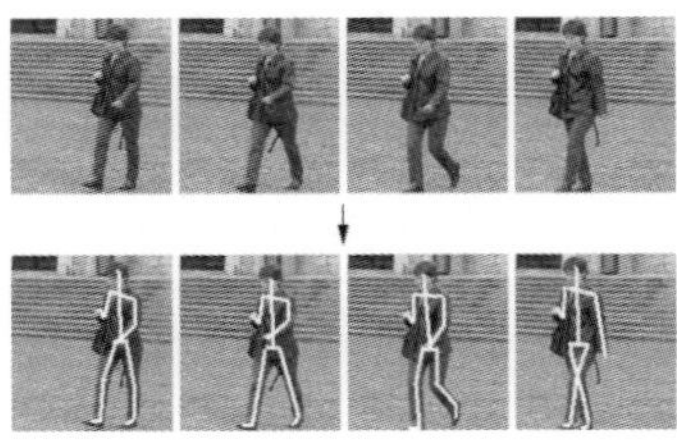

gait classification: 1) Classification of different types of human motion (such as walking, running, etc.). 2) Biometric identification of people based on their gait parameters.

Galerkin approximation: A method for determining the coefficients of a power series solution for a differential equation.

gamma: Devices such as cameras and displays that

convert between analogue (denoted a) and digital (d) images generally have a nonlinear relationship between a and d. A common model for this nonlinearity is that the signals are related by a gamma curve of the form $a = c \times d^{\gamma}$, for some constant c. For CRT displays, common values of γ are in the range 1.0–2.5.

gamma correction: The correction of brightness and color ratios so that an image has the correct dynamic range when displayed on a monitor.

gauge coordinates: A coordinate system local to the image surface itself. Gauge coordinates provide a convenient frame of reference for operators such as the gradient operator.

Gaussian convolution: See Gaussian smoothing.

Gaussian curvature: A measure of the surface curvature at a point. It is the product of the maximum and minimum of the normal curvatures in all directions through the point. See also mean curvature.

Gaussian derivative: The combination of Gaussian smoothing and a gradient filter. This results in a gradient filter that is less sensitive to noise.

Gaussian distribution: A probability density function with this distribution:

$$P(x) = \frac{1}{\sigma\sqrt{2\pi}} e^{-\frac{(x-\mu)^2}{2\sigma^2}}$$

where μ is the mean and σ is the standard deviation. If $\vec{x} \in \Re^d$, then the multivariate probability density function is $p(\vec{x}) = det(2\pi\Sigma)^{-\frac{1}{2}} exp(-\frac{1}{2}(\vec{x} - \vec{\mu})^{\top}\Sigma^{-1}(\vec{x} - \vec{\mu}))$ where $\vec{\mu}$ is the distribution mean and Σ is its covariance.

Gaussian mixture model: A representation for a distribution based on a combination of Gaussians. For instance, used to represent color histograms with multiple peaks. See expectation maximization.

Gaussian noise: Noise whose distribution is Gaussian in nature. Gaussian noise is specified by its standard deviation about a zero mean, and is often modeled as a form of additive noise.

Gaussian pyramid: A multi-resolution representation of an image formed by several images, each one a Gaussian smoothed and subsampled version of the original one at increasing standard deviation.

Gaussian smoothing: An image processing operation aimed to attenuate image noise computed by convolution with a mask sampling a Gaussian distribution.

Gaussian speckle: Speckle that has a Gaussian distribution.

Gaussian sphere: A sampled representation of a unit sphere where the surface of the sphere is defined by a number of triangular patches (often computed by dividing a dodecahedron). See also extended Gaussian image.

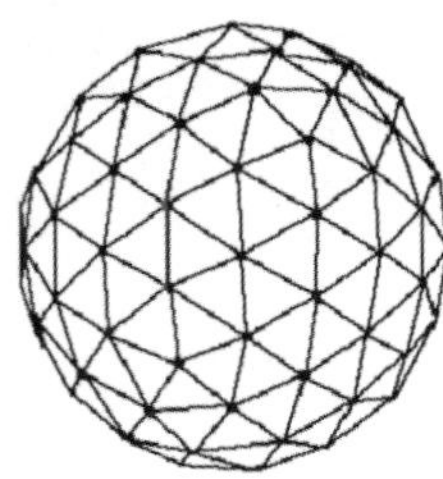

gaze control: The ability of a human subject or a robot head to control their gaze direction.

gaze direction estimation: Estimation of the direction in which a human subject is looking. Used for human–computer interaction.

gaze direction tracking: Continuous gaze direction estimation (*e.g.*, in a video sequence or a live camera feed).

gaze location: See gaze direction estimation.

generalized cone: A generalized cylinder in which the swept curve changes along the axis.

generalized curve finding: A general term referring to methods that locate arbitrary curves. For example, see generalized Hough transform.

generalized cylinder: A volumetric representation where the volume is defined by sweeping a closed curve along an axis. The axis does not need to be straight and the closed curve may vary in shape as it is moved along the axis. For example a cylinder may

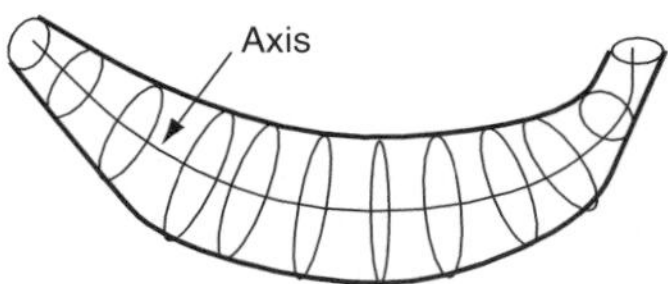

be defined by moving a circle along a straight axis, and a cone may be defined by moving a circle of changing diameter along a straight axis.

generalized Hough transform: A version of the Hough transform capable of detecting the presence of arbitrary shapes.

generalized order statistics filter: A filter in which the values within the filter mask are considered in increasing order and then combined in some fashion. The most common such filter is the median filter that selects the middle value.

generate and test: See hypothesize and verify.

generic viewpoint: A viewpoint such that small motions may cause small changes in the size or relative positions of features, but no features appear or disappear. This contrasts with a privileged viewpoint.

genetic algorithm: An optimization algorithm seeking solutions by refining iteratively a small set of candidates with a process mimicking genetic evolution. The suitability (fitness) of a set of possible solutions (population) is used to generate a new population until some conditions are satisfied (*e.g.*, the best solution has not changed for a given number of iterations).

genetic programming: Application of genetic algorithms in some programming language to evolve programs that satisfy some evaluation criteria.

genus: In the study of topology, the number of "holes" in a surface. In computer vision, sometimes used as a discriminating feature for simple object recognition.

Gestalt: German for "shape". The Gestalt school of psychology, led by the German psychologists Wertheimer, Köhler and Koffka in the first half of the twentieth century, had a profound influence on perception theories, and subsequently on computer vision. Its basic tenet was that a perceptual pattern has properties as a whole, which cannot be explained in terms of its individual components. In other words, the whole is more than the sum of its parts. This concept was captured in some basic laws (proximity, similarity, closure, "common destiny" or good form, saliency), that would apply to all mental phenomena, not just perception. Much work on low-level computer vision, most notably on perceptual grouping and perceptual organization, has exploited these ideas. See also visual illusion.

geodesic: The shortest line between two points (on a mathematically defined surface).

geodesic active contour: An active contour model similar to the snake model in that it attempts to minimize an energy function between the model and the data, but which also incorporates a geometrical model.

geodesic active region: A technique for region based segmentation that builds on geodesic active contours by adding a force that takes into account information within regions. Typically a geodesic active region will be bounded by a single geodesic active contour.

geodesic distance: The length of the shortest path between two points along some surface. This is different from the Euclidean distance that takes no account of the surface. The following example shows the geodesic distance between Calgary and London (following the curvature of the Earth).

geodesic transform: Assigns to each point the geodesic distance to some feature or class of feature.

geographic information system (GIS): A computer system that stores and manipulates geographically referenced data (such as images of portions of the Earth taken by satellite).

geometric compression: The compression of geometric structures such as polygons.

geometric constraint: A limitation on the possible physical arrangement/appearance of objects based on geometry. These types of constraints are used extensively in stereo vision (*e.g.*, the epipolar constraint), motion analysis (*e.g.*, rigid motion constraint) and object recognition (*e.g.*, focusing on specific classes of objects or relations between features).

geometric correction: In remote sensing, an algorithm or technique for correction of geometric distortion.

geometric deformable model: A deformable model in which the deformation of curves is based on the level set method and stops at object boundaries. A typical example is a geodesic active contour model.

geometric distance: In curve and surface fitting, the shortest distance from a given point to a given surface. In many fitting problems, the geometric distance is expensive to compute but yields more accurate solutions. Compare algebraic distance.

geometric distortion: Deviations from the idealized

image formation model (for example, pinhole camera) of an imaging system. Examples include radial lens distortion in standard cameras.

geometric feature: A general term describing a shape characteristic of some data, that encompasses features such as edges, corners, geons, etc.

geometric feature learning: Learning geometric features from examples of the feature.

geometric feature proximity: A measure of the distance between geometric features, *e.g.*, as by using the distance between data and overlaid model features in hypothesis verification.

geometric hashing: A technique for matching models in which some geometric invariant features are mapped into a hash table, and this hash table is used to perform the recognition.

geometric invariant: A quantity describing some geometric configuration that remains unchanged under certain transformations (*e.g.*, cross-ratio, perspective projection).

geometric model: A model that describes the geometric shape of some object or scene. A model can be 2D (*e.g.*, polycurve) or 3D (*e.g.*, surface based models), etc.

geometric model matching: Comparison of two geometric models or of a model and a set of image data shapes, for the purposes of recognition.

geometric optics: A general term referring to the description of optics from a geometrical point of view. Includes concepts such as the simple pinhole camera model, magnification, lenses, etc.

geometric reasoning: Reasoning with geometric shapes in order to address such tasks as robot motion planning, shape similarity, spatial position estimation, etc.

geometric representation: See geometric model.

geometric shape: A shape that takes a relatively simple geometric form (such as a square, ellipse, cube, sphere, generalized cylinder, etc.) or that can be described as a combination of such geometric primitives.

geometric transformation: A class of image processing operations that transform the spatial relationships in an image. They are used for the correction of geometric distortions and general image manipulation. A geometric transformation requires the definition of a pixel coordinate transformation together with an interpolation scheme. For example, a rotation does:

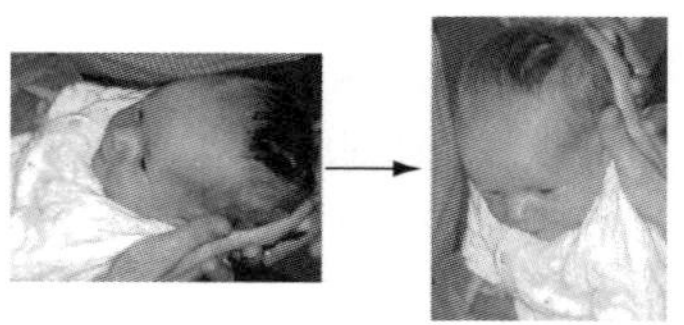

geon: GEometrical iON. A basic volumetric primitive proposed by Biederman and used in recognition by components. Some example geons are:

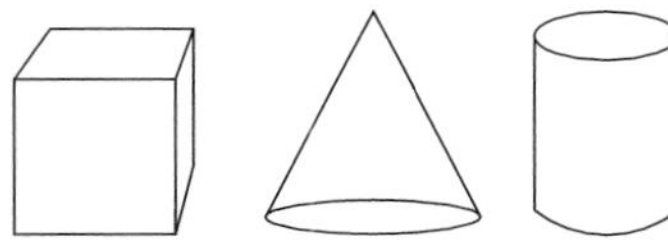

gesture analysis: Basic analysis of video data representing human gestures preceding the task of gesture recognition.

gesture recognition: The recognition of human gestures generally for the purpose of human–computer interaction. See also hand sign recognition.

Gibbs sampling: A method for probabilistic inference based on transition probabilities (between states).

GIF: Graphics Interchange Format. A common compressed image format based on the Lempel–Ziv–Welch algorithm.

GIS: See geographic information system.

glint: A specular reflection visible on a mirror-like surface.

global: A *global* property of a mathematical object is one that depends on all components of the object. For example, the average intensity of an image is a global property, as it depends on all the image pixels.

global positioning system (GPS): A system of satellites that allow the position of a *GPS receiver* to be determined in absolute Earth-referenced coordinates. Accuracy of standard civilian GPS is of the order of meters. Greater accuracy is obtainable using differential GPS.

global structure extraction: Identification of high level structures/relationships in an image (*e.g.*, symmetry detection).

global transform: A general term describing an operator that transforms an image into some other space. Sample global transforms include the discrete cosine transform, the Fourier transform, the Haar transform, the Hadamard transform, the Hartley transform, histograms, the Hough transform, the Karhunen–Loeve transform, the Radon transform, and the wavelet transform.

golden template: An image of an unflawed object/scene that is used within template matching to identify any deviations from the ideal object/scene.

gradient: Rate of change. This is frequently associated with edge detection. See also gray scale gradient.

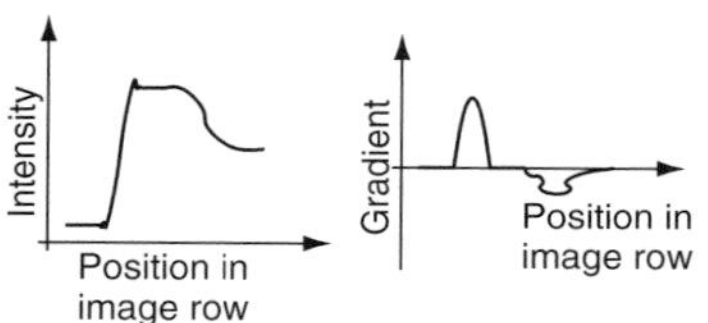

gradient based flow estimation: Estimation of the optical flow based on gradient images. This computation can be done directly through the computation of a time derivative as long as the movement between frames is quite small. See also the aperture problem.

gradient descent: An iterative method for finding the (local) minimum of a function.

gradient edge detection: Edge detection based on image gradients.

gradient filter: A filter that is convolved with an image to create an image in which every point represents the gradient in the original image in an orientation defined by the filter. Normally two orthogonal filters are used and by combining these a gradient vector can be determined for every point. Common filters include Roberts cross gradient operator, Prewitt gradient operator and the Sobel gradient operator. The Sobel horizontal gradient operator gives:

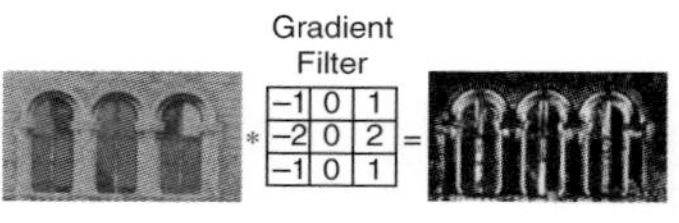

gradient image: See edge image.

gradient magnitude thresholding: Thresholding of a gradient image in order to identify 'strong' edge points.

gradient matching stereo: An approach to stereo matching in which the image gradients (or features derived from the image gradients) are matched.

gradient operator: An image processing operator that produces a gradient image from a gray scale input image I. Depending on the usage of the term, the output could be 1) the vectors ∇I of the x and y derivatives at each point or 2) the magnitudes of these gradient vectors. The usual role of the gradient operator is to locate regions of strong gradients that signals the position of an edge. The figure below shows a gray scale image and its gradient magnitude image, where darker lines indicate stronger magnitudes. The gradient was calculated using the Sobel operator.

gradient space: A representation of surface orientations in which each orientation is represented by a pair (p, q) where $p = \frac{\partial z}{\partial x}$ and $q = \frac{\partial z}{\partial y}$ (where the z axis is aligned with the optical axis of the viewing device).

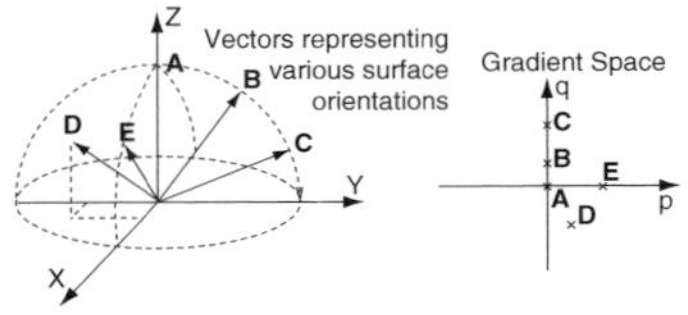

gradient vector: A vector describing the magnitude and direction of maximal change on an N-dimensional surface.

graduated non-convexity: An algorithm for finding a global minimum in a function that has many sharp local minima (a non-convex function). This is achieved by approximating the function by a convex function with just one minimum (near the global minimum of the non-convex function) and then gradually improving the approximation.

grammar: A system of rules constraining the way in which primitives (such as words) can be combined. Used in computer vision to represent objects where the primitives are simple shapes, textures or features.

grammatical representation: A representation that describes shapes using a number of primitives that can be combined using a particular set of rules (the grammar).

granulometric spectrum: The resultant distribution from a granulometry.

granulometry: The study of the size characteristics of a set (*e.g.*, the size of a set of regions). Most normally this is achieved by applying a series of morphological openings (with structured elements of increasing size) and then studying the resultant size distributions.

graph: A graph is formed by a set of vertices V and a set of edges $E \subset V \times V$ linking pairs of vertices. Vertices u and v are *neighbors* if $(u, v) \in E$ or $(v, u) \in E$. See graph isomorphism, subgraph isomorphism. This is a graph with five nodes:

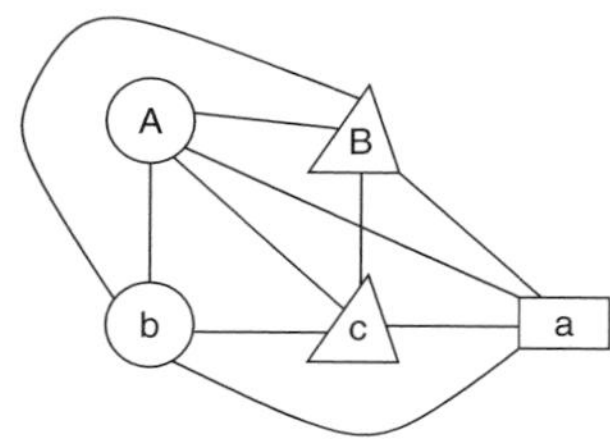

graph cut: A partition of the vertices of a directed graph V into two disjoint sets S and T. The cost of the cut is the costs of all the edges that go from a vertex in S to a vertex in T.

graph isomorphism: Two graphs are isomorphic if there exists a mapping (bijection) between their vertices that makes the edge sets identical. Determining whether two graphs are isomorphic is the graph isomorphism problem and is believed to be NP-complete. These small graphs are isomorphic with A:b, C:a, B:c:

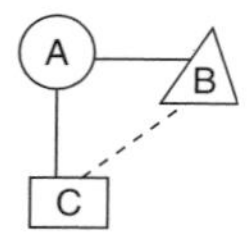

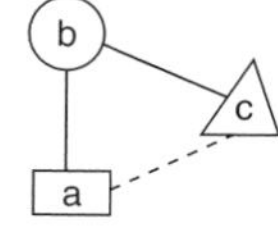

graph matching: A general term describing techniques for comparing two graph models. These techniques may attempt to find graph isomorphisms, subgraph isomorphisms, or may just try to establish similarity between graphs.

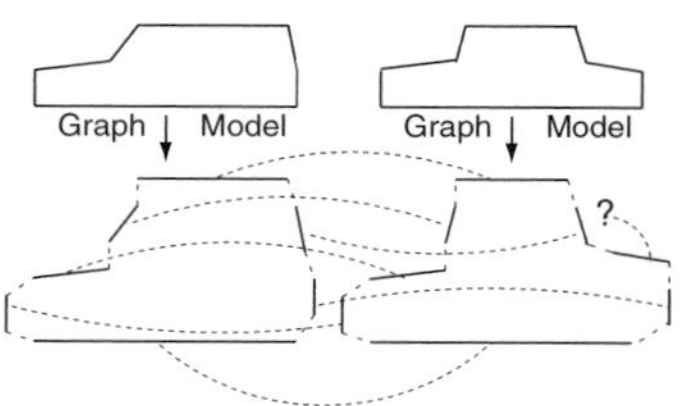

graph model: A model of data in terms of a graph. Typical uses in computer vision include object representation (see graph matching) and edge gradients (see graph searching).

graph partitioning: The operation of splitting a graph into subgraphs satisfying some criteria. For example we might want to partition a graph of all polygonal edge segments in an image into subgraphs corresponding to objects in the scene.

graph representation: See graph model.

graph searching: Search for a specific node or path through a graph. Used, among other things, for border detection (*e.g.*, in an edge gradient image) and object identification (*e.g.*, decision trees).

graph similarity: The degree to which two graph representations are similar. Typically (in computer vision) these representations will not be exactly the same and hence a double subgraph isomorphism may need to be found to evaluate similarity.

graph theoretic clustering: Clustering algorithms that use concepts from graph theory, in particular leveraging efficient graph-theoretic algorithms such as maximum flow.

grassfire algorithm: A technique for finding a region skeleton based on wave propagation. A virtual fire is lit on all region boundaries and the skeleton is defined by the intersection of the wave fronts.

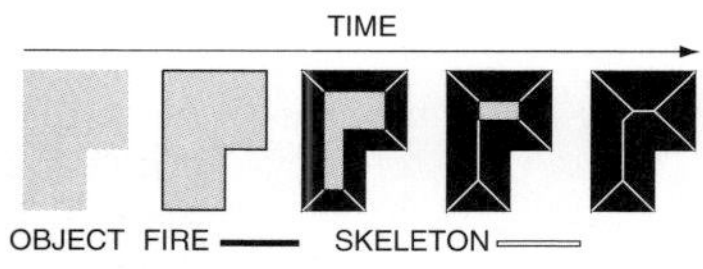

grating: See diffraction grating.

gray level . . . : See gray scale . . . .

gray scale: A monochromatic representation of the value of a pixel. Typically this represents image brightness and ranges from 0 (black) to 255 (white).

gray scale co-occurrence: The occurrence of two particular gray levels some particular distance and orientation apart. Used in co-occurrence matrices.

gray scale correlation: The cross correlation of gray scale values in image windows or full images.

gray scale distribution model: A model of how gray scales are distributed in some image region. See also intensity histogram.

gray scale gradient: The rate of change of the gray levels in a gray scale image. See also edge, gradient image and first derivative filter.

gray scale image: A monochrome image in which pixels typically represents brightness values ranging from 0 to 255. See also gray scale.

gray scale mathematical morphology: The application of mathematical morphology to gray scale images. Each quantization level is treated as a distinct set where pixels are members of the set if they have a value greater than or equal to particular quantization levels.

gray scale moment: A moment that is based on image or region gray scales. See also binary moment.

gray scale morphology: See gray scale mathematical morphology.

gray scale similarity: See gray scale correlation.

gray scale texture moment: A moment that describes texture in a gray scale image (*e.g.*, the Haralick texture operator describes image homogeneity).

gray scale transformation: A general term describing a class of image processing operations that apply to gray scale images, and simply manipulate the gray scale of pixels. Example operations include contrast stretching and histogram equalization.

gray value . . .: See gray scale . . . .

greedy search: A search algorithm seeking to maximize a local criterion instead of a global one. Greedy algorithms sacrifice generality for speed. For instance, the stable configuration of a snake is typically found by an iterative energy minimization. The snake configuration at each step of the optimization can be found globally, by searching the space of all allowed configurations of all pixels simultaneously (a large space) or locally (greedy algorithm), by searching the space of all allowed configurations of *each* pixel individually (a much smaller space).

grey . . .: See gray . . . .

grid filter: An approach to noise reduction where a nonlinear function of features (pixels or averages of a number of pixels) from the local neighborhood are used. Grid filters require a training phase where noisy data and corresponding ideal data are presented.

ground following: See ground tracking.

ground plane: The horizontal plane that corresponds to the ground (the surface on which objects stand). This concept is only really useful when the ground is roughly flat. The ground plane is highlighted here:

ground tracking: A loosely defined term describing the robot navigation problem of sensing the ground plane and following some path.

ground truth: In performance analysis, the true value, or the most accurate value achievable, of the output of a specific instrument under analysis, for instance a vision system measuring the diameter of circular holes. Ground truth values may be known theoretically, *e.g.*, from formulae, or obtained through an instrument more accurate than the one being evaluated.

grouping: 1) In human perception, the tendency to perceive certain patterns or clusters of stimuli as a coherent, distinct entity as opposed to a set of independent elements. 2) A whole class of segmentation algorithms is based on this idea. Much of this work was inspired by the Gestalt school of psychology. See also segmentation, image segmentation, supervised classification, and clustering.

grouping transform: An image analysis technique for grouping image features together (*e.g.*, based on collinearity, etc.).

H

Haar transform: A wavelet transform that is used in image compression. The basis functions used are similar to those used by first derivative edge detectors, resulting in images that are decomposed into horizontal, diagonal and vertical edges at different scales.

Hadamard transform: A transformation that can be used to transform an image to its constituent Hadamard components. A fast version of the algorithms exists that is similar to the fast Fourier transform, but all values in the basis functions are either +1 or −1. It requires significantly less computation and as such is often used for image compression.

halftoning: See dithering.

Hamming distance: The number of different bits in corresponding positions in two bit strings. For instance, the Hamming distance of 01110 and 01100 is 1, that of 10100 and 10001 is 2. A very important concept in digital communications.

hand sign recognition: The recognition of hand gestures such as those used in sign language.

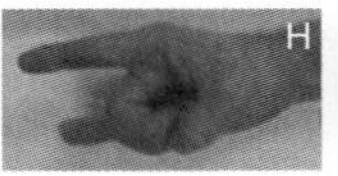

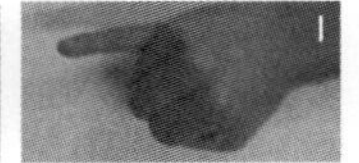

hand tracking: The tracking of a person's hand in a video sequence, often for use in human–computer interaction.

hand–eye calibration: The calibration of a manipulator (such as a robot arm) together with a visual system (such as a number of cameras). The main issue here is ensuring that both systems use the same frame of reference. See also camera calibration.

hand–eye coordination: The use of visual feedback to direct the movement of a manipulator. See also hand–eye calibration.

handwriting verification: Verification that the style of handwriting corresponds to that of some particular individual.

handwritten character recognition: The automatic recognition of characters that have been written by hand.

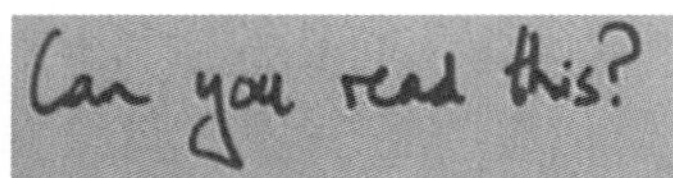

Hankel transform: A simplification of the Fourier transform for radially symmetric functions.

hat transform: See Laplacian of Gaussian (also known as Mexican hat operator) and/or top hat operator.

Harris corner detector: A corner detector where a corner is detected if the eigenvalues of the matrix **M** are large and locally maximum ($f(i,j)$ is the intensity at point (i,j)).

$$\mathbf{M} = \begin{bmatrix} \frac{\partial f}{\partial i}\frac{\partial f}{\partial i} & \frac{\partial f}{\partial i}\frac{\partial f}{\partial j} \\ \frac{\partial f}{\partial i}\frac{\partial f}{\partial j} & \frac{\partial f}{\partial j}\frac{\partial f}{\partial j} \end{bmatrix}$$

. To avoid explicit comutation of the eigenvalues, the local maxima of $det(\mathbf{M}) - 0.004 \times trace(\mathbf{M})$ can be used. This is also known as the Plessey corner finder.

Hartley transform: Similar transform to the Fourier transform, but the coefficients used are real (whereas those used in the Fourier transform are complex).

Hausdorff distance: A measure of the distance between two sets of (image) points. For every point in both sets determine the minimum distance to any point in the other set. The Hausdorff distance is the maximum of these minimum values.

HDTV: High Definition TeleVision.

height image: See range image.

Helmholtz reciprocity: An observation by Helmholtz about the bidirectional reflectance distribution function $f_r(\vec{i}, \vec{e})$ of a local surface patch, where $\vec{i}$ and $\vec{e}$ are the incoming and outgoing light rays respectively. The observation is that the reflectance is symmetric about the incoming and outgoing directions, *i.e.*, $f_r(\vec{i}, \vec{e}) = f_r(\vec{e}, \vec{i})$.

Hessian: The matrix of second derivatives of a multi-valued scalar function. It can be used to design an orientation-dependent second derivative edge detector.

$$\mathbf{H} = \begin{bmatrix} \frac{\partial^2 f(i,j)}{\partial i^2} & \frac{\partial^2 f(i,j)}{\partial i \partial j} \\ \frac{\partial^2 f(i,j)}{\partial j \partial i} & \frac{\partial^2 f(i,j)}{\partial j^2} \end{bmatrix}$$

heterarchical/mixed control: An approach to system control where control is shared amongst several systems.

heuristic search: A search process that employs commonsense rules (heuristics) to speed up search.

hexagonal image representation: An image representation where the pixels are hexagonal rather than rectangular. This representation might be used because 1) it is similar to the human retina or 2) the distances to all adjacent pixels are equal, unlike diagonally connected pixels in rectangular grids

Hexagonal Sampling Grid

hidden Markov model (HMM): A model for predicting the probability of system state on the basis of the previous state together with some observations. HMMs have been used extensively in handwritten character recognition.

hierarchical: A general term referring to the approach of considering data at a low level of detail initially and then gradually increasing the level of detail. This approach often results in better performance.

hierarchical clustering: An approach to grouping in which each item is initially put in a separate cluster, the two most similar clusters are merged and this merging is repeated until some condition is satisfied (*e.g.*, no clusters of less that a particular size remain).

hierarchical coding: Coding of (image) data at multiple layers starting with the lowest level of detail and gradually increasing the resolution. See also hierarchical image compression.

hierarchical Hough transform: A technique for improving the efficiency of the standard Hough transform. Commonly used to describe any Hough-based technique that solves a sequence of problems beginning with a low-resolution Hough space and proceeding to high-resolution space, or using low-resolution images, or operating on subimages of the input image before combining the results.

hierarchical image compression: Image compression using hierarchical coding. This leads to the concept of progressive image transmission.

hierarchical matching: Matching at increasingly greater levels of detail. This approach can be used when matching images or more abstract representations.

hierarchical model: A model formed by smaller submodels, each of which may have further smaller submodels. The model may contain multiple instances of the subcomponent models. The subcomponents may be placed relative to the model by using a coordinate system transformation or may just be listed in a set structure. This is a three-level hierarchical model with multiple usage of the subcomponents:

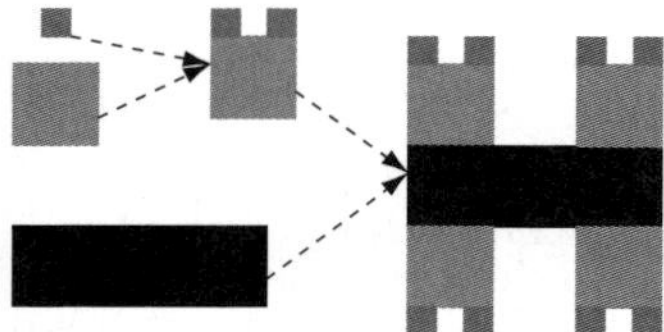

hierarchical recognition: See hierarchical matching.

hierarchical texture: A way of considering texture elements at multiple levels (*e.g.*, basic texture elements may themselves be grouped together to form a texture element at another scale, and so on).

hierarchical thresholding: A thresholding technique where an image is considered at different levels of detail in a pyramid data structure, and thresholds are identified at different levels in the pyramid starting at the highest level.

high level vision: A general term referring to image analysis and understanding tasks (*i.e.*, those tasks that address reasoning about what is seen, as opposed to basic processing of images).

high pass filter: A frequency domain filter that removes or suppresses all low-frequency components.

highlight: See specular reflection.

histogram: A representation of the frequency distribution of some values. See intensity histogram, an example of which is shown below.

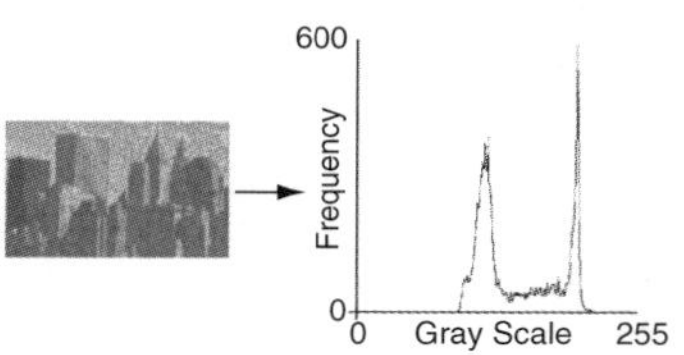

histogram analysis: A general term describing a group of techniques that abstract information from histograms (*e.g.*, determining the anti-mode/trough in a bi-modal histogram for use in thresholding).

histogram equalization: An image enhancement operation that processes a single image and results in an image with a uniform distribution of intensity levels (*i.e.*, whose intensity histogram is flat). When this technique is applied to a digital image, however, the resulting histogram will often have large values interspersed with zeros.

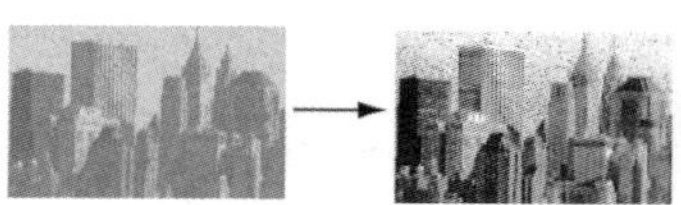

histogram modeling: A class of techniques, such as histogram equalization, modifying the dynamic range and contrast of an image by changing its intensity histogram into one with desired properties.

histogram modification: See histogram modeling.

histogram moment: A moment derived from a histogram.

histogram smoothing: The application of a smoothing filter (*e.g.*, Gaussian smoothing) to a histogram. This is often required before histogram analysis operations can be applied.

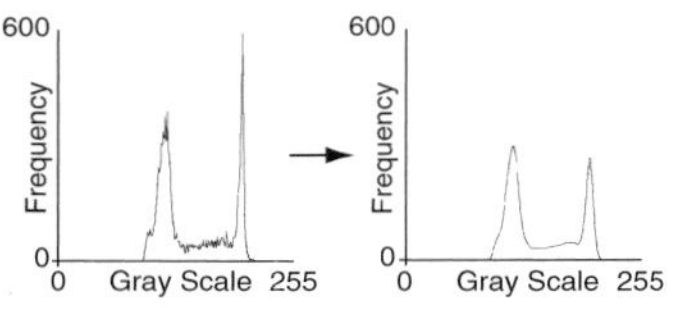

hit and miss/hit or miss operator: A morphological operation where a new image is formed by ANDing (logical AND) together corresponding bits for every pixel of an input image and a structuring element. This operator is most appropriate for binary images but may also be applied to gray scale images.

HK: See mean and Gaussian curvature shape classification.

HK segmentation: See mean and Gaussian curvature shape classification.

HMM: See hidden Markov model.

holography: The process of creating a three dimensional image (a hologram) by recording the interference pattern produced by coherent laser light that has been passed through a diffraction grating.

homogeneous, homogeneity: **1.** (Homogeneous coordinates:) In projective n-dimensional geometry, a point is represented by a $n+1$ element vector, with the Cartesian representation being found by dividing the first n components by the last one. Homogeneous quantities such as points are equal if they are scalar multiples of each other. For example a 2D point is represented as (x, y) in Cartesian coordinates and in homogeneous coordinates by the point $(x, y, 1)$ and any multiple thereof. **2.** (Homogeneous texture:) A two (or higher) dimensional pattern, defined on a space $S \subset \mathbb{R}^2$ for which some functions (*e.g.*, mean, standard deviation) applied to a window on S have values that are independent of the position of the window.

homogeneous coordinates: Points described in projective space. For example an (x, y, z) point in Euclidean space would be described as $(\lambda x, \lambda y, \lambda z, \lambda)$ for any λ in homogeneous coordinates.

homogeneous representation: A representation defined in projective space.

homography: The relationship described by a homography transformation.

homography transformation: Any invertible linear transformation between projective spaces. It is commonly used for image transfer, which maps one planar image or region to another. The transformation can be estimated using four non-collinear point pairs.

homomorphic filtering: An image enhancement technique that simultaneously normalizes brightness and enhances contrast. It works by applying a high pass filter to the original image in the frequency domain, hence reducing intensity variation (that changes slowly) and highlighting reflection detail (that changes rapidly).

homotopic transformation: A continuous deformation that preserves the connectivity of object features (*e.g.*, skeletonization). Two objects are homotopic if they can be made the same by some series of homotopic transformations.

Hopfield network: A type of neural network mainly used in optimization problems, which has been used in object recognition.

horizon line: The line defined by all vanishing points from the same plane. The most commonly used horizon line is that associated with the ground plane.

Hough transform: A technique for transforming image features directly into the likelihood of occurrence of some shape. For example see Hough transform line finder and generalized Hough transform.

Hough transform line finder: A version of the Hough transform based on the parametric equation of a line ($s = i\cos\theta + j\sin\theta$) in which a set of edge points $\{(i, j)\}$ is transformed into the likelihood of a line being present as represented in a (s, θ) space. The likelihood is quantified, in practice, by a histogram of the $\sin\theta, \cos\theta$ values observed in the images.

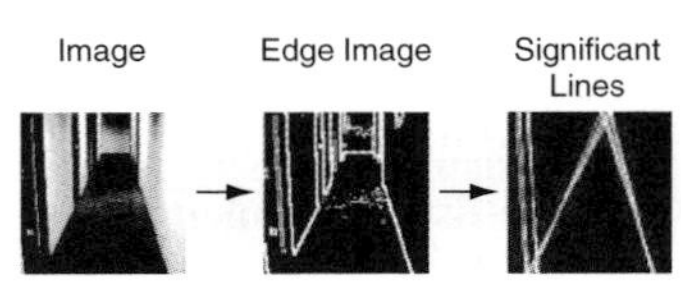

HSI: Hue-Saturation-Intensity color image format.

HSL: Hue-Saturation-Luminance color image format (see plate section for a colour version of these figures).

HSV: Hue Saturation Value color image format.

hue: Describes color using the dominant wavelength of the light. Hue is a common component of color image formats (see HSI, HSL, HSV).

Hueckel edge detector: A parametric edge detector that models an edge using a parameterized model within a circular window (the parameters are edge contrast, edge orientation and distance background mean intensity).

Huffman encoding: An optimal, variable-length encoding of values (*e.g.*, pixel values) based on the relative probability of each value. The code lengths may change dynamically if the relative probabilities of the data source change. This technique is commonly used in image compression.

human motion analysis: A general term describing the application of motion analysis to human subjects. Such analysis is used to track moving people, to recognize the pose of a person and to derive 3D properties.

HYPER: HYpothesis Predicted and Evaluated Recursively. A well known vision system developed by Nicholas Ayache and Olivier Faugeras, in which geometric relations derived from polygonal models are used for recognition.

hyperbolic surface region: A region of a 3D surface that is locally saddle-shaped. A point on a surface at which the Gaussian curvature is negative (so the signs of the principal curvatures are opposite).

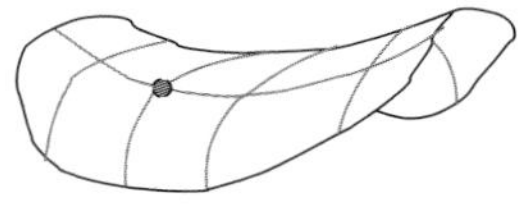

hyperfocal distance: The distance D at which a camera should be focused in order that the depth of field extends from $D/2$ to infinity. Equivalently, if a camera is focused at a point at distance D, points at $D/2$ and infinity are equally blurred.

hyperquadric: A class of volumetric shape representations that include superquadrics. Hyperquadric models can describe arbitrary convex polyhedra.

hyperspectral image: An image with a large number (perhaps hundreds) of spectral bands. An image with a lower number of spectral bands is referred to as multi-spectral image.

hyperspectral sensor: A sensor capable of collecting many (perhaps hundreds) of spectral bands simultaneously. Produces a hyperspectral image.

hypothesize and test: See hypothesize and verify.

hypothesize and verify: A common approach to object recognition in which possibilities (of object type and pose) are hypothesized and then evaluated against evidence from the images. This is done either until all possibilities are considered or until a hypothesis with a sufficiently high degree of fit is found.

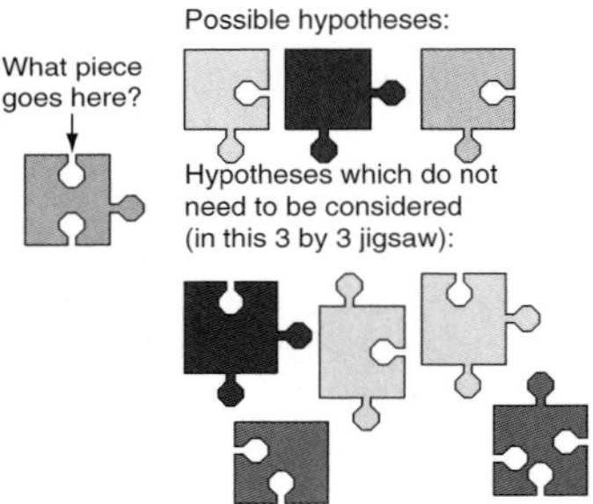
Possible hypotheses:
What piece goes here?
Hypotheses which do not need to be considered (in this 3 by 3 jigsaw):

hysteresis tracking: See thresholding with hysteresis.

I

ICA: See independent component analysis.

iconic: Having the characteristics of an image. See iconic model.

iconic model: A representation having the characteristics of an image. For example the template used in template matching.

iconic recognition: Object recognition using iconic models.

ICP: See iterative closest point.

ideal line: A line described in the continuous domain as opposed to one in a digital image, which will suffer from rasterization.

ideal point: A point described in the continuous domain as opposed to one in a digital image, which will suffer from rasterization. May also be used to refer to a vanishing point.

IDECS: Image Discrimination Enhancement Combination System. A well-known vision system developed by Haralick and Currier.

identification: The process of associating some observations with a particular instance or class of object that is already known.

identity verification: Confirmation of the identity of a person based on some biometrics (*e.g.*, face authentication). This differs from the recognition of an unknown person in that only one model has to be compared with the information that is observed.

IGS: Interpretation Guided Segmentation. A vision technique for grouping image elements into regions based on semantic interpretations in addition to raw image values. Developed by Tenenbaum and Barrow.

IHS: Intensity Hue Saturation color image format.

IIR: See infinite impulse response filter.

Dictionary of Computer Vision and Image Processing R.B. Fisher, K. Dawson-Howe, A. Fitzgibbon, C. Robertson and E. Trucco

ill-posed problem: A mathematical problem that infringes at least one of the conditions in the definition of well-posed problem. Informally, these are that the solution must (a) exist, (b) be unique, and (c) depend continuously on the data. Ill-posed problems in computer vision have been approached using regularization theory. See regularization.

illuminance: The total amount of visible light incident upon a point on a surface. Measured in lux (lumens per meter squared), or footcandles (lumens per foot squared). Illuminance decreases as the distance between the viewer and the source increases.

illuminant direction: The direction from which illuminance originates. See also light source geometry.

illumination: See illuminance.

illumination constancy: The phenomenon that allows humans to perceive the lightness/brightness of surfaces as approximately constant regardless of the illuminance.

illumination field calibration: Determination of the illuminance falling on a scene. Typically this is done by taking an image of a white object of known brightness.

illusory contour: A perceived border where there is no edge present in the image data. See also subjective contour. For example the following diagram shows the Kanizsa triangles.

image: A function describing some quantity (such as brightness) in terms of spatial layout (See image representation). Most frequently computer vision is concerned with two dimensional digital images.

image addition: See pixel addition operator.

image analysis: A general term covering all forms of analysis of image data. Generally image analysis operations result in a symbolic description of the image contents.

image acquisition: See image capture.

image arithmetic: A general term covering image processing operations that are based on the application of an arithmetic or logical operator to two images. Such operations included addition, subtraction, multiplication, division, blending, AND, NAND, OR, XOR, and XNOR.

image based: A general term describing operations or representations that are based on images.

image based rendering: The production of a new image of

a scene from an arbitrary viewpoint based on a number of images of the scene together with associated range images.

image blending: An arithmetic operation similar to image addition where a new image is formed by blending the values of corresponding pixels from two input images. Each input image is given a weight for the blending so that the total weight is 1.0.

image capture: The acquisition of an image by a recording device, *e.g.*, a camera.

image coding: The mapping or algorithm required to encode or decode an image representation (such as a compressed image).

image compression: A method of representing an image in order to reduce the amount of storage space that it occupies. Techniques can be lossless (which allows all image data to be recorded perfectly) or lossy (where some loss of quality is allowed, typically resulting in significantly better compression rates).

image connectedness: See pixel connectivity.

image coordinates: See image plane coordinates and pixel coordinates.

image database indexing: The technique of associating indices (for example, keywords) with images that allows the images to be indexed efficiently within a database.

image difference: See image subtraction.

image digitization: The process of sampling and quantizing an analogue image function to create a digital image.

image distortion: Any effect that alters an image from the ideal image. Most typically this term refers to geometric distortions, although it can also refer to other types of distortion such as image noise and effects of sampling and quantization.

Correct Image

Distorted Image

image encoding: The process of converting an image into a different representation. For example see image compression.

image enhancement: A general term covering a number of image processing operations, that alter an image in order to make it easier for humans to perceive. Example operations include contrast stretching and histogram equalization. For example, the following shows a histogram equalization operation:

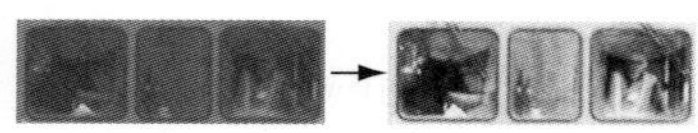

image feature: A general term for an interesting image structure that could arise from a corresponding interesting scene structure. Features can be single points such as interest points, curve vertices, image edges, lines or curves or surfaces, etc.

image feature extraction: A group of image processing techniques concerned with the identification of particular features in an image. Examples include edge detection and corner detection.

image flow: See optic flow.

image formation: A general term covering issues relating to the manner in which an image is formed. For example in the case of a digital camera this term would include the camera geometry as well as the process of sampling and quantization.

image grid: A geometric map describing the image sampling in which every image point is represented by a vertex (or hole) in the map/grid.

image indexing: See image database indexing.

image intensifier: A device for amplifying an image, so that the resultant sensed luminous flux is significantly higher.

image interleaving: Describes the way in which image pixels are organized. Different possibilities include pixel interleaving (where the image data is ordered by pixel position), and band interleaving (where the image data is ordered by band, and is then ordered by pixel position within each band).

image interpolation: A method for computing a value for a pixel in an output image based on non-integer coordinates in some input image. The computation is based on the values of nearby pixels in the input image. This type of operation is required for most geometric transformations and computations requiring sub-pixel resolution. Types of interpolation scheme include nearest-neighbor interpolation, bilinear interpolation, bi cubic interpolation, etc. This figure shows the result of interpolation in image enlargement:

image interpretation: A general term for computer vision processes that extract descriptions from images (as opposed to processes that produce output images for human viewing).

There is often the assumption that the descriptions are very high-level, *e.g.*, "the boy is walking to the store carrying a book" or "these cells are cancerous". A broader definition would also allow processes that extract information needed by a subsequent (usually non-image processing) activity, *e.g.*, the position of a bright spot in an image.

image invariant: An image feature or measurement image that is invariant to some properties. For example invariant color features are often used in image database indexing.

image irradiance equation: Usually expressed as $E(x, y) = R(p, q)$, this equality (up to a constant scale factor to account for illumination strength, surface color and optical efficiency) says that the observed brightness E at pixel (x, y) is equal to the reflectance R of the surface for surface normal $(p, q, -1)$. Usually there is a one-degree-of-freedom family of surface normals with the same reflectance value so the observed brightness only partially constrains local surface orientation and thus shape.

image magnification: The extent to which an image is expanded for viewing. If the image size is actually changed then image interpolation must be used. Normally quoted relative to the original size (*e.g.*, $\times 2$, $\times 10$, etc.).

Magnified image (x4)

image matching: The comparison of two images, often evaluated using cross correlation. See also template matching.

Image 1

Image 2

Locations where Image 2 matches Image 1

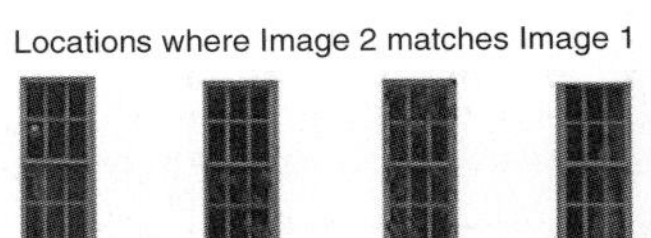

image memory: See frame store.

image morphing: A gradual transformation from one image to another image.

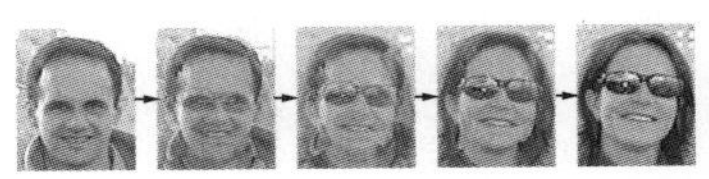

image morphology: An approach to image processing that considers all operations in terms of set operations. See mathematical morphology.

image mosaic: A composition of several images, to provide a single larger image with covering a wider field of view. For example, the following is a mosaic of three images:

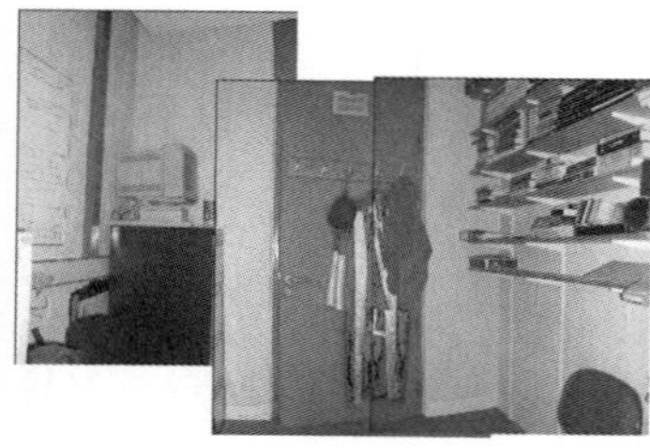

image motion estimation: Computation of optical flow for all pixels/features in an image.

image multiplication: See pixel multiplication operator.

image noise: Degradation of an image where pixels have values which are different from the ideal values. Often noise is modeled as having a Gaussian distribution with a zero mean, although it can take on different forms such as salt-and-pepper noise depending upon the cause of the noise (*e.g.*, the environment, electrical inference, etc.). Noise is measured in terms of the signal-to-noise ratio.

Original image

Image with gaussian noise

Image with salt-and-pepper noise

image modality: A general term for the sensing technique used to capture an image, *e.g.*, a visible light, infrared or X-ray image.

image normalization: The purpose of image normalization is to reduce or eliminate the effects of different illumination on the same or similar scenes. A typical approach is to subtract the mean of the image and divide by the standard deviation, which produces a zero mean, unit variance image. Since images are not Gaussian random samples, this approach does not completely solve the problem. Further, light source placement can also cause variations in shading that are not corrected by this approach. This figure shows an original image (left) and its normalization (right):

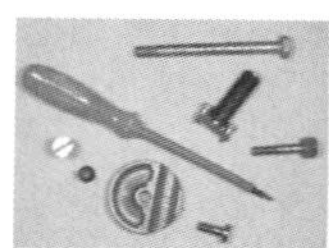

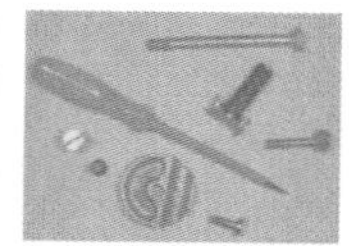

image of absolute conic: See absolute conic.

image pair rectification: See image rectification.

image plane: The mathematical plane behind the lens onto which an image is focused. In practice, the physical sensing surface aims to be placed here, but its position will vary slightly due to minor variations in sensor shape and placement. The term is also used to describe the geometry of the image recorded at this location. See:

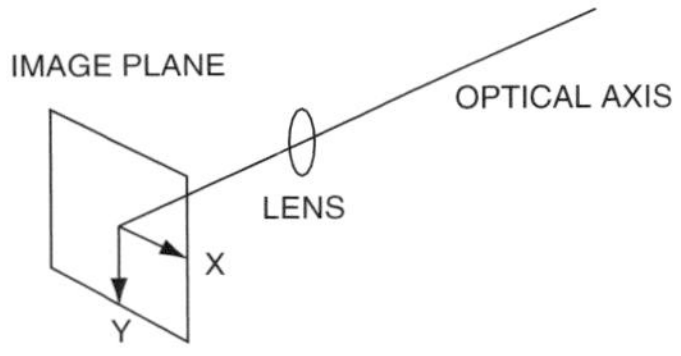

image plane coordinates: The position of points in the physical image sensing plane. These have physically meaningful values, such as centimeters. These can be converted to pixel coordinates, which are in pixels. The two meanings are sometimes used interchangeably.

image processing: A general term covering all forms of processing of captured image data. It can also mean processing that starts from an image and results in an image, as contrasted to ending with symbolic descriptions of the image contents or scene.

image processing operator: A function that may be applied to an image in order to transform it in some way. See also image processing.

image pyramid: A hierarchical image representation in which each level contains a smaller version of the image at the previous level. Often pixel values are obtained by a smoothing process. Usually the reduction is by a power of two (*i.e.*, 2 or 4). The figure below shows four levels of a pyramid in which each level is formed by averaging together two pixels from the previous layer. The levels are enlarged to the original image size for inspection of the effect of the compression.

image quality: A general term, usually referring to the extent to which the image data records the observed scene faithfully. The specific issues that are important to image quality are problem specific, but may include low image noise, high image contrast, good image focus, low motion blur, etc.

image querying: A shorthand term for indexing into image databases. This is often done based on color, texture or shape indices. The database keys could be based on global or local measures.

image reconstruction: A term used in image compression to describe the process of recreating a digital image from some compressed form.

image rectification: A warping of a stereo pair of images such that conjugate epipolar lines (defined by the two cameras' epipoles and any 3D scene point) are collinear. Usually the lines are transformed to be parallel to the horizontal axis so that corresponding image features can be found on the same raster line. This reduces the computational complexity of the stereo correspondence problem.

image registration: See registration.

image representation: A general term for how the image data is represented. Image data can be one, two, three or more dimensional. Image data is often stored in arrays where the spatial layout of the array reflects the spatial layout of the data. The figure below shows a small 10×10 pixel image patch with the gray scale values for the corresponding pixels.

123	123	123	123	123	123	123	123	96	96
123	123	112	96	96	123	123	123	123	96
123	123	96	96	112	123	137	123	123	96
123	123	96	96	123	214	234	178	123	96
123	100	72	109	178	230	230	137	123	96
125	78	51	142	218	178	96	76	96	96
92	100	92	92	81	76	76	96	123	123
81	109	129	129	100	81	92	123	123	123
51	109	142	137	123	123	123	123	123	123
33	76	123	123	137	137	123	123	123	123

image resolution: Usually used to record the number of pixels in the horizontal and vertical directions in the image, but may also refer to the separation between pixels (*e.g.*, $1\,\mu$m) or the angular separation between the lines of sight corresponding to adjacent pixels.

image restoration: The process of removing some known (and modelled) distortion from an image, such as blur in an out-of-focus image. The process may not produce a perfect image, but may remove an undesired distortion (*e.g.*, motion blur) at the cost of another ignorable distortion (*e.g.*, phase distortion).

image sampling: The process of measuring some pixel values from the physical image focused onto the image plane. The sampling could be monochrome, color or multi-spectral, such as RGB. The sampling usually results in a rectangular array of pixels sampled at nearly equally spacing, but other sampling could be used such as space variant sensing.

image scaling: The operation of increasing or reducing the size of an image by some scale factor. This operation may require the use of some type of image interpolation method. See also image magnification.

image segmentation: The grouping of image pixels into meaningful, usually connected, structures such as curves and regions. The term is applied to a variety of image modalities, such as intensity data or range data and properties, such as similar feature orientation, feature motion, surface shape or texture.

image sequence: A series of images generally taken at regular intervals in time. Typically the camera and/or objects in the scene will be moving.

image sequence fusion: The integration of information from the many images in an image sequence. Different types of fusion include 3D structure recovery, production of a mosaic of the scanned scene, tracking of a moving object, improved scene imaging due to image averaging, etc.

image sequence matching: Computing the correspondence between pixels or image features in frames of the image sequence. With the correspondences, one can construct image mosaics, stabilize image jitter or recover scene structure.

image sequence stabilization: Normal hand-held video camera recordings contain some image motion due to the jitter of the human operator. Image stabilization attempts to estimate the random portion of the camera motion jitter and translate the images in the sequence to reduce or remove the jitter. A similar application would be to remove systematic camera motions to produce a motionless image. See also feature stabilization.

image sharpening operator: An image enhancement operator that increases the high spatial frequency component of the image, so as to make the edges of objects appear sharper or less blurred. See also edge enhancement. These images show a raw image (left) and an image sharpened with the unsharp operator (right).

image size: The number of pixels in an image, for example, 768 horizontally by 494 vertically.

image smoothing: See noise reduction.

image stabilization: See image sequence stabilization

image storage devices: See frame store.

image subtraction operator: See pixel subtraction operator.

image transfer: 1) See novel view synthesis. 2) Alternatively, a general term describing the movement of an image from one device to another, or alternatively from one representation to another.

image understanding: A general term referring to the derivation of high-level (abstract) information from an image or series of images. This term is often used to refer to the emulation of human visual capabilities.

Image Understanding Environment (IUE): A C++ based collection of data-types (classes) and standard computer vision algorithms. The motivation behind the development of the IUE was to reduce the independent re-invention of basic computer vision code in government funded computer vision research.

image warping: A general term for transforming the positions of pixels in an image, usually while maintaining image topology (*i.e.*, neighboring original pixels remain neighbors in the warped image). This results in an image with a new shape. This operation might be done, for example, to correct some geometric distortion, align two images (see image rectification), or transform shapes into a more easily processed form (*e.g.*, circles into straight lines).

imaging geometry: A general term referring to the relative placement of sensors, structured light sources, point light sources, etc.

imaging spectroscopy: The acquisition and analysis of surface composition by using image data from multiple spectral channels. A typical sensor (AVIRIS) records 224 measurements at 10 nm increments from 400 to 2500 nm. The term might refer to the raw multi-dimensional signal or to the classification of that signal into surface types (*e.g.*, vegetation or mineral types).

imaging surface: The surface within a camera on which the image is projected by the lens. This surface in a digital camera is comprised of photosensitive elements that record the incident illumination. See also image plane.

implicit curve: A curve that is defined by an equation of the form $f(\vec{x}) = 0$. Then the curve is the set of points $S = \{\vec{x} | f(\vec{x}) = 0\}$.

implicit surface: The representation of a surface as the set of points that makes a function have the value zero. For example, the sphere $x^2 + y^2 + z^2 = r^2$ of radius r at the origin could be represented by the function $f(x, y, z) = x^2 + y^2 + z^2 - r^2$. The set of points where $f(x, y, z) = 0$ is the implicit surface.

impossible object: An object that cannot physically exist, such as:

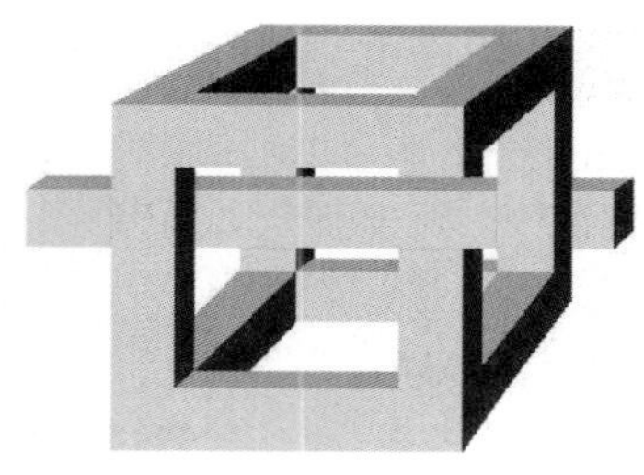

impulse noise: A form of image corruption where image pixels have their value replaced by the maximum value (*e.g.*, 255). See also salt-and-pepper noise. This figure shows impulse noise on an image:

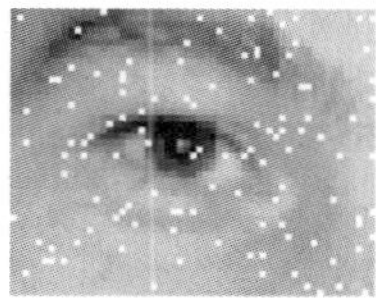

incandescent lamp: A light source whose light arises from the glowing of a very hot structure, such as a tungsten filament in the common light bulb.

incident light: A general term referring to the light that strikes or illuminates a surface.

incremental learning: Learning that is incremental in nature. See continuous learning.

independent component analysis: A multi-variate data analysis method. It finds a linear transformation that makes each component of the transformed data vectors independent of each other. Unlike principal component analysis, which considers only second order properties (covariances) and transforms onto basis vectors that are orthogonal to each other, ICA considers properties of the whole distribution and transforms onto basis vectors that need not be orthogonal.

index of refraction: The absolute index of refraction in a material is the ratio of the speed of an electromagnetic wave in a vacuum to the speed in the material. More commonly used is the relative index of refraction of two media, which is the ratio of their absolute indices of refraction. This ratio is used in lens design and explains the bending of light rays as the light passes into a new material (Snell's Law).

indexing: The process of retrieving an element from a data structure using a key. A powerful concept imported into computer vision from programming. For example, the problem of establishing the identity of an object given an image and a set of candidate models is typically approached by locating some characterizing elements in the image, or features, then using the features' properties to index a data base of models. See also model base indexing.

industrial vision: A general term covering uses of machine vision technology to industrial processes. Applications include product inspection, process feedback, part or tool alignment. A large range of lighting and sensing techniques are used. A common feature of industrial vision systems is fast processing rates (*e.g.*, several times a second), which may require limiting the rate at which targets are analyzed or limiting the types of processing.

infinite impulse response filter (IIR): A filter that produces an output value (y_n) based on the current and past input values (x_i) together with past output values (y_j). $y_n = \sum_{i=0}^{p} a_i x_{n-i} + \sum_{j=1}^{q} b_j y_{n-j}$ where a_i and b_j are weights.

inflection point: A point at which the second derivative of a curve changes its sign, corresponding to a change in concavity. See also curve inflection.

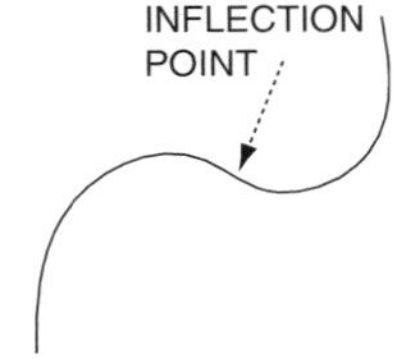

influence function: A function describing the effect of an individual observations on a statistical model. This allows us to evaluate whether the observation is having an undue influence on the model.

information fusion: Fusion of information from multiple sources. See sensor fusion.

infrared: See infrared light.

infrared imaging: Production of a image through use of an infrared sensor.

infrared light: Electromagnetic energy with wavelengths approximately in the range 700 nm to 1 mm. Immediately shorter wavelengths are visible light and immediately longer wavelengths are microwave radio. Infrared light is often used in machine vision systems because: 1) it is easily observed by most semiconductor image sensors yet is not visible by humans or 2) it is a measure of the heat emitted by the observed scene.

infrared sensor: A sensor capable of observing or measuring infrared light.

inlier: A sample that falls within an assumed probability distribution (*e.g.*, within the 95 percentile). See also outlier.

inspection: A general term for visually examining a target to detect defects. Common practical inspection examples include printed circuit boards for breaks or solder joint failures, paper production for holes or discolorations, and food for irregularities.

integer lifting: A method used to construct wavelet representations.

integer wavelet transform: An integer version of the discrete wavelet transform.

integral invariant: An integral (of some function) that is invariant under a set of transformations. For example, local integrals along a curve of curvature or arc length are invariant to rotation and translation. Integral invariants potentially have greater stability to noise than, *e.g.*, differential invariants, such as curvature itself.

integration time: The length of time that a light-sensitive sensor medium is exposed to the incident light (or other stimulus). Shorter times reduce the signal strength and possible motion blur (if the sensor or objects in the scene are moving).

intensity: 1) The brightness of a light source. 2) Image data that records the brightness of the light that comes from the observed scene.

intensity based database indexing: This is a form of image database indexing that uses intensity descriptors such as histograms of pixel (monochrome or color) values or vectors of local derivative values.

intensity cross correlation: Cross correlation using intensity data.

intensity data: Image data that represents the brightness of the measured light. There is not usually a linear mapping between the brightness of the measured light and the stored values. The term can refer to the intensity of observed visible light as well.

intensity gradient: The mathematical gradient operation ∇ applied to an intensity image I gives the intensity gradient ∇I at each image point. The intensity gradient direction shows the local image direction in which the maximum change in intensity occurs. The intensity gradient magnitude gives the magnitude of the local rate of change in image intensity. These terms are illustrated below. At each of the two designated points, the length of the vector shows the magnitude of the change in intensity and the direction of the vector shows the direction of greatest change.

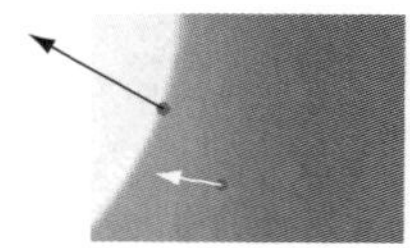

intensity gradient direction: The local image direction in which the maximum change in intensity occurs. See also intensity gradient.

intensity gradient magnitude: The magnitude of the local rate of change in image intensity. See also intensity gradient. The image that follows shows the raw image and its intensity gradient magnitude (contrast enhanced for clarity).

intensity histogram: A data structure that records the number of pixels of each intensity value. A typical gray scale image will have pixels with values in [0,255]. Thus the histogram will have 256 entries recording the number of pixels that had value 0, the number having value 1, etc. A dark object against a lighter background and its histogram are shown here

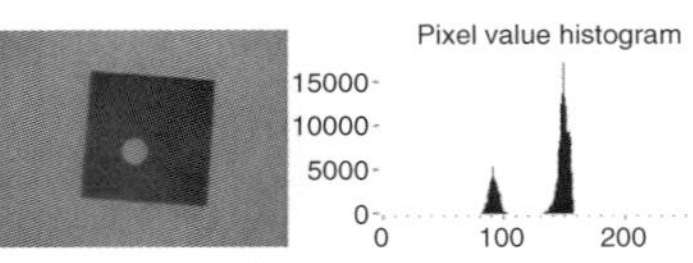

intensity image: An image that records the measured intensity data.

intensity level slicing: An image processing operation in which pixels with values other than the selected value (or range of values) are set to zero. If the image is viewed as a landscape, with height proportional to brightness, then the slicing operator takes a cross section through the height surface. The right image below shows (in black) the intensity level 80 of the left image.

intensity matching: This approach finds corresponding points in a pair of images by matching the gray scale intensity patterns. The goal is to find image neighborhoods that have nearly identical pixel intensities. All image points could be considered for matching or only feature or interest points. An algorithm where intensity matching is used is correlation based stereo matching.

intensity sensor: A sensor that measures intensity data.

interest point: A general term for pixels that have some interesting property. Interest points are often used for making feature point correspondences between images. Thus, the points usually have some identifiable property. Further, because of the need to limit the combinatorial explosion that matching can produce, interest points are often expected to be infrequent in an image. Interest points are often points of high variation in pixel values. See also point feature. Example interest points from the Harris corner detector(courtesy of Marc Pollefeys) are seen here:

interest point feature detector: An operator applied to an image to locate interest points. Well-known examples are the Moravec and the Plessey interest point operators.

interference: When 1) ordinary light interacts with matter that has dimensions similar to the wavelength of the light or 2) coherent light interacts with itself, then interference occurs. The most notable effect from a computer vision perspective is the production of interference fringes and the speckle of laser illumination. May alternatively refer to electrical interference which can affect an image when it is being transmitted on an electrical medium.

interference fringe: When optical interference occurs, the most noticeable effect it has is the production of interference fringes where the light illuminates a surface. These are parallel roughly equally spaced lighter and darker bands of brightness. One important consequence of these bands is blurring of the edge positions.

interferometric SAR: An enhancement of synthetic aperture radar (SAR) sensing to incorporate phase information from the reflected signal, increasing accuracy.

interior orientation: A photogrammetry term for the calibration of the intrinsic parameters of a camera, including its focal length, principal point, lens distortion, etc. This allows transformation of measured image coordinates into camera coordinates.

interlaced scanning: A technique arising from television engineering, whereby alternate rows of an image are scanned or transmitted instead of consecutive rows. Thus, one television frame is transmitted by sending first the odd rows, forming the odd field, and then the even rows, forming the even field.

intermediate representation: A representation that is created as a stage in the derivation of some other representation from some input representation. For example the raw primal sketch, full primal sketch, and 2.5D sketch were intermediate representation between input images and a

3D model in Marr's theory. In the following example a binary image of the notice board is an intermediate representation between the input image and the textual output.

internal energy (or force): A measure of the stability of a shape (such as smoothness) of an active shape or deformable contour model which is part of the deformation energy. This measure is used to constrain the appearance of the model.

internal parameters (of camera): See intrinsic parameters.

inter-reflection: The reflection caused by light reflected off a surface and bouncing off another surface of the same object. See also mutual illumination.

interval tree: An efficient structure for searching in which every node in the tree is a parent to nodes in a particular interval of values.

interpolation: A mathematical process whereby a value is inferred from other nearby values or from a mathematical function linking nearby values. For example, dense values along a curve can be linearly interpolated between two known curve points by fitting a line connecting the two curve points. Image, surface and volume values can be interpolated, as well as higher dimensional structures. Interpolating functions can be curved as well as linear.

interpretation tree search: An algorithm for matching between members between two discrete sets. For each feature from the first set, it builds a depth-first search tree considering all possible matching features from the second set. After a match is found for one feature (by satisfying a set of consistency tests), then it tries to match the remaining features. The algorithm can cope when no match is possible for a given feature by allowing a given number of skipped features. Here we see an example of a partial interpretation tree that is matching model features to data features:

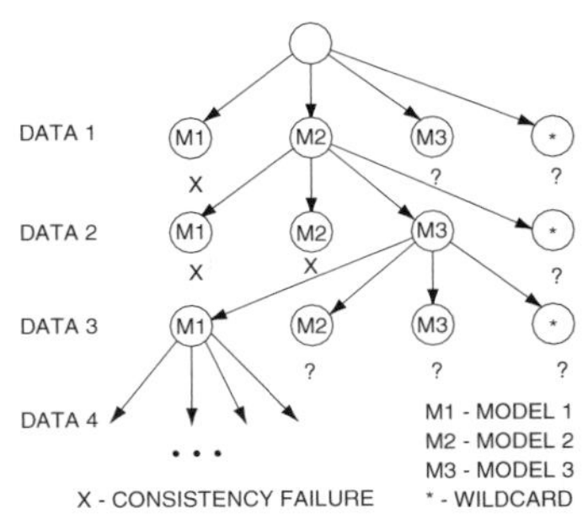

intrinsic camera parameters: Parameters such as focal length, coefficients of radial lens distortion, and the position of the principal point, that describe the mapping from image pixels to world rays in a camera. Determining the parameters of this mapping is the task of camera calibration. For a pinhole camera, world rays $\vec{r}$ are mapped to homogeneous image coordinates $\vec{x}$ by $\vec{x} = \mathbf{K}\vec{r}$ where $\mathbf{K}$ is the upper triangular 3×3 matrix

$$\mathbf{K} = \begin{pmatrix} \alpha_u f & s & u_0 \\ 0 & \alpha_v f & v_0 \\ 0 & 0 & 1 \end{pmatrix}$$

In this form, f represents the focal length, s is the skew angle between the image coordinate axes, (u_0, v_0) is the principal point, and α_u and α_v are the the aspect ratios (*e.g.*, pixels/mm) in the u and v image directions.

intrinsic dimensionality: The number of dimensions (degrees of freedom) inherent in a data set, independent of the dimensionality of the space in which it is represented. For example, a curve in 3D is intrinsically 1D although its points are represented in 3D.

intrinsic image: A term describing one of a set of images registered with the input intensity image that describe properties intrinsic to the scene, instead of properties of the input image. Example intrinsic images include: distance to scene points, scene surface orientations, surface reflectance, etc. The right image below shows a depth image registered with the intensity image on the left.

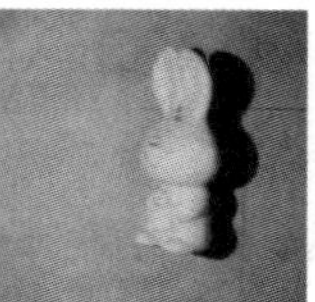

intruder detection: An application of machine vision, usually analyzing a video sequence to detect the appearance of an unwanted person in a scene.

invariant: Something that does not change under specified operations (*e.g.*, translation invariant).

invariant contour function: The contour function characterizes the shape of a planar figure based on the external boundary. Values invariant to position, scale or orientation can be computed from the contour functions. These invariants can be used for recognition of instances of the planar figure.

inverse convolution: See deconvolution.

inverse Fourier transform: A transformation that allows a signal to be recreated from its Fourier coefficients. See Fourier transform.

inverse square law: A physical law that says the illumination power received at distance d from a point light source is inversely proportional to the square of d, *i.e.*, is proportional to $\frac{1}{d^2}$.

invert operator: A low-level image processing operation where a new image is formed by replacing each pixel by an inverted value. For binary images, this is 1 if the input pixel is 0 or 0 if the input pixel is 1. For gray level images, this depends on the maximum range of intensity values. If the range of intensity values is [0,255] then the inverse inverse of a pixel with value x is $256-x$. The result is like a photographic negative. Below is a gray level image and its inverted image:

IR: See infrared.

irradiance: The amount of energy received at a point on a surface from the corresponding scene point.

isometry: A transformation that preserves distances. Thus the transformation $T: x \mapsto u$ is an isometry if, for all pairs (x, y), we have $|x-y| = |T(x) - T(y)|$.

isophote curvature: Isophotes are curves of constant image intensity. Isophote curvature is defined at any given pixel as: $-\frac{L_{vv}}{L_w}$, where L_w is magnitude of the gradient perpendicular to the isophote and L_{vv} is the curvature of the intensity surface along the isophote at that point.

iso-surface: A surface in a 3D space where the value of some function is constant. *i.e.*, $f(x, y, z) = C$ where C is some constant.

isotropic gradient operator: A gradient operator that computes the scalar magnitude of the gradient, *i.e.*, a value that is independent of edge direction.

isotropic operator: An operator that produces the same output irrespective of the local orientation of the pixel neighborhood where the operator is applied. For example, a mean smoothing operator produces the same output value, even if the image data is rotated at the point where the operator is being applied. On the other hand, a directional derivative operator would produce different values if the image were rotated. This concept is particularly relevant to feature detectors, some of which are sensitive to the local orientation of the image pixel values and some of which are not (isotropic).

iterated closest point: See iterative closest point.

iterative closest point (ICP): A shape alignment algorithm that works by iterating its two-stage process until some termination point: step 1) given an estimated transformation of the first shape onto the second, find the closest feature from the second shape for each feature of the first shape, and step 2) given the new set of closest features, re-estimate the transformation that maps the first feature set onto the second. Most variations of the algorithm need a good initial estimate of the alignment.

IUE: See Image Understanding Environment.

J

Jacobian: The matrix of derivatives of a vector function. Typically if the function $\vec{f}(\vec{x})$ is written in component form as

$$\vec{f}(\vec{x}) = \vec{f}(x_1, x_2, \ldots, x_p) = \begin{pmatrix} f_1(x_1, x_2, \ldots, x_p) \\ f_2(x_1, x_2, \ldots, x_p) \\ \vdots \\ f_n(x_1, x_2, \ldots, x_p) \end{pmatrix}$$

then the Jacobian $\mathbf{J}$ is the $n \times p$ matrix

$$\mathbf{J} = \begin{pmatrix} \frac{\partial f_1}{\partial x_1} & \cdots & \frac{\partial f_1}{\partial x_p} \\ \vdots & & \vdots \\ \frac{\partial f_n}{\partial x_1} & \cdots & \frac{\partial f_n}{\partial x_p} \end{pmatrix}$$

joint entropy registration: Registration of data using joint entropy (a measure of the degree of uncertainty) as a criterion.

JPEG: A common format for compressed image representation designed by the Joint Photographic Experts Group (JPEG).

junction label: A symbolic label for the pattern of edges meeting at the junction. This approach is mainly used in blocks world scenes where all objects are polyhedra, and thus all lines are straight and meet at only a limited number of configurations. Example "Y" (*i.e.*, corner of a block seen front on) and "arrow" (*i.e.*, corner of a block seen from the side) junctions are shown here. See also line label.

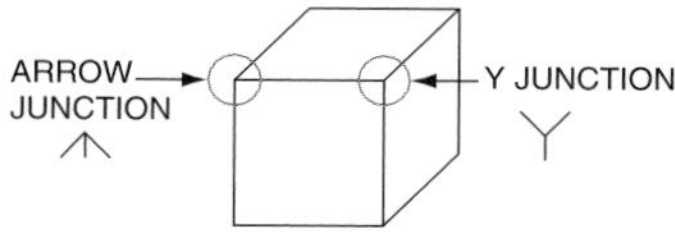

Dictionary of Computer Vision and Image Processing R.B. Fisher, K. Dawson-Howe, A. Fitzgibbon, C. Robertson and E. Trucco

K

k-means: An iterative squared error clustering algorithm. Input is a set of points $\{\vec{x}_i\}_{i=1}^n$, and initial guess at the locations $\vec{c}_1, \ldots, \vec{c}_k$ of k cluster centers. The algorithm alternates two steps: points are assigned to the cluster center closest to them, and then the cluster centers are recomputed as the mean of the associated points. Iterating yields an estimate of the k cluster centers that is likely to minimize $\sum_{\vec{x}} \min_{\vec{c}} |\vec{x} - \vec{c}|^2$.

k-means clustering: See k-means.

k-medians (also k-medoids): A variant of k-means clustering in which multi-dimensional medians are computed instead of means. The definition of multi-dimensional median varies, but options for the median $\vec{m}$ of a set of points $\{\vec{x}^i\}_{i=1}^n$, *i.e.*, $\{(x_1^i, \ldots, x_d^i)\}_{i=1}^n$ include the componentwise definition $\vec{m} = (\text{median}\{x_1^i\}_{i=1}^n, \ldots, \text{median}\{x_d^i\}_{i=1}^n)$ and the analogue of the one dimensional definition $\vec{m} = \text{argmin}_{\vec{m} \in R^d} \sum_{i=1}^n |\vec{m} - \vec{x}^i|$.

k-nearest-neighbor algorithm: A nearest neighbor algorithm that uses the classifications of the nearest k neighbors when making a decision.

Kalman filter: A recursive linear estimator of a varying state vector and associated covariance from observations, their associated covariances and a dynamic model of the state evolution. Improved estimates are calculated as new data is obtained.

Karhunen–Loève transformation: The projection of a vector (or image when treated as a vector) onto an orthogonal space that has uncorrelated components constructed from the autocorrelation (scatter) matrix of a set of example vectors. An advantage is the orthogonal components have a natural ordering (by the largest eigenvalues of the covariance of the original vector space) so that one can select the most significant variation in the dataset. The transformation can be used as a basis for image compression,

Dictionary of Computer Vision and Image Processing R.B. Fisher, K. Dawson-Howe, A. Fitzgibbon, C. Robertson and E. Trucco

for estimating linear models in high dimensional datasets and estimating the dominant modes of variation in a dataset, etc. It is also known as the principal component transformation. The following image shows a dataset before and after the KL transform was applied.

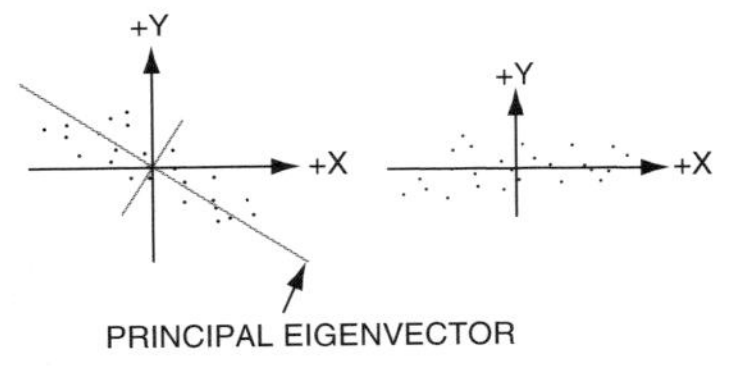

kernel: 1) A small matrix of numbers that is used in image convolutions. 2) The structuring element used in mathematical morphology. 3) The mathematical transformation used kernel discriminant analysis.

kernel discriminant analysis: A classification approach based on three key observations: 1) some problems need curved classification boundaries, 2) the classification boundaries should be defined locally by the classes rather than globally and 3) a high dimensional classification space can be avoided by using the kernel method. The method provides a transformation via a kernel so that linear discriminant analysis can be done in the input space instead of the transformed space.

kernel function: 1) A function in an integral transformation (*e.g.*, the exponential term in the Fourier transform); 2) a function applied at every point in an image (see convolution).

kernel principal component analysis: An extension of the principal component analysis (PCA) method that allows classification with curved region boundaries. The kernel method is equivalent to a nonlinear mapping of the data into a high dimensional space from which the global axes of maximum variation are extracted. The method provides a transformation via a kernel so that PCA can be done in the input space instead of the transformed space.

key frames: Primarily a computer graphics animation technique, where key frames in a sequence are drawn by more experienced animators and then intermediate interpolating frames are drawn by less experienced animators. In computer vision motion sequence analysis, key frames are the analogous video frames, typically displaying motion discontinuities between which the scene motion can be smoothly interpolated.

KHOROS: An image processing development environment with a large set of operators. The system comes with a pull-down interactive development workspace where operators can be instantiated and connected by click and drag operations.

kinetic depth: A technique for estimating the depth at image feature points (usually edges) by exploiting a controlled sensor motion. This technique generally does not work at all points of the image because of insufficient image structure or sensor precision in smoothly varying regions, such as walls. See also shape from motion. A typical motion case is for the camera to rotate on a circular trajectory while fixating on a point in front of the camera, as seen here:

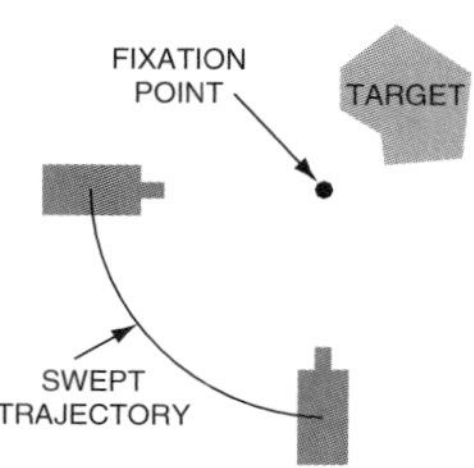

Kirsch compass edge detector: A first derivative edge detector that computes the gradient in different directions according to which calculation mask is used. Edges have high gradient values, so thresholding the intensity gradient magnitude is one approach to edge detection. A Kirsch mask that detects edges at 45° is:

$$\begin{bmatrix} -3 & 5 & 5 \\ -3 & 0 & 5 \\ -3 & -3 & -3 \end{bmatrix}$$

knowledge-based vision: A style of image interpretation that relies on multiple processing components capable of different image analysis processes, some of which may solve the same task in different ways. Linking the components together is a reasoning algorithm that knows about the capabilities of the different components, when they might be usable or might fail. An additional common component is some form of task dependent knowledge encoded in a knowledge representation that is used to help guide the reasoning algorithm. Also common is some uncertainty mechanism that records the confidence that the system has about the outcomes of its processing. For example, a knowledge-based vision system might be used for aerial analysis of road networks, containing specialized detection modules for straight roads, road junctions, forest roads as well as survey maps, terrain type classifiers, curve linking, etc.

knowledge representation: A general term for methods of computer encoding knowledge. In computer vision systems, this is usually knowledge about recognizable objects and visual processing methods. A common knowledge representation scheme is the geometric model that records the 2D or 3D shape of objects. Other commonly used vision knowledge representation schemes are graph models and frames.

Koenderink's surface shape classification: An alternative to the more common mean curvature and Gaussian curvature 3D surface shape classification labels. Koenderink's scheme decouples the two intrinsic shape parameters into one parameter (S) that represents the local surface shape (including cylindrical, hyperbolic, spherical and planar) and a second parameter (C) that encodes the magnitude of the curvedness of the shape. The shape classes represented in Koenderink's classification scheme are illustrated:

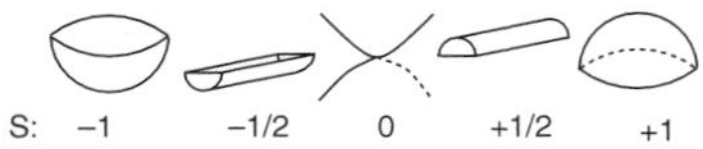

Kohonen network: A multivariate data clustering and analysis method that produces a topological organization of the input data. The response of the whole network to a given data vector can be used as a lower dimensional signature of the data vector.

KTC noise: A type of noise associated with Field Effect Transistor (FET) image sensors. The "KTC" term is used because the noise is proportional to $\sqrt{kTC}$ where T is the temperature, C is the capacitance of the image sensor and k is Boltzmann's constant. This noise arises during image capture at each pixel independently and is also independent of integration time.

Kullback–Leibler distance/ divergence: A measure of the relative entropy or distance between two probability densities $p_1(\vec{x})$ and $p_2(\vec{x})$, defined as

$$D(p_1 \| p_2) = \int p_1(\vec{x}) \log \frac{p_1(\vec{x})}{p_2(\vec{x})} d\vec{x}$$

kurtosis: A measure of the flatness of a distribution of grayscale values. If n_g is the number of pixels out of N with gray scale value g, then the fourth histogram moment is $\mu_4 = \frac{1}{N}\Sigma_g n_g (g - \mu_1)^4$, where μ_1 is the mean pixel value. The kurtosis is $\mu_4 - 3$.

Kuwahara: An edge-preserving noise reduction filter. The filter uses four regions surrounding the pixel being smoothed. The smoothed value for that pixel is the mean value of the region with smallest variance.

L

label: A description associated with something for the purposes of identification. For example see region labeling.

labeling problem: Given a set S of image structures (which may be pixels as well as more structured objects like edges) and a set of labels L, the labeling problem is the question of how to assign a label $l \in L$ for each image structure $s \in S$. This process is usually dependent on both the image data and neighboring labels. A typical remote sensing application is to label image pixels by their land type, such as water, snow, sand, wheat field, forest, etc. A range image (below left) has its pixels labeled by the sign of their mean curvature (white: negative, light gray: zero, dark gray: positive, black: missing data).

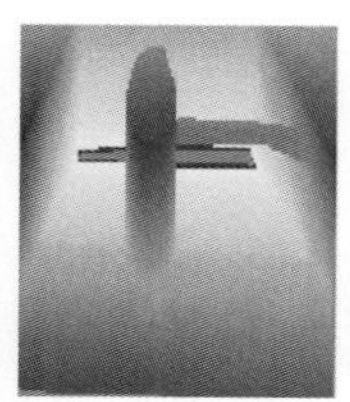

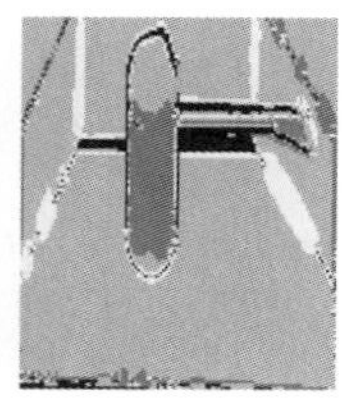

lacunarity: A scale dependent measure of translational invariance based on the size distribution of holes within a set. High lacunarity indicates that the set is heterogeneous and low lacunarity indicates homogeneity.

LADAR: LAser Detection And Ranging or Light Amplification for Detection and Ranging. See laser radar.

Lagrange multiplier technique: A method of constrained optimization to find a solution to a numerical problem that includes one or more constraints. The classical form of the Lagrange multiplier technique finds the parameter vector $\vec{v}$ minimizing (or maximizing) the function $f(\vec{v}) = g(\vec{v}) + \mu h(\vec{v})$, where $g()$ is the function being minimized and $h()$ is a constraint function that has value zero when its argument satisfies the constraint. The Lagrange multiplier is μ.

Laguerre formula: A formula for computing the directed angle between two 3D lines based on the cross ratio of four points. Two points arise where the two image lines intersect the ideal line (*i.e.*, the line through the vanishing points) and the other two points are the ideal line's absolute points (intersection of the ideal line and the absolute conic).

Lambert's law: The observed shading on ideal diffuse reflectors is independent of observer position and varies with the angle θ between the surface normal and source direction:

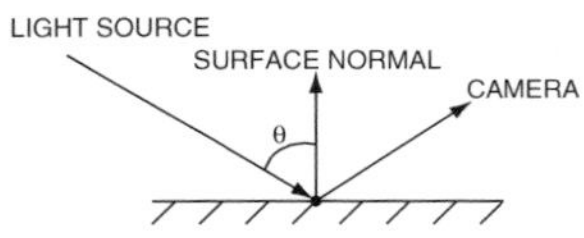

Lambertian surface: A surface whose reflectance obeys Lambert's law, more commonly known as a matte surface. These surfaces have equally bright appearance from all viewpoints. Thus, the shading of the surface depends only on the relative direction of the incident illumination.

landmark detection: A general term for detecting an image feature that is commonly used for registration. The registration might be between a model and the image or it might be between two images, etc. Landmarks might be task specific, such as components on an electronic circuit card or an anatomical feature such as the tip of the nose, or might be a more general image feature such as interest points.

LANDSAT: A series of satellites launched by the United States of America that are a common source of satellite images of the Earth. LANDSAT 7 for example was launched in April 1999 and provides complete coverage of the Earth every 16 days.

Laplacian: Loosely, the Laplacian of a function is the sum of its second order partial derivatives. For example the Laplacian of $f(x, y, z) : \mathbb{R}^3 \mapsto \mathbb{R}$ is $\nabla^2 f(x, y, z) = \frac{\partial^2 f}{\partial x^2} + \frac{\partial^2 f}{\partial y^2} + \frac{\partial^2 f}{\partial z^2}$. In computer vision, the Laplacian operator may be applied to an image, by convolution with the Laplacian kernel, one definition of which is given by the sum of second derivative kernels $[-1, 2, -1]$ and $[-1, 2, -1]^\top$, with zero padding to make the result 3×3:

$$\begin{pmatrix} 0 & -1 & 0 \\ -1 & 4 & -1 \\ 0 & -1 & 0 \end{pmatrix}$$

Laplacian of Gaussian operator: A low-level image operator that applies the second derivative Laplacian operator (∇^2) after a Gaussian smoothing operation everywhere in an image. It is an isotropic operator. It is often used as part of a zero crossing edge detection operator because the locations where the value changes sign (positive to negative or vice versa) of the output image are located near the edges in the input image, and the detail of the detected edges can be controlled by use of the scale parameter of the Gaussian smoothing. An example mask that implements the Laplacian of Gaussian operator with smoothing parameter $\sigma = 1.4$ is:

0	1	1	2	2	2	1	1	0
1	2	4	5	5	5	4	2	1
1	4	5	3	0	3	5	4	1
2	5	3	-12	-24	-12	3	5	2
2	5	0	-24	-40	-24	0	5	2
2	5	3	-12	-24	-12	3	5	2
1	4	5	3	0	3	5	4	1
1	2	4	5	5	5	4	2	1
0	1	1	2	2	2	1	1	0

Laplacian pyramid: A compressed image representation in which a pyramid of Laplacian images is created. At each level of the scheme, the current gray scale image has the Laplacian applied to it. The next level gray scale image is formed by Gaussian smoothing and subsampling. At the final level, the smoothed and subsampled image is kept. The original image can be approximately reconstructed level by level through expanding and smoothing the current level image and then adding the Laplacian.

laser: Light Amplification by Stimulated Emission of Radiation. A very bright light source often used for machine vision applications because of its properties: most light is at a single spectral frequency, the light is coherent, so various interference effects can be exploited and the light beam can be processed so that divergence is slight. Two common applications are for structured light triangulation and range sensing.

laser illumination: A very bright light source useful because of its limited spectrum, bright power and coherence. See also laser.

laser radar: (LADAR) A LIDAR range sensor that uses laser light. See also laser range sensor.

laser range sensor: A laser-based range sensor records the distance from the sensor to a target or target scene by detecting the image of a laser spot or stripe projected onto the scene. These sensors are commonly based on structured light triangulation, time of flight or phase difference technologies.

laser speckle: A time-varying light pattern produced by interference of the light reflected from a surface illuminated by a laser.

laser stripe triangulation: A structured light triangulation system that uses laser light. For example, a projected plane of light that would normally result in a straight line in the camera image is distorted by any objects in the scene where the distortion is proportional to the height of the object. A typical triangulation geometry is illustrated here:

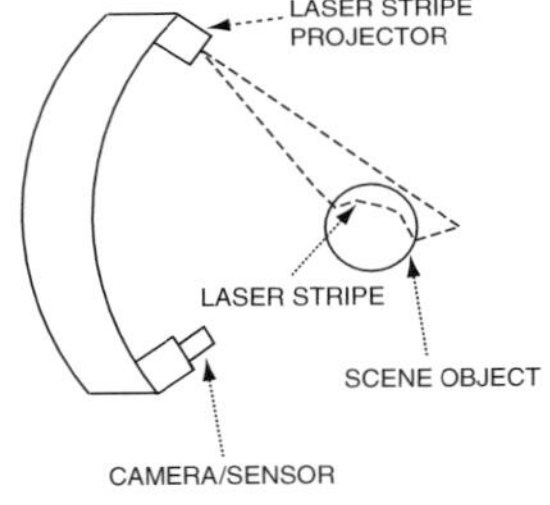

lateral inhibition: A process whereby a given feature weakens or eliminates nearby features. An example of this appears in the Canny edge detector where locally maximal intensity gradient magnitudes cause adjacent gradient values that lie across (as contrasted with along) the edge to be set to zero.

Laws' texture energy measure: A measure of the amount of image intensity variation at a pixel. The measure is based on 5 one dimensional finite difference masks convolved orthogonally to give 25 2D masks. The 25 masks are then convolved with the image. The outputs are smoothed nonlinearly and then combined to give 14 contrast and rotation invariant measures.

least mean square estimation: Also known as least square estimation or mean square estimation. Let $\vec{v}$ be the parameter vector that we are searching for and $e_i(\vec{v})$ be the error measure associated with the ith of N data items. The error measure often used is the Euclidean, algebraic or Mahalanobis distance between the ith data item and a curve or surface being fit, that is parameterized by $\vec{v}$. Then the mean square error is:

$$\frac{1}{N}\sum_{i=1}^{N} e_i(\vec{v})^2$$

The desired parameter vector $\vec{v}$ minimizes this sum.

least median of squares estimation: Let $\vec{v}$ be the parameter vector that we are searching for and $e_i(\vec{v})$ be the error associated with the ith of N data items. The error measure often used is the Euclidean, algebraic or Mahalanobis distance between the ith data item and a curve or surface being fit that is parameterized by $\vec{v}$. Then the median square error is the median or middle value of the sorted set $\{e_i(\vec{v})^2\}$. The desired parameter vector $\vec{v}$ minimizes this median value. This estimator usually requires more computation for the iterative and sorting algorithms but can be more robust to outliers than the least mean square estimator.

least square curve fitting: A least mean square estimation process that fits a parametric curve model or a line to a collection of data points, usually 2D or 3D. Fitting often uses the Euclidean, algebraic or Mahalanobis distance to evaluate the goodness of fit. Here is an example of least square ellipse fitting:

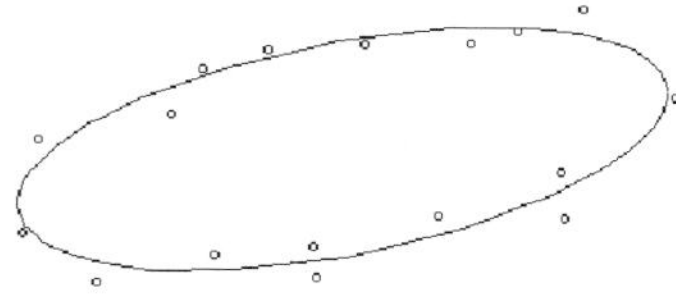

least square estimation: See least mean square estimation.

least squares fitting: A general term for a least mean square estimation process that fits some parametric shape, such as a curve or surface, to a collection of data. Fitting often uses the Euclidean, algebraic or Mahalanobis distance to evaluate the goodness of fit.

least square surface fitting: A least mean square estimation process that fits a parametric surface model to a collection of data points, usually range data. Fitting often uses the Euclidean, algebraic or Mahalanobis distance to evaluate the goodness of fit. The range image (above left) has planar and cylindrical surfaces fitted to the data (above right).

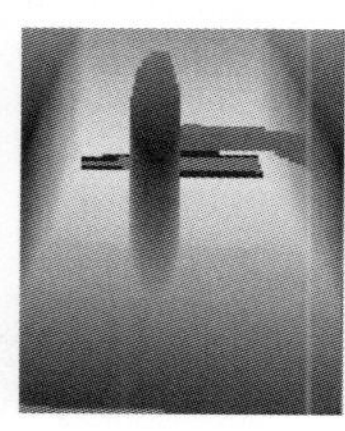

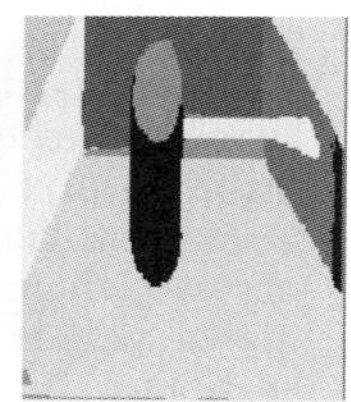

leave-one-out test: A method for testing a solution in which one sample is left out of the training set and used instead for testing. This can be done for every sample.

LED: Light Emitting semiconductor Diode. Often used as detectable point light source markers or controllable illumination.

left-handed coordinate system: A 3D coordinate system with the XYZ axes arranged as shown below. The alternative is a right-handed coordinate system.

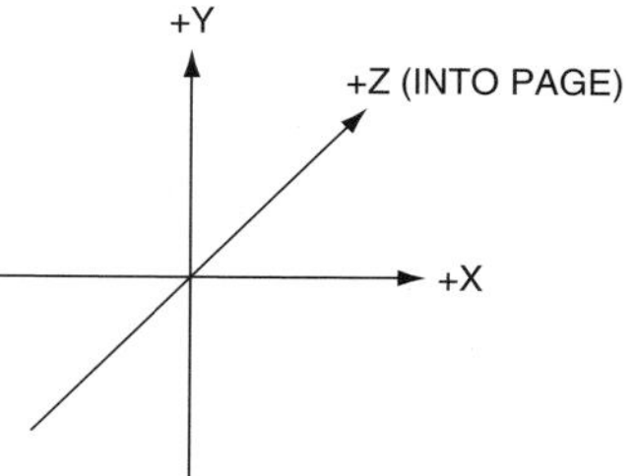

Legendre moment: The Legendre moment of a piecewise continuous function $f(x, y)$ with order (m, n) is $\frac{1}{4}(2m+1)(2n+1) \int_{-1}^{+1} \int_{-1}^{+1} P_m(x) P_n(y) f(x, y) \mathrm{d}x \mathrm{d}y$ where $P_m(x)$ is the mth order Legendre polynomial. These moments can be used for characterizing image data and images can be reconstructed from the infinite set of moments.

Lempel–Ziv–Welch (LZW): A form of file compression based on encoding commonly occurring byte sequences. This form of compression is used in the common GIF image file format.

lens: A physical optical device for focusing incident light onto an imaging surface, such as photographic film or an electronic sensor. Lenses can also be used to change magnification, enhance or modify a field of view.

lens distortion: Unexpected variation in the light field passing through a lens. Examples are radial lens distortion or chromatic aberration and usually arise from how the lens differs from the ideal lens.

lens equation: The simplest case of a convex converging lens with focal length f perfectly focused on a target at distance D has distance d between the lens and the image plane as related by the lens equation $\frac{1}{f} = \frac{1}{D} + \frac{1}{d}$ and illustrated here:

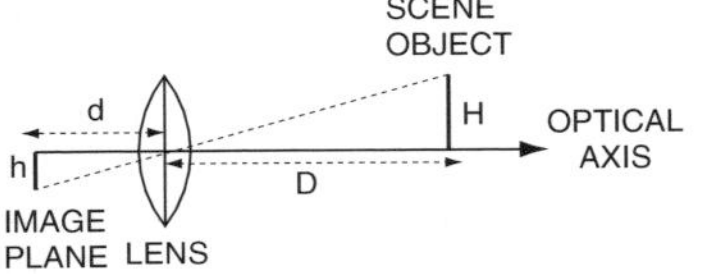

lens type: A general term for lens shapes and functions, such as convex or half-cylindrical, converging, magnifying, etc.

level set: The set of data points $\vec{x}$ that satisfy a given equation of the form: $f(\vec{x}) = c$. Varying the value of c gives different sets of usually closely related points. A visual analogy is of a geographic surface and the ocean rising. If the function $f()$ is the sea level, then the level sets are the shore lines for different sea levels c. The figure below shows an intensity image and the pixels at level (brightness) 80.

Levenberg–Marquardt optimization: A numerical multivariate optimization method that switches smoothly between gradient descent when far from a (local) optimum and a second-order inverse Hessian (quadratic) method when nearer.

license plate recognition: A computer vision application that aims to identify a vehicle's license plate from image data. Image data is often acquired from automatic cameras at places where vehicles slow down such as bridges and toll barriers.

LIDAR: LIght Detection And Ranging. A range sensor using (usually) laser light. It can be based on the time of flight of a pulse of laser light or the phase shift of a waveform. The measurement could be of a single point or an array of measurements if the light beam is swept across the scene/object.

Lie groups: A group that can be represented as a continuous and differentiable manifold of a space, such that group operations are also continuous. An example of a Lie group is the orthogonal group $SO(3) = \{\mathbf{R} \in \mathcal{R}^{3\times3} : \mathbf{R}^\top \mathbf{R} = I, det(\mathbf{R}) = 1\}$ of rigid 3D rotations.

light: A general term for the electromagnetic radiation used in many computer vision applications. The term could refer to the illumination in the scene or the irradiance coming from the scene onto the sensor. Most computer vision applications use light that is visible, infrared or ultraviolet.

light source: A general term for the source of illumination in a scene, whether deliberate or accidental. The light source might be a point light source or an extended light source.

light source detection: The process of detecting the position of or direction to the light sources in the scene, even if not observable. The light sources are usually assumed to be point light sources for this process.

light source geometry: A general term referring to the shape and placement of the light sources in a scene.

light source placement: A general term for the positions of the light sources in a scene. It may also refer to the care that machine vision applications engineers take when placing the light sources so as to minimize unwanted lighting effects, such as shadows and specular reflections, and to enhance the visibility of desired scene structures, *e.g.*, by back lighting or oblique lighting.

light stripe ranging: See structured light triangulation.

lightfield: A function that encodes the radiance on an empty point in space as a function of the point's position and the direction of the illumination. A lightfield allows image based rendering of new (unoccluded) scene views from arbitrary positions within the lightfield.

lighting: A general term for the illumination in a scene, whether deliberate or accidental.

lightness: The estimated or perceived reflectance of a surface, when viewed in monochrome.

lightpen: A user-interface device that allows people to indicate places on a computer screen by touching the screen at the desired place with the pen. The computer can then draw items, select actions, etc. It is effectively a type of mouse that acts on the display screen instead of on a mat.

likelihood ratio: The ratio of probabilities of observing data D with and without condition C: $\frac{P(D|C)}{P(D|\neg C)}$.

limb extraction: A process of image interpretation that extracts 1) the arms or legs of people or animals, *e.g.*, for tracking or 2) the barely visible edge of a curved surface as it curves away from an observer (derived from an astronomical term). See figure below. See also occluding contour.

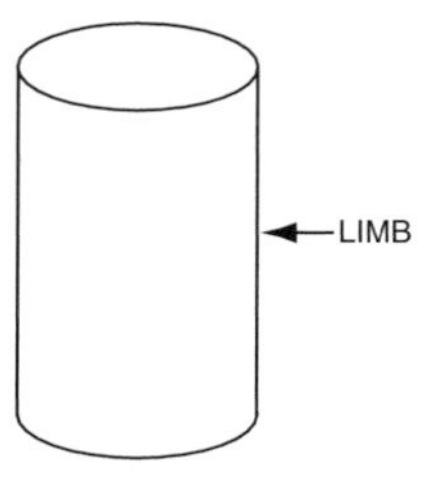

line: Usually refers to a straight ideal line that passes through two points, but may also refer to a general curve marking, *e.g.*, on paper.

line cotermination: When two lines have endpoints in exactly or nearly the same location. See examples:

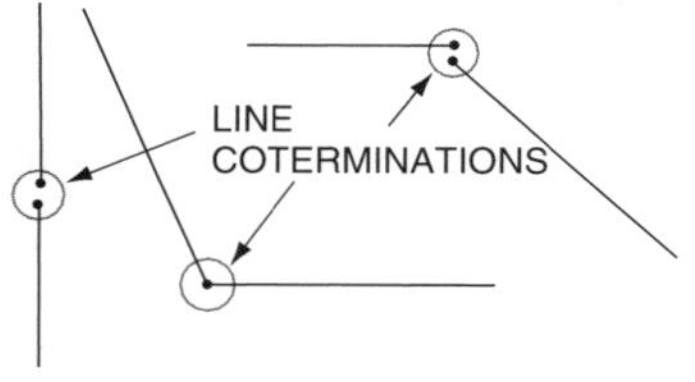

line detection operator: A feature detection process that detects lines. Depending on the specific operator, locally linear line segments may be detected or straight lines might be globally detected. Note that this detects lines as contrasted with edges.

line drawing analysis: 1) Analysis of hand-made or CAD drawings to extract a symbolic description or shape description. For example, research has investigated extracting 3D building models from CAD drawings. Another application is the analysis of hand-drawn circuit sketches to form a circuit description. 2) Analysis of the line junctions in a polyhedral blocks world scene, in order to understand the 3D structure of the scene.

line fitting: A curve fitting problem where the objective is to estimate the parameters of a straight line that best interpolates given point data.

line following: See line grouping.

line grouping: Generally refers to the process of creating a longer curve by grouping together shorter fragments found by line detection. These might be short connecting locally detected line fragments, or might be longer straight line segments separated by a gap. May also refer to the grouping of line segments on the basis of grouping principles such as parallelism.

See also edge tracking, perceptual organization, Gestalt.

line intersection: Where two or more lines intersect at a point. The lines cross or meet at a line junction. See:

LINE INTERSECTIONS

line junction: The point at which two or more lines meet. See junction labeling.

line label: In an ideal polyhedral blocks world scene, lines arise from only a limited set of physical situations such as convex or concave surface shape discontinuities (fold edges), occluding edges where a fold edge is seen against the background (blade edge), crack edges where two polyhedra have aligned edges or shadow edges. Line labels identify the type of line (*i.e.*, one of these types). Assigning labels is one step in scene understanding that helps deduce the 3D structure of the scene. See also junction label. Here is an example of the usual line labels for convex(+), concave(−) and occluding (>) edges.

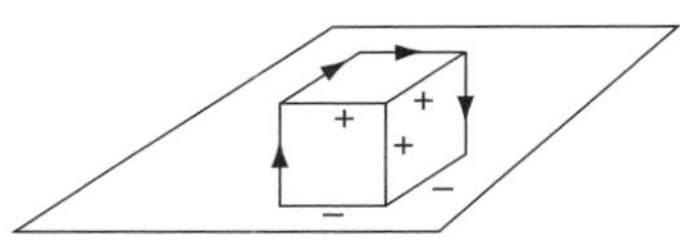

line linking: See line grouping.

line matching: The process of making a correspondence between the lines in two sets. One set might be a geometric model such as used in model based recognition or model registration or alignment. Alternatively, the lines may have been extracted from different images, as when doing feature based stereo or estimating the epipolar geometry between the two lines.

line moment: A line moment is similar to the traditional area moment but is calculated only at points $(x(s), y(s))$ along the object contour. The pq^{th} moment is: $\int x(s)^p y(s)^q \mathrm{d}s$. The infinite set of line moments uniquely determine the contour.

line moment invariant: A set of invariant values computable from the line moments. These may be invariant to translation, scaling and rotation.

line of sight: A straight line from the observer or camera into the scene, usually to some target. See:

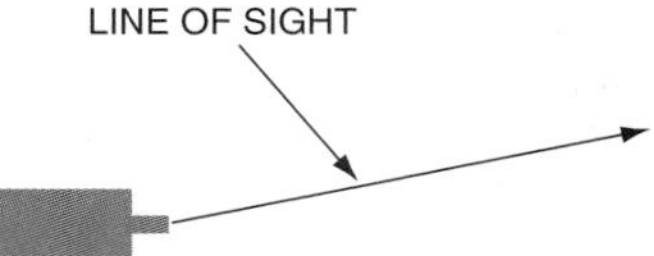

line scan camera: A camera that uses a solid-state or semiconductor (*e.g.*, CMOS) linear array sensor, in which all of the photosensitive elements are in a single 1D line. Typical line scan cameras have between 32 and 8192 elements. These sensors are used for a variety of machine vision applications such as scanning, flow process control and position sensing.

line segmentation: See curve segmentation.

line spread function: The line spread function describes how an ideal infinitely thin line would be distorted after passing through an optical system. Normally, this can be computed by integrating the point spread functions of an infinite number of points along the line.

line thinning: See thinning.

linear: 1) Having a line-like form. 2) A mathematical description for a process in which the relationship between some input variables $\vec{x}$ and some output variables $\vec{y}$ is given by $\vec{y} = \mathbf{A}\vec{x}$ where $\mathbf{A}$ is a matrix.

linear array sensor: A solid-state or semiconductor (*e.g.*, CMOS) sensor in which all of the photosensitive elements are in a single 1D line. Typical linear array sensors have between 32 and 8192 elements and are used in line scan cameras.

linear discriminant analysis: See linear discriminant function.

linear discriminant function: Assume a feature vector $\vec{x}$ based on observations of some structure. (Assume that the feature vector is augmented with an extra term with value 1.) A linear discriminant function is a basic classification process that determines which of two classes or cases the structure belongs to based on the sign of the linear function $l = \vec{a} \cdot \vec{x} = \sum a_i x_i$, for a given coefficient vector $\vec{a}$. For example, to discriminate between unit side squares and unit diameter circles based on the area A, the feature vector is $\vec{x} = (A, 1)'$ and the coefficient vector $\vec{a} = (1, -0.89)'$. If $l > 0$, then the structure is a square, otherwise a circle.

linear features: A general term for features that are locally or globally straight, such as lines or straight edges.

linear filter: A filter whose output is a weighted sum of its inputs, *i.e.*, all terms in the filter are either constants or variables. If $\{x_i\}$ are the inputs (which may be pixel values from a local neighborhood or pixel values from the same position in different images of the same scene, etc.), then the linear filter output would be $\sum a_i x_i + a_0$, for some constants a_i.

linear regression: Estimation of the parameters of a linear relationship between two random variables X and Y given sets of samples $\vec{x}_i$ and $\vec{y}_i$. The objective is to estimate the matrix $\mathbf{A}$ and vector $\vec{a}$ that minimize the residual $r(\mathbf{A}, \vec{a}) = \sum_i \|\vec{y}_i - \mathbf{A}\vec{x}_i - \vec{a}\|^2$. In this form, the $\vec{x}_i$ are assumed to be noise-free quantities. When both variables are subject to error, orthogonal regression is preferred.

linear transformation: A mathematical transformation of a set of values by addition and multiplication by constants. If the set of values is a vector $\vec{x}$, the general linear transformation produces another vector $\vec{y} = \mathbf{A}\vec{x}$, where $\vec{y}$ need not have the same dimension as $\vec{x}$ and $\mathbf{A}$ is a constant matrix (*i.e.*, is not a function of $\vec{x}$).

lip shape analysis: An application of computer vision to understanding the position and shape of human lips as part of face analysis. The goal might be face recognition or expression understanding.

lip tracking: An application of computer vision to following the position and shape of human lips in a video sequence. The goal might be for lip reading, augmentation of deaf sign analysis or focusing of resolution during image compression.

local: A *local property* of a mathematical object is one that is defined in terms only of a small neighborhood of the object, for instance, curvature. In image processing, a local operator operates on a small number of nearby pixels at a time.

local binary pattern: Given a local neighborhood about a point, use the value of the central pixel to threshold the neighborhood. This creates a local descriptor of the gray scale structure that is invariant to lightness and contrast transformations, that can be used to create local texture primitives.

local contrast adjustment: A form of contrast enhancement that adjusts pixel intensities based on the values of nearby pixels instead of the values of all pixels in the image. The right image has the eye area's brightness (from original image at the left) enhanced while maintaining the background's contrast:

local curvature estimation: A part of surface or curve shape estimation that estimates the curvature at a given point based on the position of nearby parts of the curve or surface. For example, the curve $y = sin(x)$ has zero local curvature at the point $x = 0$ (*i.e.*, the curve is locally uncurved or straight), although the curve has nonzero local curvature at other other points (*e.g.*, at $\frac{\pi}{4}$). See also differential geometry.

Local Feature Focus (LFF) method: A 2D part identification and pose estimation algorithm that can cope with large amounts of occlusion of the parts. The algorithm uses a mixture of property-based classifiers, graph models and geometric models. The key identification process is based around local configurations of image features that is more robust to occlusion.

local invariant: See local point invariant.

local operator: An image processing operator that computes its output at each pixel from the values of the nearby pixels instead of using all or most of the pixels in the image.

local point invariant: A property of local shape or intensity that is invariant to, *e.g.*, translation, rotation, scaling, contrast or brightness changes, etc. For example, a surface's Gaussian curvature is invariant to change in position.

local surface shape: The shape of a surface in a "small" region around a point, often classified into one of a small number of surface shape classifications. Computed as a function of the surface curvatures.

local variance contrast: The variance of the pixel values computed in a neighborhood about each pixel. Contrast is the difference between the larger and smaller values of this variance. Large values of this property occurs in highly textured or varying areas.

log-polar image: An image representation in which the pixels are not in the standard Cartesian layout but instead have a space varying layout. In the log-polar case, the image is parameterized by a polar coordinate θ and a radial coordinate r. However, unlike polar coordinates, the radial distance increases exponentially as r grows. The mapping from position (θ, r) to Cartesian coordinates is $(\beta^r cos(\theta), \beta^r sin(\theta))$, where β is some design parameter. Further, the amount of area of the image plane represented by each pixel grows exponentially with r, although the precise pixel size depends on factors like amount of pixel overlap, etc. See also foveal image. The receptive fields of a log-polar image (courtesy of Herman Gomes) can be seen in the outer rings of:

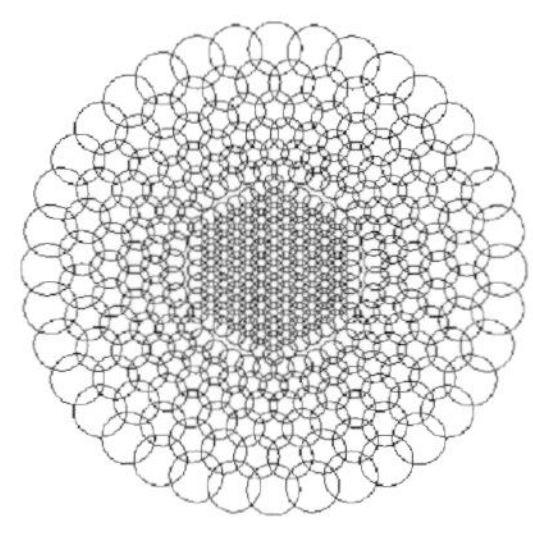

log-polar stereo: A form of stereo vision in which the input images come from log-polar sensors instead of the standard Cartesian layout.

logarithmic transformation: See pixel logarithm operator.

logical object representation: An object representation based on some logical formalism such as the predicate calculus.

For example, a square can be defined as:

$$square(s) \iff polygon(s) \\ \& \; number\_of\_sides(s, 4) \\ \& \; \forall e_1 \; \forall e_2 (e_1 \neq e_2 \; \& \\ side\_of(s, e_1) \& \; side\_of(s, e_2) \\ \& \; length(e_1) = length(e_2) \\ \& \; (parallel(e_1, e_2) \\ \mid perpendicular(e_1, e_2))).$$

long baseline stereo: See wide baseline stereo.

long motion sequence: A video sequence of more than just a few frames in which there is significant camera or scene motion. The essential idea is that the 3D scene structure can be inferred by effectively a stereo vision process. Here the matched image features can be tracked through the sequence, instead of having to solve the stereo correspondence problem. If a long sequence is not available, then analysis could use optical flow or short baseline stereo.

look-up table: Given a finite set of input values $\{x_i\}$ and a function on these values, $f(x)$, a look-up table records the values $\{(x_i, f(x_i))\}$ so that the value of the function $f()$ can be looked up directly rather than recomputed each time. Look-up tables can be easily used for color remapping or standard functions of integer pixel values (*e.g.*, the logarithm of a pixel's value).

lossless compression: A category of image compression in which the original image can be exactly reconstructed from the compressed image. This contrasts with lossy compression.

lossy compression: A category of image compression in which the original image cannot be exactly reconstructed from the compressed image. The goal is to lose insignificant image details (*e.g.*, noise) while limiting perception of changes to the image appearance. Lossy algorithms generally produce greater compression than lossless compression.

low angle illumination: A machine vision technique, often used for industrial vision, where a light source (usually a point light source) is placed so that a ray of light from the source to the inspection point is almost perpendicular to the surface normal at that point. The situation can also arise naturally, *e.g.*, from the sun position at dawn or dusk. One consequence of this low angle is that shallow surface shape defects and cracks cast strong shadows that may simplify the inspection process. See:

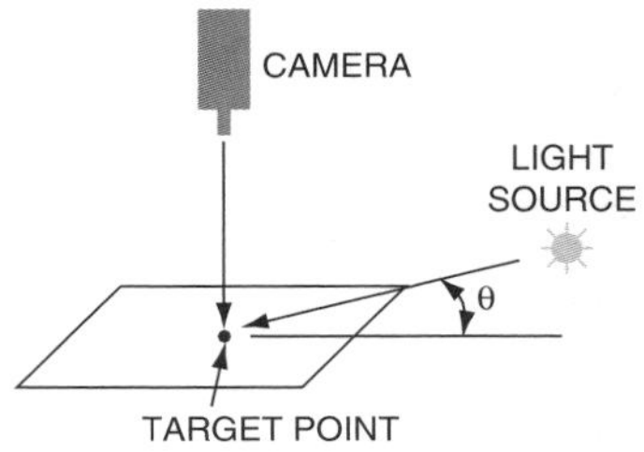

low frequency: Usually referring to low spatial frequency in the context of computer vision. The low-frequency components of an image are the slowly changing intensity components of the image, such as large regions of bright and dark pixels. If low temporal frequency is the intended meaning, then low frequency refers to slowly changing patterns of brightness or darkness at the same pixel in a video sequence. This image shows the low-frequency components of an image.

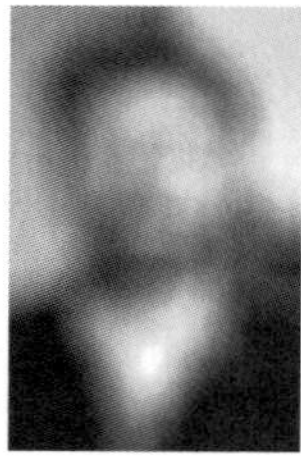

low level vision: A general and somewhat imprecisely (*i.e.*, contentiously) defined term for the initial stages of image analysis in a vision system. It can also be used for the initial stages of processing in biological vision systems. Roughly, low level vision refers to the first few stages of processing applied to intensity images. Some authors use this term only for operations that result in other images. So, edge detection is about where most authors would say that low-level vision ends and middle-level vision starts.

low pass filter: This term is imported from 1D signal processing theory into image processing. The term "low" is a shorthand for "low frequency", that, in the context of a single image, means low spatial frequency, *i.e.*, intensity patterns that change over many pixels. Thus a low pass filter applied to an image leaves the low spatial frequency patterns, or large, slowly changing patterns, and removes the high spatial frequency components (sharp edges, noise). Low pass filters are a kind of smoothing or noise reduction filter. Alternatively, filtering is applied to the changing values of a given pixel over an image sequence. In this case the pixel values can be treated as a sampled time sequence and the original signal processing definition of "low pass filter" is appropriate. Filtering this way removes rapid temporal changes. See also high pass filter. Here is an image and a low-pass filtered version:

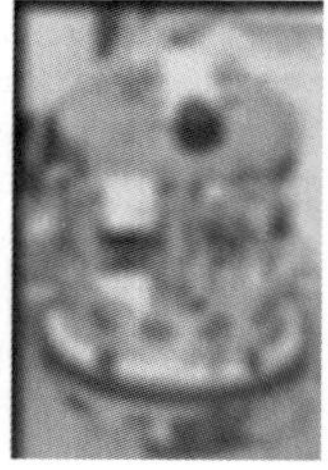

Lowe's curve segmentation method: An algorithm that tries to split a curve into a sequence of straight line segments. The algorithm has three main stages: 1) a recursive splitting of segments into two shorter, but more line-like segments, until all remaining segments are very short. This forms a tree of segments. 2) Merging segments in the tree in a bottom-up fashion according to a straightness measure. 3) Extracting the remaining unmerged segments from the tree as the segmentation result.

luma: The luminance component of light. Color can be divided into luma and chroma.

luminance: The measured intensity from a portion of a scene.

luminance efficiency: The sensor specific function $V(\lambda)$ that determines how the observed light $I(x, y, \lambda)$ at sensor position (x, y) of wavelength λ contributes to the measured luminance $l(x, y) = \int I(\lambda)V(\lambda)d\lambda$ at that point.

luminous flux: The amount of light at all wavelengths that passes through a given region in space. Proportional to perceived brightness.

luminosity coefficient: A component of tristimulus color theory. The luminosity coefficient is the amount of luminance contributed by a given primary color to the total perceived luminance.

M

M-estimation: A robust generalization of least square estimation and maximum likelihood estimation.

Mach band effect: An effect in the human visual system in which a human observer perceives a variation in brightness at the edges of a region of constant brightness. This variation makes the region appear slightly darker when it is beside a brighter region and appear slightly brighter when it is beside a darker region.

machine vision: A general term for processing image data by a computer and often synonymous with computer vision. There is a slight tendency to use "machine vision" for practical vision systems, such as for industrial vision, and "computer vision" for more exploratory vision systems or for systems that aim at some of the competences of the human vision system.

macrotexture: The intensity pattern formed by spatially organized texture primitives on a surface, such as a tiling. This contrasts with microtexture.

magnetic resonance imaging (MRI): See NMR.

magnification: The process of enlargement (*e.g.*, of an image). The amount of enlargement applied.

magnitude-retrieval problem: The reconstruction of a signal based on only the phase (not the magnitude) of the Fourier transform.

Mahalanobis distance: The distance between two N-dimensional points scaled by the statistical variation in each component of the point. For example, if $\vec{x}$ and $\vec{y}$ are two points from the same distribution that has covariance matrix $\mathbf{C}$ then the Mahalanobis distance is given by

$$((\vec{x}-\vec{y})'\mathbf{C}^{-1}(\vec{x}-\vec{y}))^{\frac{1}{2}}$$

Dictionary of Computer Vision and Image Processing R.B. Fisher, K. Dawson-Howe, A. Fitzgibbon, C. Robertson and E. Trucco

The Mahalanobis distance is the same as the Euclidean distance if the covariance matrix is the identity matrix. A common usage in computer vision systems is for comparing feature vectors whose elements are quantities having different ranges and amounts of variation, such as a 2-vector recording the properties of area and perimeter.

mammogram analysis: A mammogram is an X-ray of the human female breast. The main purpose of analysis is the detection of potential signs of cancerous growths.

Manhattan distance: Also called the Manhattan metric. Motivated by the problem of only being able to walk along city blocks in dense urban environments, the distance between points (x_1, y_1) and (x_2, y_2) is $\mid x_1 - x_2 \mid + \mid y_1 - y_2 \mid$.

many view stereo: See multi-view stereo.

MAP: See maximum a *posteriori* probability.

map analysis: Analyzing an image of a map (*e.g.*, obtained with a flat-bed scanner) in order to extract a symbolic description of the terrain described by the map. This is now a largely obsolete process given digital map databases.

map registration: The registration of a symbolic map to (usually) aerial or satellite image data. This may require identifying roads, buildings or land features. This image shows a road model (black) overlaying an aerial image.

marching cubes: An algorithm for locating surfaces in volumetric datasets. Given a function $f()$ on the voxels, the algorithm estimates the position of the surface $f(\vec{x}) = c$ for some c. This requires estimating where the surface intersects each of the twelve edges of a voxel. Many implementations propagate from one voxel to its neighbors, hence the "marching" term.

marginal distribution: A probability distribution of a random variable X derived from the joint probability distribution of a number of random variables integrated over all variables except X.

Markov Chain Monte Carlo: Markov Chain Monte Carlo (MCMC) is a statistical inference method useful for estimating the parameters of complex distributions. The method generates samples from the distribution by running the Markov Chain that models the problem for a long time (hopefully to equilibrium) and then uses the ensemble of samples to estimate the distribution. The states of the Markov Chain are the possible configurations of the problem.

Markov random field (MRF): An image model in which the value at a pixel can be expressed as a linear weighted sum of the values of pixels in a finite neighborhood about the original pixel plus an additive random noise value.

Marr's theory: A shortened term for "Marr's theory of the human vision system". Some of the key stages in this integrated but incomplete theory are the raw primal sketch, full primal sketch, 2.5D sketch and 3D object recognition.

Marr–Hildreth edge detector: An edge detector based on multi-scale analysis of the zero-crossings of the Laplacian of Gaussian operator.

mask: A term for an $m \times n$ array of numbers or symbolic labels. A mask can be the smoothing mask used in a convolution, the target in a template matching or the kernel used in a mathematical morphology operation, etc. Here is a simple mask for computing an approximation to the Laplacian operator:

0	1	0
1	–4	1
0	1	0

matched filter: A matched filter is an operator that produces a strong result in the output image when it processes a portion of the input image containing a pattern for which it is "matched". For example, the filter could be tuned for the letter "e" in a given font size and type style, or a particular face viewed at the right scale. It is similar to template matching except the matched filter can be tuned for spatially separated patterns. This is a signal processing term imported into image processing.

matching function: See similarity metric.

matching method: A general term for finding the correspondences between two structures (*e.g.*, surface matching) or sets of features (*e.g.*, stereo correspondence).

mathematical morphology operation: A class of mathematically defined image processing operations in which the result is based on the spatial pattern of the input data values rather than values themselves. For example, a morphological line thinning algorithm would identify places in an image where a line description was represented by data more than 1 pixel wide (*i.e.*, the pattern to match). As this is redundant, the thinning algorithm would chose one of the redundant pixels to be set to 0. Mathematical morphology operations can apply to both binary and gray scale images. This figure shows a small image patch image before and after a thinning operation.

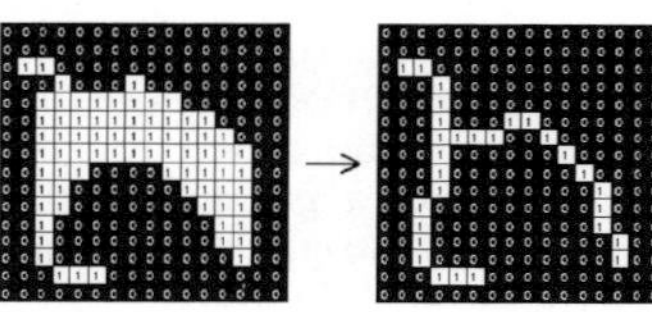

matrix: A mathematical structure of a given number of rows and columns with each entry usually containing a number. A matrix can be used to represent a transformation between two coordinate systems, record the covariance of a set of vectors, etc. A matrix for rotating a 2D vector by $\frac{\pi}{6}$ radians is:

$$\begin{bmatrix} cos(\frac{\pi}{6}) & sin(\frac{\pi}{6}) \\ -sin(\frac{\pi}{6}) & cos(\frac{\pi}{6}) \end{bmatrix}$$

$$= \begin{bmatrix} 0.866 & 0.500 \\ -0.500 & 0.866 \end{bmatrix}$$

matrix array camera: A 2D solid state imaging sensor, such as those found in typical current video, webcam and machine vision cameras.

matte surface: A surface whose reflectance follows the Lambertian model.

maximal clique: A clique (all nodes are connected to all other nodes in the clique) where no further nodes exist that are connected to all nodes in the clique. Maximal cliques may have different sizes – the issue is maximality, not size. Maximal cliques are used in association graph matching algorithms to represent maximally matched structures. The graph below has two maximal cliques: BCDE and ABD.

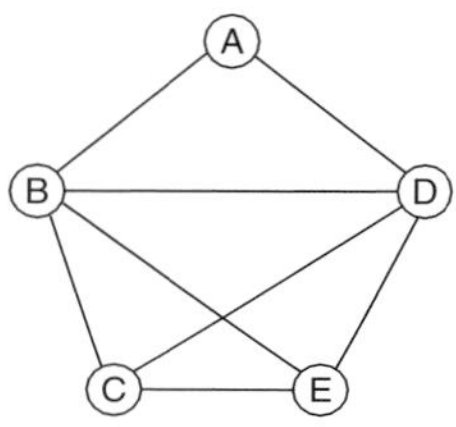

maximum *a posteriori* probability: The highest probability after some event or observations. This term is often used in the context of parameter estimation, pose estimation or object recognition problems, in which case we wish to estimate the parameters, position or identity respectively that have highest probability given the observed image data.

maximum entropy: A method for extracting the maximum amount of information (entropy) from a measurement (such as an image) in the presence of noise. This method will always give a conservative result; only presenting structure where there is evidence for it.

maximum entropy restoration: An image restoration technique based on maximum entropy.

maximum likelihood estimation: Estimating the parameters of a problem that has the highest likelihood or probability, *i.e.*, given the observed data. For example, the maximum likelihood estimate of the mean of a Gaussian distribution is the average of the observed samples drawn from that distribution.

MCMC: See Markov Chain Monte Carlo.

MDL: See minimum description length.

mean and Gaussian curvature shape classification: A classification of a local (*i.e.*, very small) surface patch (often at single pixels from a range image) into one of a set of simple surface shape classes based on the signs of the mean and Gaussian curvatures. The standard set of shape classes is: {plane, concave cylinder, convex cylinder, concave ellipsoid, convex ellipsoid, saddle valley, saddle ridge, minimal}. Sometimes the classes {saddle valley, saddle ridge, minimal} are conflated into the single class "hyperbolic". This table summarizes the classifications based on the curvature signs:

MEAN CURVATURE
– 0 +
GAUSSIAN CURVATURE
–
0
+
IMPOSSIBLE

mean curvature: A mathematical characterization for a component of local surface shape at a point on a smooth surface. Each point can be uniquely described by a pair of principal curvatures. The mean curvature is the average of the principal curvatures.

mean filter: See mean smoothing operator.

mean shift: An adaptive gradient ascent technique that operates by iteratively moving the center of a search window to the average of certain points within the window.

mean smoothing operator: A noise reduction operator that can be applied to a gray scale image or to separate components of a multispectral image. The output value at each pixel is the average of the values of all pixels in a neighborhood of the input pixel. The size of the neighborhood determines how much smoothing (or noise reduction) is done, but also how much blurring of fine detail also occurs. A image with Gaussian noise with $\sigma = 13$ and its mean smoothing are:

measurement resolution: The degree to which two differing quantities can be distinguished by measurement. This may be the minimum spatial distance that two adjacent pixels represent (spatial resolution) or the minimum time difference between visual observations (temporal resolution), etc.

medial axis skeletonization: See medial axis transform.

medial axis transform: An operation on a binary image that transforms regions into sets of pixels that are the centers of circles that are bitangent to the boundary and that fit entirely within the region. The value of each point on the axis is the radius of the bitangent circle. This can be used to represent the region by a simpler axis-like structure and is most effective on elongated regions. A region and its medial axis are below.

medial line: A curve going through the middle of an elongated structure. See also medial axis transform. This figure shows a region and its medial line.

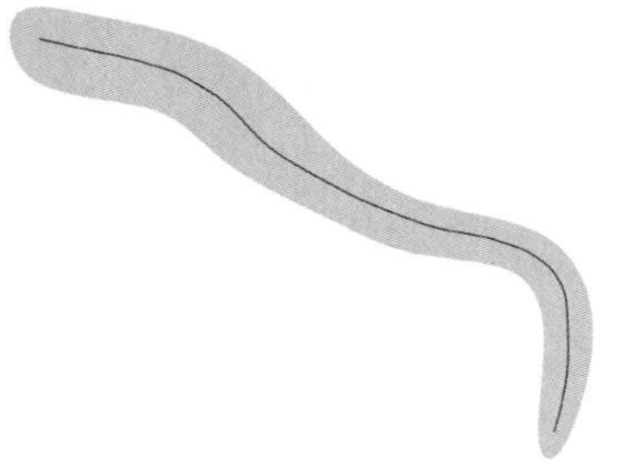

medial surface: The medial surface of a volume is the 3D generalization of the medial axis of a planar region. It is the locus of centers of spheres that touch the surface of the volume at three or more points.

median filter: See median smoothing.

median flow filtering: A noise reduction operation on vector data that generalizes the median filter on image data. The assumption is that the vectors in a spatial neighborhood about the current vector should be similar. Dissimilar vectors are rejected. The term "flow" arose through the filter's development in the context of image motion.

median smoothing: An image noise reduction operator that replaces a pixel's value by the median (middle) of the sorted pixel values in its neighborhood. An image with salt-and-pepper noise and the result of applying median smoothing are:

medical image registration: A general term for registration of two or more medical image types or an atlas with some image data. A typical registration would align X-ray CAT and NMR images.

membrane model: A surface fitting model that minimizes a combination of the smoothness of the fit surface and the closeness of the fit surface to the original data. The surface class must have C^0 continuity and thus it differs from the smoother thin plate model that has C^1 continuity.

mesh model: A tessellation of an image or surface into polygonal patches, much used in computer aided design (CAD). The vertices of the mesh are called nodes, or nodal points. A popular class of meshes is based on triangles, for instance the Delaunay triangulation. Meshes can be uniform, *i.e.*, all polygons are the same, or non-uniform. Uniform meshes can be represented by small sets of parameters. Surface meshes have been used for modeling free-form surfaces (*e.g.*, faces, landscapes). See also surface fitting. This icosahedron is a mesh model of a nearly spherical object:

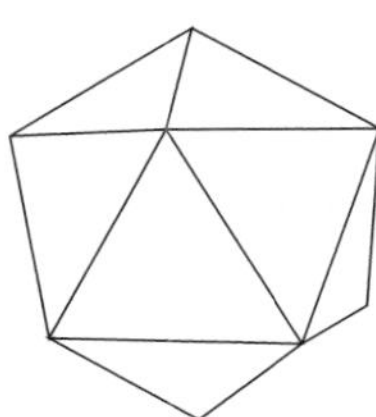

mesh subdivision: Methods for subdividing cells in a mesh model into progressively smaller cells. For example see Delaunay triangulation.

metameric colors: Colors that are defined by a limited number of channels each of which integrates a range of the spectrum. Hence the same metameric color can be caused by a variety of spectral distributions.

metric determinant: The metric determinant is a measure of curvature. For surfaces, it is the square root of the determinant of the first fundamental form matrix of the surface.

metric property: A visual property that is a measurable quantity, such as a distance or area. This contrasts with logical properties such as image connectedness.

metric reconstruction: Reconstruction of the 3D structure of a scene with correct spatial dimensions and angles. This contrasts with projective reconstruction. Two views of a metrical and projective

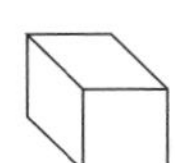

OBSERVED VIEW

RECONSTRUCTED VIEW

METRICAL RECONSTRUCTION

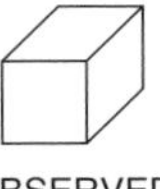

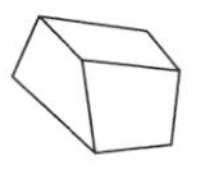

OBSERVED VIEW

RECONSTRUCTED VIEW

PERSPECTIVE RECONSTRUCTION

reconstruction of a cube are below. The metrical projection looks "correct" from all views, but the perspective projection may look "correct" only from the views where the data was acquired.

metric stratum: These are the set of similarity transformations (*i.e.*, rigid transformations with a scaling). This is what can be recovered from image data without external information such as some known length.

metrical calibration: Calibration of intrinsic and extrinsic camera parameters to enable metric reconstruction of a scene.

Mexican hat operator: A convolution operator that implements either a Laplacian of Gaussian or difference of Gaussians operator (which produce very similar results). The mask that can be used to implement this convolution has a shape similar to a Mexican hat (sombrero), as seen here:

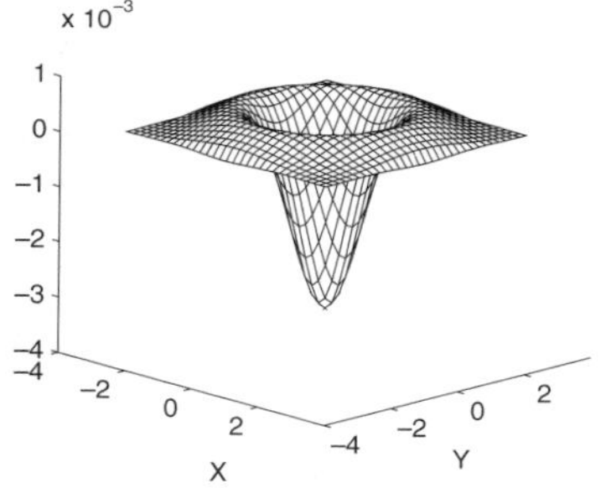

micron: One millionth of a meter; a micrometer.

microscope: An optical device observing small structures such as organic cells, plant fibers or integrated circuits.

microtexture: See statistical texture.

mid-sagittal plane: The plane that separates the body (and brain) into left and right halves. In medical imaging (*e.g.*, NMR), it usually refers to a view of the brain sliced down the middle between the two hemispheres.

middle level vision: A general term referring to the stages of visual data processing between low level and high level vision. There are many variations of the definition of this term but a usable rule of thumb is that middle level vision starts with descriptions of the contents of an image and results in descriptions of the features of the scene. Thus, binocular stereo would be a middle level vision process because it acts on image edge fragments to produce 3D scene fragments.

MIMD: See multiple instruction multiple data.

minimal point: A point on a hyperbolic surface where the two principal curvatures are equal in magnitude but opposite in sign, *i.e.*, $\kappa_1 = -\kappa_2$.

minimal spanning tree: Consider a graph G and a subset

T of the arcs in G such that all nodes in G are still connected in T and there is exactly one path joining any two nodes. T is a spanning tree. If each arc has a weight (possibly constant), the minimal spanning tree is the tree T with smallest total weight. This is a graph and its minimal spanning tree:

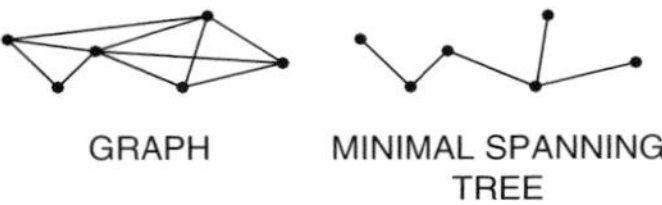

minimum bounding rectangle: The rectangle of smallest area that surrounds a set of image data.

minimum description length (MDL): A criterion for comparing descriptions usually based on the implicit assumption that the best description is the one that is shortest (*i.e.*, takes the fewest number of bits to encode). The minimum description usually requires several components: 1) the models observed (*e.g.*, whether lines or circular arcs), 2) the parameters of the models (*e.g.*, the line endpoints), 3) how the image data varies from the models (*e.g.*, explicit deviations or noise model parameters) and 4) the remainder of the image that is not explained by the models.

minimum distance classifier: Given an unknown sample with feature vector $\vec{x}$, select the class c with model vector $\vec{m}_c$ for which the distance $|| \vec{x} - \vec{m}_c ||$ is smallest.

minimum spanning tree: See minimal spanning tree.

MIPS: millions of instructions per second.

mirror: A specularly reflecting surface for which incident light is reflected only at the same angle and in the same plane as the surface normal.

miss-one-out test: See leave-one-out test.

missing data: Data that is unavailable, hence requiring it to be estimated. For example a moving person may become occluded resulting in missing position data for a number of frames.

missing pixel: A pixel for which no value is available (*e.g.*, if there was a problem with a sensing element in the image sensor).

mixed pixel: A pixel whose measurement arises from more than one scene phenomena. For example, a pixel that observes the edge between two regions. This pixel has a gray level that lies between the different gray levels of the two regions.

mixed reality: Image data that contains both original image data and overlaid computer graphics. See also augmented reality. This image shows an example of mixed reality, where the butterfly is a graphical object added to the image of the small robot:

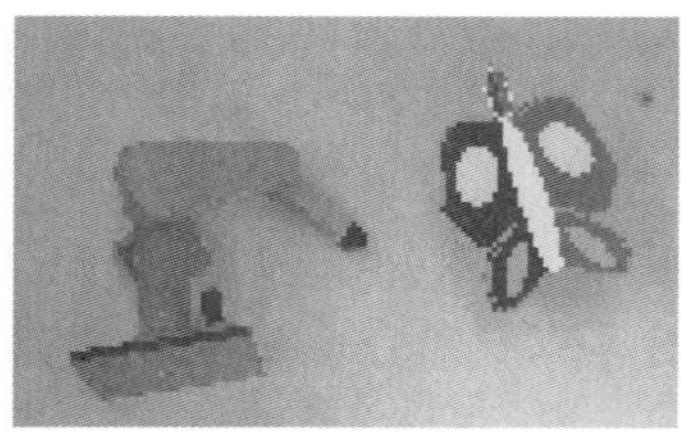

mixture model: A probabilistic representation in which more than one distribution is combined, modeling a situation where the data may arise from different sources or have different behaviors, each with different probability distributions.

MLE: See maximum likelihood estimation.

modal deformable model: A deformable model based on modal analysis (*i.e.*, study of the different shapes that an object can assume).

mode filter: A noise reduction filter that, for each pixel, outputs the mode (most common) value in its local neighborhood. The figure below shows a raw image with salt-and-pepper noise and the filtered version at the right.

model: An abstract representation of some object or class of objects.

model acquisition: The process of learning a model, usually based on observed instances or examples of the structure being modeled. This may be simply learning the parameters of a distribution from examples. For example, one might learn the image texture properties that distinguish tumorous cells from normal cells. Alternatively, the structure of the object might be learned as well, such as constructing a model of a building from a video sequence. Another type of model acquisition is learning the properties of an object, such as what properties and relations define a square as compared to other geometric shapes.

model base: A database of models usually used as part of an identification process.

model base indexing: Selecting one or more candidate models from a model database of structures known by the system. This is usually to eliminate exhaustive testing with every member of the model base.

model based coding: A method of encoding the contents of an image (or video sequence) using a pre-defined or learned set of models. This could be for producing a more compact description of the image data (see model based compression) or for producing a symbolic description. For example, a Mondrian style image could be encoded by the positions, sizes and colors of the colored rectangular regions.

model based compression: An application of model based coding for the purpose of reducing the amount of memory required to describe the image while still allowing reconstruction of the original image.

model based feature detection: Using a parametric model of a feature to locate instances of the feature in an image. For example, a parametric edge detector uses a parameterized model of a step edge that encodes edge direction and edge magnitude.

model based recognition: Identification of the structures in an image by using some internally represented model of the objects known to the computer system. The models are usually geometric models. The recognition process finds image features that match the model features with the right shape and position. The advantage of model based recognition is that the model encodes the object shape thus allowing predictions of image data and less chance of coincidental features being falsely recognized.

model based segmentation: An image segmentation process that uses geometric models to partition the image into different regions. For example, aerial images could have the visible roads segmented by using a geographic information system model of the road network.

model based tracking: An image tracking process that uses models to locate the position of moving targets in an image sequence. For example, the estimated position, orientation and velocity of a modeled vehicle in one image allows a strong prediction of its location in the next image in the sequence.

model based vision: A general term for using models of the objects expected to be seen in the image data to help with the image analysis. The model allows, among other things, prediction of additional model feature positions, verification that a set of features could be part of the model and understanding of the appearance of the model in the image data.

model building: See also model acquisition. The process of constructing a geometric model usually based on observed instances or examples of the structure being modeled, such as from a video sequence.

model fitting: See model registration.

model invocation: See model base indexing.

model reconstruction: See model acquisition.

model registration: A general term for aligning a geometric model to a set of image data. The process may require estimating the rotation, translation and scale that maps a model onto the image data. There may also be shape parameters, such as model length, that need to be estimated. The fitting may need to account for perspective distortion. This figure shows a 2D model registered on an intensity image of the same part.

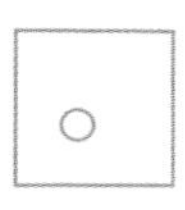

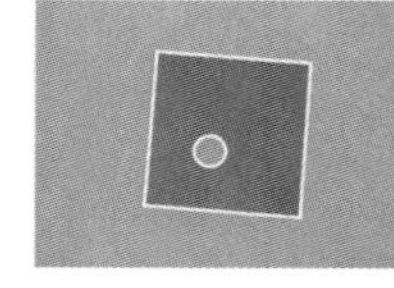

model selection: See model base indexing.

modulation transfer function (MTF): Informally, the MTF is a measure of how well spatially varying patterns are observed by an optical system. More formally, in a 2D image, let $X(f_h, f_v)$ and $Y(f_h, f_v)$ be the Fourier transforms of the input $x(h, v)$ and output $y(h, v)$ images. Then, the MTF of a horizontal and vertical spatial frequency pair (f_h, f_v) is $\mid H(f_h, f_v) \mid / \mid H(0, 0) \mid$, where $H(f_h, f_v) = Y(f_h, f_v)/X(f_h, f_v)$. This is also the magnitude of the optical transfer function.

Moiré fringe: An interference pattern that is observed when spatially sampling, at a given spatial frequency, a signal that has a slightly different spatial frequency. The result is a set of light and dark bands in the observed image. As well as causing image degradation, this effect can also be used in range sensors, where the fringe positions give an indication of surface depth. An example of typical observed fringe patterns is:

Moiré interferometry: A technique for contouring surfaces that works by projecting a fringe pattern (*e.g.*, of straight lines) and observing this pattern through another grating. This effect can be acheieved in other ways as well. The technique is useful for measuring extremely small stress and distortion movements.

Moiré pattern: See moiré fringe.

Moiré topography: A method for measuring the local shape of a surface by analyzing the spacing of moiré fringes on the target surface.

moment: A method for summarizing the distribution of pixel positions or values. Moments are a parameterized family of values. For example, if $I(x, y)$ is a binary image then $\Sigma_{x,y} I(x, y) x^p y^q$ computes its pq^{th} moment m_{pq}. (See also gray level moments and moments of intensity.)

moment characteristic: See moment invariant.

moment invariant: A function of image moment values that keeps the same value even if the image is transformed in some manner. For example, the value $\frac{1}{A^2}((\mu_{20})^2 + (\mu_{02})^2)$ is invariant where μ_{pq} are central moments of a binary image region and A is the area of the region. This value is a constant even if the image data is translated, rotated or scaled.

moments of intensity: An image moment value that takes account of the gray scales of the image pixels as well as their positions. For example, if $G(x, y)$ is a gray scale image, then $\Sigma_{x,y} G(x, y) x^p y^q$ computes its pq^{th} moment of intensity g_{pq}. See also gray level moment.

Mondrian: A famous visual artist from the Netherlands, whose later paintings were composed of adjacent rectangular blocks of constant (*i.e.*, without shading) color. This style of image has been used for much color vision research and, in particular, color constancy because of its simplified image structure without shading, specularities, shadows or light sources.

monochrome: Containing only different shades of a single color. This color is usually different shades of gray, going from pure black to pure white.

monocular: Using a single camera, sensor or eye. This contrasts with binocular and multi-ocular stereo where more than one sensor is used. Sometimes there is also the implication that the image data is acquired from only a single viewpoint as a single camera taking images over time is mathematically equivalent to multiple cameras.

monocular depth cue: Image evidence that indicates that one surface may be closer to the viewer than another. For example, motion parallax or occlusion relationships give evidence of relative depths.

monocular visual space: The visual space behind the lens in an optical system. This space is commonly assumed to be without structure but scene depth can be recovered from the defocus blurring that occurs in this space.

monotonicity: A sequence of values or function that is either continuously increasing (monotone increasing) or continuously decreasing (monotone decreasing).

Moravec interest point operator: An operator that locates interest points at pixels where neighboring intensity values change greatly in at least one direction. These points can be used for stereo matching or feature point tracking. The operator computes the sum of the squares of pixel differences in a line vertically, horizontally and both diagonal directions in a 5×5 window about the given pixel. The minimum of these four values is selected and then all values that are not local maxima or are below a given threshold are suppressed. This image shows the interest points found by the Moravec operator as white dots on the original image.

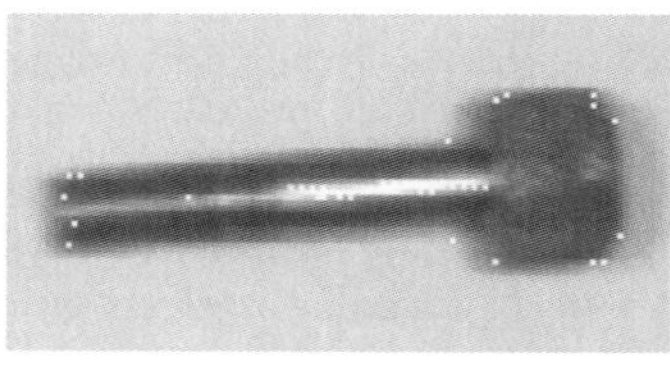

morphological gradient: A gray scale mathematical morphology operation applied to gray scale images that results in an output image similar to the standard intensity gradient. The gradient is calculated by $\frac{1}{2}(D_G(A, B) - E_G(A, B))$ where $D_G()$ and $E_G()$ are the gray scale dilate and erode respectively of image A by kernel B.

morphological segmentation: Using mathematical morphology operations applied to binary images to extract isolated regions of the desired shape. The desired shape is specified by the morphological kernel. The process could also be used to separate touching objects.

morphological smoothing: A gray scale mathematical morphology operation applied to gray scale images that results in an output image similar to that produced by standard noise reduction. The smoothing is calculated by $C_G(\,O_G(A, B), B)$ where $C_G()$ and $O_G()$ are the gray scale close and open operations respectively of image A by kernel B.

morphological transformation: One of a large class of binary and gray scale image trans formations whose primary characteristic is they react to the pattern of the pixel values rather than the values themselves. Examples include dilation, erosion, skeletonizing, thinning, etc. The right figure below is the opening of the left figure, when using a disk shaped structuring element 11 pixels in diameter.

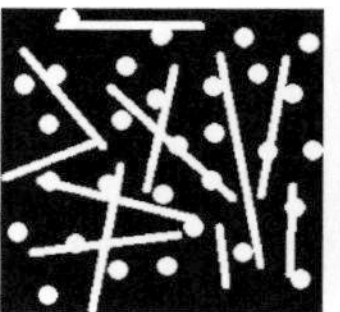

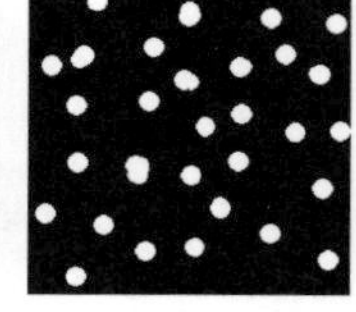

morphology: The shape of a structure. See also mathematical morphology.

morphometry: Techniques for the measurement of shape.

mosaic: The construction of a larger image from a collection of partially overlapping images taken from different view points. The reconstructed image could have different geometries, *e.g.*, as if seen from a single perspective viewpoint, or as if seen from an orthographic viewpoint. See also image mosaic.

motion: A general language term, but, in the context of computer vision, refers to analysis of an image sequence where the camera position or scene structure changes over time.

motion analysis: Analysis of an image sequence in order to extract useful information. Examples of information routinely extracted include: shape of observed scene, figure–ground separation, egomotion estimation, and estimates of a target's position and motion.

motion blur: The blurring of an image that arises when either the camera or something in the scene moves while the image is being acquired. The image below shows the blurring that occurs when an object moves during image capture.

motion coding: 1) A component of video sequence compression in which efficient methods are used for representing movement of image regions between video frames. 2) A term for neural cells tuned to respond for direction and speeds of image motion.

motion detection: Analysis of an image sequence to determine if or when something in the observed scene moves. See also change detection.

motion discontinuity: When the smooth motion of either the camera or something in the scene changes, such as the speed or direction of motion. Another form of motion discontinuity is between two groups of adjacent pixels that have different motions.

motion estimation: Estimating the motion direction and speed of the camera or something in the scene.

motion factorization: Given a set of tracked feature points through an image sequence, a measurement matrix can be constructed. This matrix can be factored into component matrices that represent the shape and 3D motion of the structure up to an 3D affine transform (which is removable using knowledge of the intrinsic camera parameters).

motion field: The projection of the relative motion vector for each scene point onto the image plane. In many circumstances this is closely related to the optical flow, but may differ as image intensities can also change due to illumination changes. Similarly, motion of a uniformly shaded region is not observable locally because there is no changes in image intensity values.

motion layer segmentation: The segmentation of an image into different regions where the motion is locally consistent. The layering effect is most noticeable when the observer is moving through a scene with objects at different depths (causing different amounts of parallax) some of which might also be moving. See also motion segmentation.

motion model: A mathematical model of types of motion allowable for the target object or camera, such as only linear motion along the optical axis with constant velocity. Another example might allow velocities and accelerations in any direction, but occasionally discontinuities, such as for a bouncing ball.

motion representation: See motion model.

motion segmentation: See motion layer segmentation.

motion sequence analysis: The class of computer vision algorithms that process sequences of images captured close together in space and time, typically by a moving camera. These analyses are often characterized by assumptions on temporal coherence that simplify computation.

motion smoothness constraint: The assumption that nearby points in the image have similar motion directions and speeds, or similar optical flow. This constraint is based on the fact that adjacent pixels generally record data from the projection of adjacent surface patches from the scene. These scene components will have similar motion relative to the observer. This assumption can help reduce motion estimation errors or constrain the ambiguity in optical flow estimates arising from the aperture problem.

motion tracking: Identification of the same target feature points through an image sequence. This could also refer to tracking complete objects as well as feature points, including estimating the trajectory or motion parameters of the target.

movement analysis: A general term for analyzing an image sequence of a scene where objects are moving. It is often used for analysis of human motion such as for people walking or using sign language.

moving average smoothing: A form of image noise reduction that occurs over time by averaging the most recent images together. It is based on the assumption that variations in time of the observed intensity at a pixel are random. Thus, averaging the values will produce intensity estimates closer to the true (mean) value.

moving light display: An image sequence of a darkened scene containing objects with attached point light sources. The light sources are observed as a set of moving bright spots. This sort of image sequence was used in the early research on structure from motion.

moving object detection: Analyzing an image sequence, usually with a stationary camera, to detect whether any objects in the scene move.

moving observer: A camera or other sensor that is moving. Moving observers have been extensively used in recent research on structure from motion.

MPEG: Moving Picture Experts Group. A group developing standards for coding digital audio and video, as used in video CD, DVD and digital television. This term is often used to refer to media that is stored in the MPEG 1 format.

MPEG 2: A standard formulated by the ISO Motion Pictures Expert Group (MPEG), a subset of ISO Recommendation 13818, meant for transmission of studio-quality audio and video. It covers four levels of video resolution.

MPEG 4: A standard formulated by the ISO Motion Pictures Expert Group (MPEG), originally concerned with similar applications as H.263 (very low bit rate channels, up to 64 kbps). Subsequently extended to encompass a large set of multimedia applications, including over the Internet.

MPEG 7: A standard formulated by the ISO Motion Pictures Expert Group (MPEG). Unlike MPEG 2 and MPEG 4, that deal with compressing multimedia contents within specific applications, it specifies the structure and features of the compressed multimedia content produced by the different standards, for instance to be used in search engines.

MRF: See Markov random field.

MRI: Magnetic Resonance Imaging. See nuclear magnetic res- onance.

MSRE: Mean Squared Reconstruction Error.

MTF: See modulation transfer function.

multi-dimensional edge detection: variation on standard edge detection of gray scale images in which the input image is multi-spectral (*e.g.*, a RGB color image). The edge detection operator may detect edges in each dimension independently and then combine the edges or may use all information at each pixel directly. The following image shows edges detected from red, green and blue components of an RGB image.

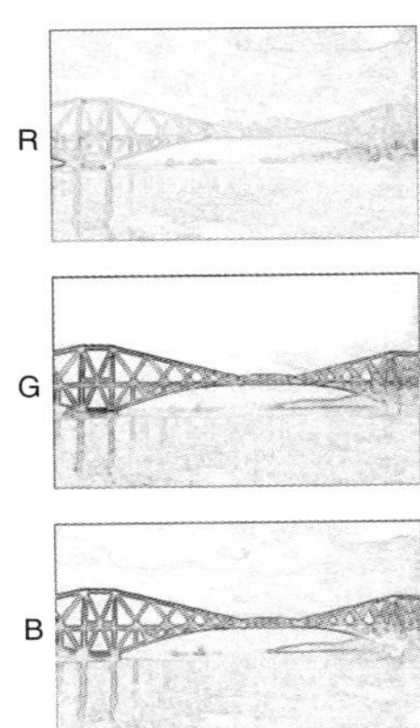

multi-dimensional histogram: A histogram with more than one dimension. For example consider measurements as vectors, *e.g.*, from a multi-spectral image, with N dimensions in the vector. Then one could create a histogram represented by an array with dimension N. The N components in each vector are used to index into the array. Accumulating counts or other evidence values in the array makes it a histogram.

multi-grid method: An efficient algorithm for solving systems of discretized differential (or other) equations. The term "multi-grid" is used because the system is first solved at a coarse sampling level, which is then used to initialize a higher-resolution solution.

multi-image registration: A general term for the geometric alignment of two or more image datasets. Alignment allows pixels from the different source images to lie on top of each other or to be combined. (See also sensor fusion.) For example, two overlapping intensity images could be registered to help create a mosaic. Alternatively, the images need not be from the same type of sensor. (See multi-modal fusion.) For example, NMR and CAT images of the same body part could be registered to provide richer information, *e.g.*, for a doctor. This image shows two unregistered range images on the left and the registered datasets on the right.

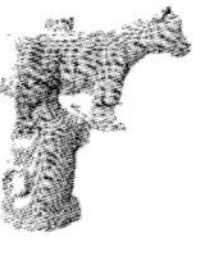

multi-level: See multi-scale method.

multi-modal analysis: A general term for image analysis using image data from more than one sensor type. There is often the assumption that the data is registered so that each pixel records data of two or more types from the same portion of the observed scene.

multi-modal fusion: See sensor fusion.

multi-modal neighborhood signature: A description of a feature point based on the image data in its neighborhood. The data comes several registered sensors, such as X-ray and NMR.

multi-ocular stereo: A stereo triangulation process that uses more than one camera to infer 3D information. The terms binocular stereo and trinocular stereo are commonly used when there are only two or three cameras respectively.

multi-resolution method: See multi-scale method.

multi-scale description: See multi-scale method.

multi-scale integration: 1) Combining information extracted by using operators with different scales. 2) Combining information extracted from registered images with different scales. These two definitions could just be two ways of considering the same process if the difference in operator scale is only a matter of the amount of smoothing. An example of multi-scale integration occurs combining edges extracted from images with different amounts of smoothing to produce more reliable edges.

multi-scale method: A general term for a process that uses information obtained from more than one scale of image. The different scales might be obtained by reducing the image size or by Gaussian smoothing of the image. Both methods reduce the spatial frequency of the information. The main reasons for multi-scale methods are: 1) some structures have different natural scales (*e.g.*, a thick bar could also be considered to be two back-to-back edges) and 2) coarse scale information is generally more reliable in the presence of image noise, but the spatial accuracy is better in finer scale information (*e.g.*, an edge detector might use a coarse scale to reliably detect the edges and a finer scale to locate them more accurately). Below is an image with two scales of blurring.

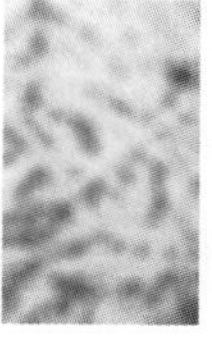
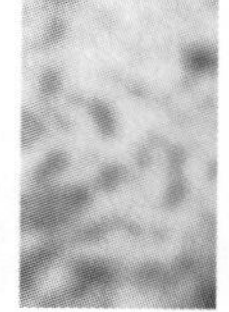

multi-scale representation: A representation having image features or descriptions that belong to two or more scales. An example might be zero crossings detected from intensity images that have received increasing amounts of Gaussian smoothing. A multi-scale model representation might represent an arm as a single generalized cylinder at a coarse scale, two generalized cylinders at an intermediate scale and with a surface triangulation at a fine scale. The representation might have results from several discrete scales or from a more continuous range of scales, as in a scale space. Below are zero crossings found at two scales of Gaussian blurring.

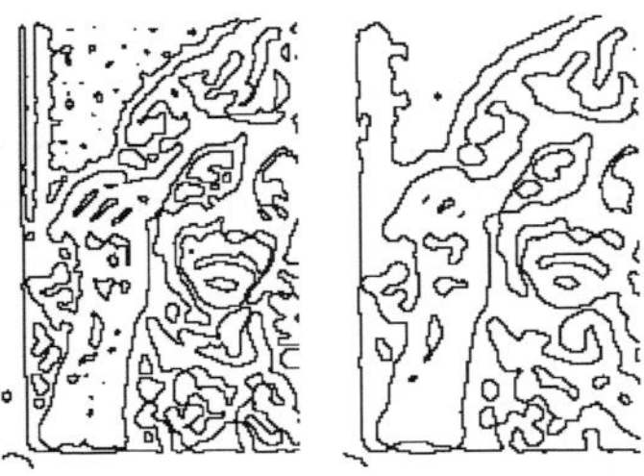

multi-sensor geometry: The relative placement of a set of sensors or multiple views from a single sensor but from different positions. One key consequence of the different placements is ability to deduce the 3D structure of the scene. The sensors need not be the same type but usually are for convenience.

multi-spectral analysis: Using the observed image brightness at different wavelengths to aid in the understanding of the observed pixels. A simple version uses RGB image data. Seven or more bands, including several infrared wavelengths are often used for satellite remote sensing analysis. Recent hyperspectral sensors can give measurements at 100–200 different wavelengths.

multi-spectral image: An image containing data measured at more than one wavelength. The number of wavelengths may be as low as two (*e.g.*, some medical scanners), three (*e.g.*, RGB image data), or seven or more bands, including several infrared wavelengths (*e.g.*, satellite remote sensing). Recent hyperspectral sensors can give measurements at 100–200 different wavelengths. The typical image representation uses a vector to record the different spectral measurements at each pixel of an image array. The following image shows the red, green and blue components of an RGB image.

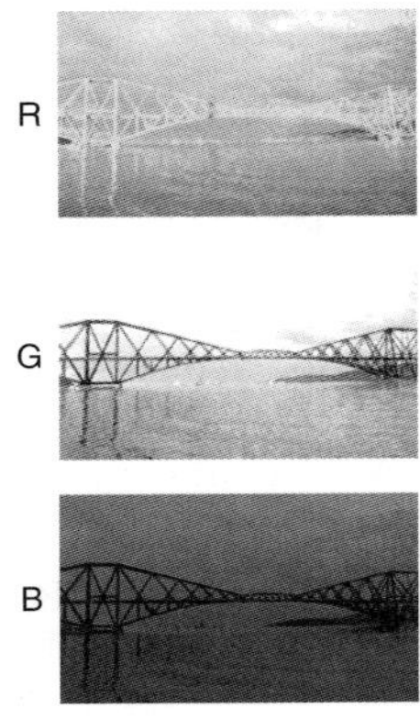

multi-spectral segmentation: Segmentation of a multi-spectral image. This can be addressed by segmenting the image channels individually and then combining the results, or alternatively the segmentation can be based on some combination of the information from the channels.

multi-spectral thresholding: A segmentation technique for multi-spectral image data. A common approach is to threshold each spectral channel independently and then logically AND together the resulting images. An alternative is to cluster pixels in a multi-spectral space and choose thresholds that select desired clusters. The images below show a colored image first thresholded in the blue channel (0–100 accepted) and then ANDed with the thresholded green channel (0–100 accepted). (See plate section for a colour version of these figures.)

multi-tap camera: A camera that provides multiple outputs.

multi-thresholding: Thresholding using a number of thresholds giving a result that has a number of gray scales or colors. In the following example the image has been thresholded with two thresholds (113 and 200).

multi-variate normal distribution: A Gaussian distribution for a variable that is a vector rather than as a scalar. Let $\vec{x}$ be the vector variable with dimension N. Assume that this variable has mean value $\vec{\mu}_x$ and covariance matrix $\mathbf{C}$. Then the probability of observing the particular value $\vec{x}$ is given by:

$$\frac{1}{(2\pi)^{\frac{N}{2}}\,|\mathbf{C}|^{\frac{1}{2}}}e^{-\frac{1}{2}(\vec{x}-\vec{\mu}_x)^\top \mathbf{C}^{-1}(\vec{x}-\vec{\mu}_x)}$$

multi-view geometry: See multi- sensor geometry.

multi-view image registration: See multi-image registration.

multi-view stereo: See multi-sensor geometry.

multiple instruction multiple data (MIMD): A form of parallelism in which, at any given time, each processor might be executing a different instruction or program on a different dataset or pixel. This contrasts with single instruction multiple data parallelism where all processors execute the same instruction simultaneously although on different pixels.

multiple motion segmentation: See motion segmentation.

multiple target tracking: A general term for tracking multiple objects simultaneously in an image sequence. Example applications include tracking football players and automobiles on a road.

multiple view interpolation: A technique for creating (or recognizing) new unobserved views of a scene from example images captured from other viewpoints.

multiplicative noise: A model for the corruption of a signal where the noise is proportional to the signal strength. $f(x, y) = g(x, y) + g(x, y).v(x, y)$ where $f(x, y)$ is the observed signal, $g(x, y)$ is the ideal (original) signal and $v(x, y)$ is the noise.

Munsell color notation system: A system for precisely specifying colors and their relationships, based on hue, value (brightness) and chroma (saturation). The "Munsell Book of Color" contains colored chips indexed by these three attributes. The color of any unknown surface can be identified by comparison with the colors in the book under specified lighting and viewing conditions.

mutual illumination: When light reflecting from one surface illuminates another surface and vice versa. The consequence of this is that light observed coming from a surface is a function of not only the light source spectrum and the reflectance of the target surface, but also the reflectance of the nearby surface (through the spectrum of the light reflecting from the nearby surface onto the first surface). The following diagram shows how mutual illumination can occur.

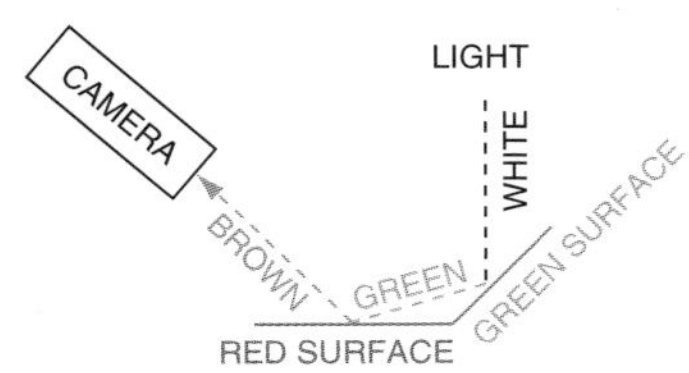

mutual information: The amount of information two pieces of data (such as images) have in common. In other words given a data item A and an unknown data item B, the mutual information $MI(A, B) = H(B) - H(B|A)$ where $H(x)$ is the entropy.

mutual interreflection: See mutual illumination.

N

NAND operator: An arithmetic operation where a new image is formed by NANDing (logical AND followed by NOT) together corresponding bits for every pixel of the two image images. This operator is most appropriate for binary images but may also be applied to gray scale images. For example the following shows the NAND operator applied to two binary images:

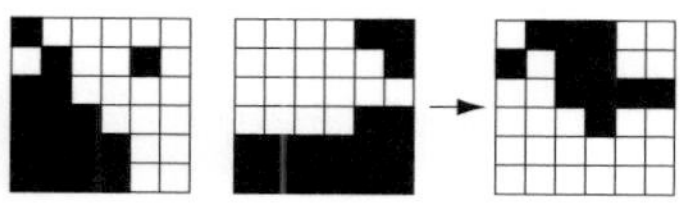

narrow baseline stereo: A form of stereo triangulation in which the sensor positions are close together. The baseline is the distance between the sensor positions. Narrow baseline stereo often occurs when the image data is from a video sequence taken by a moving camera.

near infrared: Light wavelengths approximately in the range 750–5000 nm.

nearest neighbor: A classification, labeling or grouping principle in which a data item is associated with or takes the same label as the previously classified data item that is nearest to the first data item. This distance might be based on spatial distance or a distance in a property space. In this figure the unknown square is classified with the label of the nearest point, namely a circle.

Necker cube: A line drawing of a cube drawn under orthographic projection, which as a

Dictionary of Computer Vision and Image Processing R.B. Fisher, K. Dawson-Howe, A. Fitzgibbon, C. Robertson and E. Trucco

result can be interpreted in two ways.

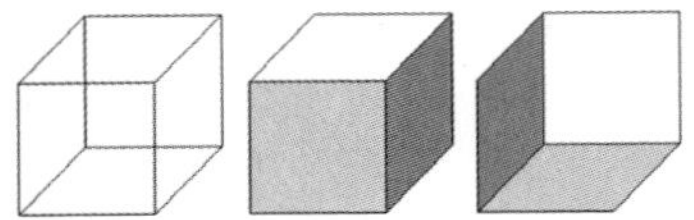

Necker reversal: An ambiguity in the recovery of 3D structure from multiple images. Under affine viewing conditions, the sequence of 2D images of a set of rotating 3D points is the same as the sequence produced by the rotation in the opposite direction of a different set of points, so that two solutions to the structure and motion problem are possible. The different set of points is the reflection of the first set about any plane perpendicular to the optical axis of the camera.

needle map: An image representation used for displaying 2D and 3D vector fields, such as surface normals. Each pixel has a vector. Diagrams showing these use little lines with the magnitude and direction of the vector projected onto the image of a 3D vector. To avoid overcrowding the image, the pixels where the lines are drawn are a subset of the full image. This image shows a needle map of the surface normals on the block sides.

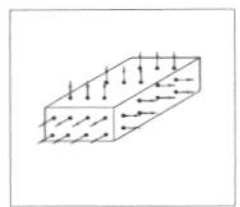

negate operator: See invert operator.

neighborhood: 1) The neighborhood of a vertex v in a graph is the set of vertices that are connected to v by an arc. 2) The neighborhood of a point (or pixel) x is a set of points "near" x. A common definition is the set of points within a certain distance of x, where the distance metric may be Manhattan distance or Euclidean distance. 3) The *4 connected neighborhood* of a 2D location (x, y) is the set of image locations $\{(x+1, y), (x-1, y), (x, y+1), (x, y-1)\}$. The 8 connected neighborhood is the set of pixels $\{(x+i, y+j) | -1 \leq i, j \leq 1\}$. The 26 connected neighborhood of a 3D point (x, y, z) is defined analogously.

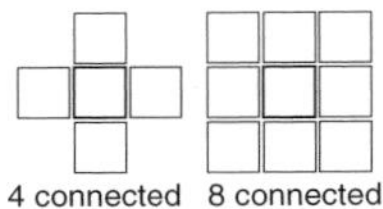

neural network: A classifier that maps input data $\vec{x}$ of dimension n to a space of outputs $\vec{y}$ of dimension m. As a black box, the network is a function $f: \mathbb{R}^n \mapsto [0, 1]^m$. The most commonly used form of neural network is the multi-layer perceptron (MLP). An MLP is characterized by a $m \times n$ matrix of weights $\mathbf{W}$, and a transfer function σ that maps the reals to $[0, 1]$. The output of the single-layer network is $\vec{f}(\vec{x}) = \sigma(\mathbf{W}\vec{x})$ where σ is applied elementwise to vector arguments. A multi-layer network is a cascade of single-layer networks, with different weights matrices at each layer. For example, a two-layer network with k hidden nodes is defined by weights matrices $\mathbf{W}_1 \in \mathbb{R}^{k \times n}$ and $\mathbf{W}_2 \in \mathbb{R}^{m \times k}$, and written $f(\vec{x}) = \sigma(\mathbf{W}_2 \sigma(\mathbf{W}_1 \vec{x}))$. A common choice for σ is the *sigmoid* function $\sigma(t) = (1 + e^{-st})^{-1}$ for some value of s. When we make it explicit that $\vec{f}$ is a function of the weights as well as the input vector, it is written $\vec{f}(\mathbf{W}; \vec{x})$.Typically, a neural network is trained to predict the relationship between the $\vec{x}$'s and $\vec{y}$'s of a given collection of training examples. Training means setting the weights matrices to minimize the *training error* $e(\mathbf{W}) = \sum_i d(\vec{y}_i, \vec{f}(\mathbf{W}; \vec{x}_i))$ where d measures distance between the network output and a training example. Common choices for $d(\vec{y}, \vec{y}')$ include the 2-norm $\|\vec{y} - \vec{y}'\|^2$.

Newton's optimization method: To find a local minimum of function $f: \mathbb{R}^n \mapsto \mathbb{R}$ from starting position $\vec{x}_0$. Given the function's gradient ∇f and Hessian H evaluated at $\vec{x}_k$, the Newton update is $\vec{x}_{k+1} = \vec{x}_k - \mathsf{H}^{-1} \nabla f$. If f is a quadratic form then a single Newton step will directly yield the global minimum. For general f, repeated Newton steps will generally converge to a local optimum.

next view planning: When inspecting an object or obtaining a geometric or appearance-based model, it may be necessary to observe the object from several places. Next view planning determines where to next place the camera (by moving either the object or the camera) based on either what was observed (in the case of unknown objects) or a geometric model (in the case of known objects).

next view prediction: See next view planning.

NMR: See nuclear magnetic resonance.

node of graph: A symbolic representation of some entity or feature. It is connected to other nodes in a graph by arcs, that represent relationships between the different entities.

noise: A general term for the deviation of a signal away from its "true" value. In the case of images, this leads to pixel values (or other measurements) that are different from their expected values. The causes of noise can be random factors, such as thermal noise in the sensor, or minor scene events,

such as dust or smoke. Noise can also represent systematic, but unmodeled, events such as short term lighting variations or quantization. Noise might be reduced or removed using a noise reduction method. Above are images without and with salt-and-pepper noise.

noise model: A way to model the statistical properties of noise without having to model the causes of the noise. One general assumption about noise is that it has some underlying, but perhaps unknown, distribution. A Gaussian noise model is a commonly used for random factors and a uniform distribution is often used for unmodeled scene effects. Noise could be modeled with a mixture model. The noise model typically has one or more parameters that control the magnitude of the noise. The noise model can also specify how the noise affects the signal, such as additive noise (which offsets the true value) or multiplicative noise (which rescales the true value). The type of noise model can constrain the type of noise reduction method.

noise reduction: An image processing method that tries to reduce the distortion of an image that has been caused by noise. For example, the images from a video sequence taken with a stationary camera and scene can be averaged together to reduce the effect of Gaussian noise because the average value of a signal corrupted with this type of noise converges to the true value. Noise reduction methods often introduce other distortions, but these may be less significant to the application than the original noise. An image with salt-and-pepper noise and its noise reduced by median smoothing are shown in the figure.

noise removal: See noise reduction.

noise source: A general term for phenomena that corrupt image data. This could be systematic unmodeled processes (*e.g.*, 60 Hz electromagnetic noise) or random processes (*e.g.*, electronic shot noise). The sources could be in the scene (*e.g.*, chaff), in the medium (*e.g.*, dust), in the lens (*e.g.*, imperfections) or in the sensor (*e.g.*, sensitivity variations).

noise suppression: See noise reduction.

noise-whitening filter: A noise modifying filter that outputs images whose pixels have noise that is independent of 1) other pixels' noise (spatial noise) or 2) other values of that pixel at other times (temporal noise). The resulting image's noise is white noise.

non-accidentalness: A general principle that can be used to improve image interpretation based on the concept that when regularities appear in an image, they are most likely to result from regularities in the scene. For example, if two straight lines end near to each other, then this could have arisen from a coincidental alignment of the line ends and the observer. However, it is much more probable that the two lines end at the same point in the observed scene. This figure shows line terminations and orientations that are unlikely to be coincidental.

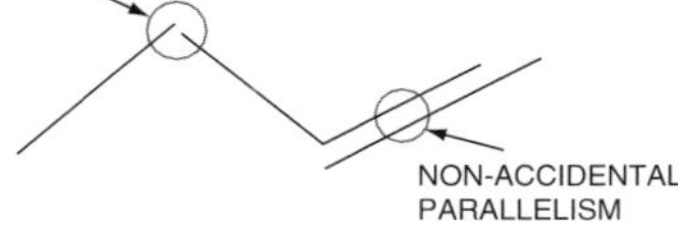

non-hierarchical control: A way of structuring the sequence of actions in an image interpretation system. Non-hierarchical control is when there is no master process that orders the sequence of actions or operators applied. Instead, typically, each operator can observe the current results and decide if it is capable of executing and if it is desirable to do so.

nonlinear filter: A process where the outputs are a nonlinear function of the inputs. This covers a large range of algorithms. Examples of nonlinearity might be: 1) doubling the values of all input data does not double the values of the output results (*e.g.*, a filter that reports the position at which a given value appears), 2) applying an operator to the sum of two images gives different results from adding the results of the operator applied to the two original images (*e.g.*, thresholding).

non-maximal suppression: A technique for suppressing multiple responses (*e.g.*, high values of gradient magnitude) representing a single edge or other feature. The resulting edges should be a single pixel wide.

non-parametric clustering: A data clustering process such as k-nearest neighbor that does not assume an underlying probability distribution.

non-parametric method: A probabilistic method used when the form of the underlying probability distribution is unknown or multi-modal. Typical applications are to estimate the *a posteriori* probability of a classification given an observation. Parzen windows or k-nearest neighbor classifiers are often used.

non-rigid model representation: A model representation where the shape of the model can change, perhaps under the control of a few parameters. These models are useful for representing objects whose shape can change, such as moving humans or biological specimens. The differences in shape may occur over time or be between different instances. Changes in apparent shape due to perspective projection and observer viewpoint are not relevant here. By contrast, a rigid model would have the same actual shape irrespective of the viewpoint of the observer.

non-rigid motion: A motion of an object in the scene in which the shape of the object also changes. Examples include: 1) the position of a walking person's limbs and 2) the shape of a beating heart. Changes in apparent shape due to perspective projection and viewpoint are not relevant here.

non-rigid registration: The problem of registering, or aligning, two shapes that can take on a variety of configurations (unlike rigid shapes). For instance, a walking person, a fish, and facial features like mouth and eyes are all non-rigid objects, the shape of which changes in time. This type of registration is frequently needed in medical imaging as many human body parts deform. Non-rigid registration is considerably more complex than rigid registration. See also alignment, registration, rigid registration.

non-rigid tracking: A tracking process that is designed to track non-rigid objects. This means that it can cope with changes in actual object shape as well as apparent shape due to perspective projection and observer viewpoint.

non-symbolic representation: A model representation in which the appearance is described by a numerical or image-based description rather than a symbolic or mathematical description. For example, non-symbolic models of a line would be a list of the coordinates of the points in the line or an image of the line. Symbolic object representations include the equation of the line or the endpoints of the line.

normal curvature: A plane that contains the surface normal $\vec{n}$ at point $\vec{p}$ to a surface intersects that surface to form a planar curve Γ that passes through $\vec{p}$. The normal curvature is the curvature of Γ at $\vec{p}$. The intersecting plane can be at any specified orientation about the surface normal. See:

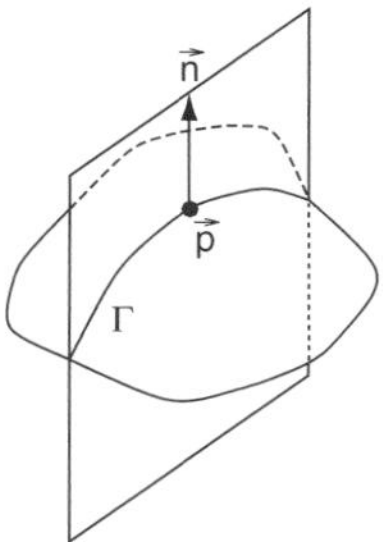

normal distribution: See Gaussian distribution.

normal flow: The component of optical flow in the direction of the intensity gradient. The orthogonal component is not locally observable because small motions orthogonally do not change the appearance of local neighborhoods.

normalized correlation: 1) An image or signal similarity measure that scales the differences between the signals by a measure of the average signal strength:

$$\frac{\sum_i (x_i - y_i)^2}{\sqrt{(\sum_i x_i^2)(\sum_i y_i^2)}}$$

This scales the difference so that it is less significant if the inputs are larger. The similarities lie in the range [0,1], where 0 is most similar. 2) A statistical cross correlation process where the correlation coefficient is normalized to lie in the range $[-1, 1]$, where 1 is most similar. In the case of two scalar variables, this means dividing by the standard deviations of the two variables.

NOT operator: See invert operator.

novel view synthesis: A process whereby a new view of an object is synthesized by combining information from several images of the object from different viewpoints. One method is by 3D reconstruction, *e.g.*, from binocular stereo, and then rendering the reconstruction using computer graphics. However, the main approaches to novel view synthesis use epipolar geometry and the pixels of two or more images of the object to directly synthesize a new image without creating a 3D reconstruction.

NP-complete: A concept in computational complexity covering a special set of problems. All of these problems currently can be solved, in the worst case, in time exponential $O(e^N)$ in the number or size N of their input data. For the subset of exponential problems called NP-complete, if an algorithm for one could be found that executes in polynomial time $O(N^p)$ for some p, then a related algorithm could be found for any other NP-complete algorithm.

NTSC: National Television System Committee. A television signal recording system used for encoding video data at approximately 60 video fields per second. Used in the USA, Japan and other countries.

nuclear magnetic resonance (NMR): An imaging technique based on magnetic properties of the atomic nuclei. Protons and neutrons within atomic nuclei generate a magnetic dipole that can respond to an external magnetic field. Several properties related to the relaxation of that magnetic dipole give rise to values that depend on the tissue type, thus allowing identification or at least visualization of the different soft tissue types. The measurement of the signal is a way of measuring the density of certain types of atoms, such as hydrogen in the case of biological NMR scanners. This technology is used for medical body scanning, where a detailed 3D volumetric image can be produced. Signal levels are highly correlated with different biological structures so one can easily observe different tissues and their positions. Also called MRI/magnetic resonance imaging.

NURBS: Non-Uniform Rational B-Splines: a type of shape modeling primitive based on ratios of b-splines. Capable of accurately representing a wide range of geometric shapes including freeform surfaces.

Nyquist frequency: The minimum sampling frequency for which the underlying true image (or signal) can be reconstructed from the samples. If sampling at a lower frequency, then aliasing will occur, creating apparent image structure that does not exist in the original image.

Nyquist sampling rate: See Nyquist frequency.

O

object: 1) A general term referring to a group of features in a scene that humans consider to compose a larger structure. In vision it is generally thought of as that to which attention is directed. 2) A general system theory term, where the object is what is of interest (unlike the background). Resolution or scale may determine what is considered the object.

object centered representation: A model representation in which the position of the features and components of the model are described relative to the position of the object itself. This might be a relative description (the nose is 4 cm from the mouth) or might use a local coordinate system (*e.g.*, the right eye is at position (0,25,10) where (0,0,0) is the nose.) This contrasts with, for example, a viewer centered representation. Here is a rectangular solid defined in its local coordinate system:

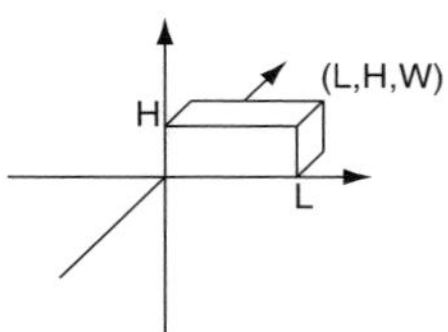

object contour: See occluding contour.

object grouping: A general term meaning the clustering of all of the image data associated with a distinct observed object. For example, when observing a person, object grouping could cluster all of the pixels from the image of the person.

object plane: In the case of convex simple lenses typically used in laboratory TV cameras, the object plane is the 3D scene plane where all points are exactly in focus on the image plane (assuming a perfect lens and the optical axis is perpendicular to the image

Dictionary of Computer Vision and Image Processing R.B. Fisher, K. Dawson-Howe, A. Fitzgibbon, C. Robertson and E. Trucco

plane). The object plane is illustrated here:

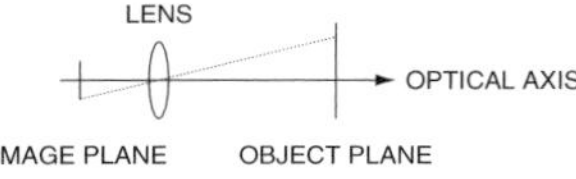

object recognition: A general term for identifying which of several (or many) possible objects is observed in an image. The process may also include computing the object's image or sceneposition, or labeling the image pixels or image features that belong to the object.

object representation: An encoding of an object into a form suitable for computer manipulation. The models could be geometric models, graph models or appearance models, as well as other forms.

object verification: A component of an object recognition process that attempts to verify a hypothesized object identity by examining evidence. Commonly, geometric object models are used to verify that object features are observed in the correct image positions.

objective function: 1) The cost function used in an optimization process. 2) A measure of the misfit between the data and the model.

oblique illumination: See low angle illumination.

observer: The individual (or camera) making observations. Most frequently this refers to the camera system from which images are being supplied. See also observer motion estimation.

observer motion estimation: When an observer is moving, image data of the scene provides optical flow or trackable scene feature points. These allow an estimate of how the observer is moving relative to the scene, which is useful for navigation control and position estimation.

obstacle detection: Using visual data to detect objects in front of the observer, usually for mobile robotics applications.

Occam's razor: An argument attributed to William of Occam (Ockham), an English nominalist philosopher of the early fourteenth century, stating that assumptions must not be needlessly multiplied when explaining something (*entia non sunt multiplicanda praeter necessitatem*). Often used simply to suggest that, other conditions being equal, the simplest solution must be preferred. Notice variant spelling *Ockham*. See also minimum description length.

occluding contour: The visible edge of a smooth curved surface as it bends away from an observer. The occluding contour defines a 3D space curve on the surface, such that a line of sight from the observer to a point on the space curve is perpendicular to the surface normal at that point. The 2D image of this curve may also be called the occluding contour. The contour can often be found by an edge detection process. The cylinder boundaries on both the left and right are occluding contours from our viewpoint:

occluding contour analysis: A general term that includes 1) detection of the occluding contour, 2) inference of the shape of the 3D surface at the occluding contour and 3) determining the relative depth of the surfaces on both sides of the occluding contour.

occluding contour detection: Determining which of the image edges arise from occluding contours.

occlusion: Occlusion occurs when one object lies between an observer and another object. The closer object occludes the more distant one in the acquired image. The occluded surface is the portion of the more distant object hidden by the closer object. Here, the cylinder occludes the more distant brick:

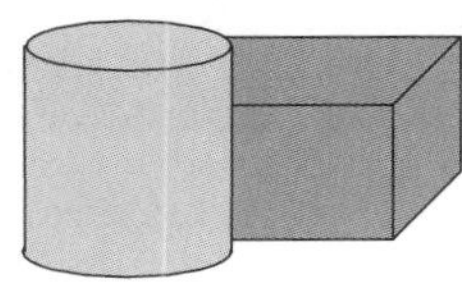

occlusion recovery: The process of attempting to infer the shape and appearance of a surface hidden by occlusion. This recovery helps improve completeness when reconstructing scenes and objects for virtual reality. This image shows two occluded pipes and an estimated recovery:

occlusion understanding: A general term for analyzing scene occlusions that may include occluding contour detection, determining the relative depths of the surfaces on both sides of an occluding contour, searching for tee junctions as a cue for occlusion and depth order, etc.

occupancy grid: A map construction technique used mainly for autonomous vehicle navigation. The grid is a set of squares or cubes representing the scene, which are marked according to whether the observer believes the corresponding scene region is empty (hence navigable) or full. A probabilistic measure could also be used. Visual evidence from range, binocular stereo or sonar sensors are typically used to construct and update the grid as the observer moves.

OCR: See optical character recognition.

octree: A volumetric representation in which 3D space is recursively divided into eight (hence "oct") smaller volumes by planes parallel to the XY, YZ,

XZ coordinate system planes. A tree is formed by linking the eight subvolumes to each parent volume. Additional subdivision need not occur when a volume contains only object or empty space. Thus, this representation can be more efficient than a pure voxel representation. Here are three levels of a pictorial representation of an octree, where one octant and the largest (leftmost) level is expanded to give the middle figure, and similarly an octant of the middle:

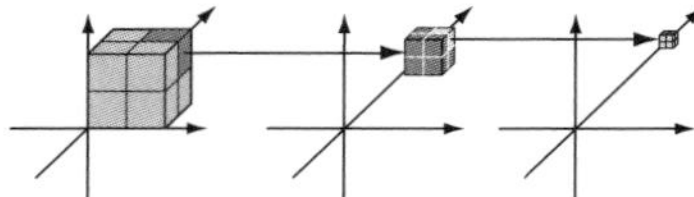

odd field: Standard interlaced video transmits all of the even scan lines in an image frame first and then all of the odd lines. The set of odd lines is the odd field.

O'Gorman edge detector: A parametric edge detector. A decomposition of the image and model by orthogonal Walsh function masks was used to compute the step edge parameters (contrast and orientation). One advantage of the parametric model was a goodness of model fit as well as the edge contrast that increased the reliability of the detected edges.

omnidirectional sensing: Literally, sensing all directions simultaneously. In practice, this means using mirrors and lenses to project most of the lines of sight at a point onto a single camera image. The space behind the mirrors and camera(s) is typically not visible. See also catadioptric optics. Here a camera using a spherical mirror achieves a very wide field of view:

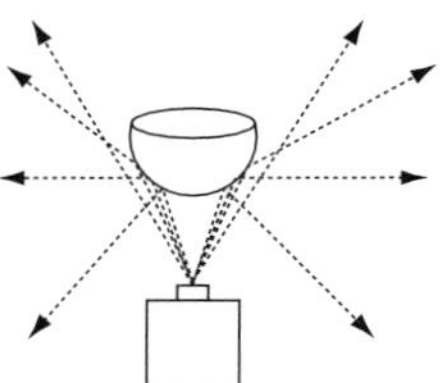

opaque: When light cannot pass through a structure. This causes shadows and occlusion.

open operator: A mathematical morphology operator applied to a binary image. The operator is a sequence of N erodes followed by N dilates, both using a specified structuring element. The operator is useful for separating touching objects and removing small regions. The right image was created by opening the left image with an 11-pixel disk kernel:

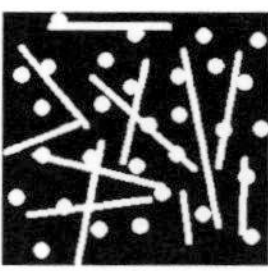
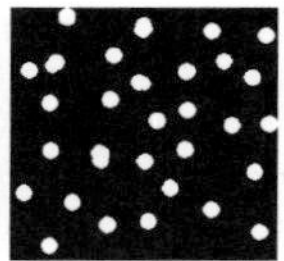

operator: A general term for a function that is applied to some data in order to transform it in some way. For example see image processing operator.

opponent color: A color representation system originally developed by Hering in which an image is represented by three channels with contrasting colors: Red–Green, Yellow–Blue, and Black–White.

optical: A process that uses light and lenses is an optical process.

optical axis: The ray, perpendicular to the lense and through the optical center, around which the lense is symmetrical.

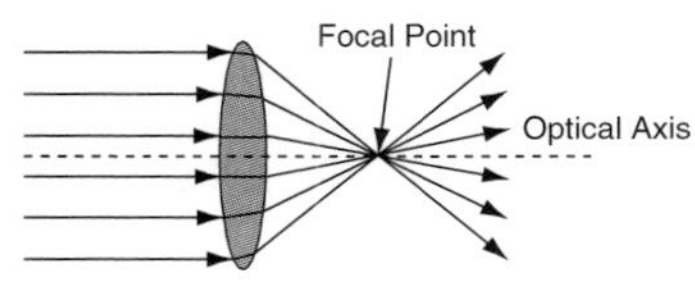

optical center: See focal point.

optical character recognition (OCR): A general term for extracting an alphabetic text description from an image of the text. Common specialisms include bank numerals, handwritten digits, handwritten characters, cursive text, Chinese characters, Arabic characters, etc.

optical flow: An instantaneous velocity measurement for the direction and speed of the image data across the visual field. This can be observed at every pixel, creating a field of velocity vectors. The set of apparent motions of the image pixel brightness values.

optical flow boundary: The boundary between two regions where the optical flow is different in direction or magnitude. The regions can arise from objects moving in different directions or surfaces at different depths. See also optical flow field segmentation. The dashed line in this image is the boundary between optical flow moving left and right:

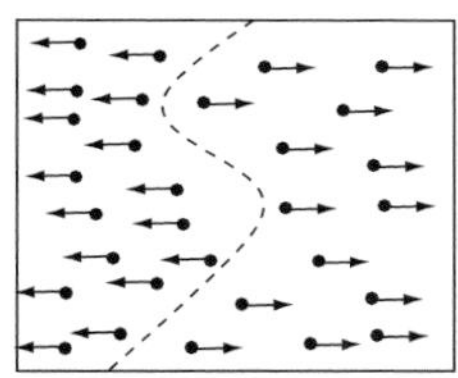

optical flow constraint equation: The equation $\frac{\partial I}{\partial t} + \nabla I \cdot \vec{u}_x = 0$ that links the observed change in image I's intensities over time $\frac{\partial I}{\partial t}$ at image position $\vec{x}$ to the spatial change in pixel intensities at that position ∇I and the velocity $\vec{u}_x$ of the image data at that pixel. The constraint does not completely determine the image motion, as this has two degrees of freedom. The equation provides only one constraint, thus leading to an aperture problem.

optical flow field: The field composed of the optical flow vector at each pixel in an image.

optical flow field segmentation: The segmentation of an optical flow image into regions where the optical flow has a similar direction or magnitude. The regions can arise from objects moving in different directions or surfaces at different depths. See also optical flow boundary.

optical flow region: A region where the optical flow has a similar direction or magnitude. Regions can arise from objects moving in different directions, or surfaces at different depths. See also optical flow boundary.

optical flow smoothness constraint: The constraint that nearby pixels in an image usually have similar optical flow because they usually arise from projection of adjacent surface patches having similar motions relative to the observer. The constraint can be relaxed at optical flow boundaries.

optical image processing: An image processing technique in which the processing occurs by use of lenses and coherent light instead of by a computer. The key principle is that a coherent light beam that passes through a transparency of the target image and is then focused produces the Fourier transform of the image at the focal point where frequency domain filtering can occur. A typical processing arrangement is:

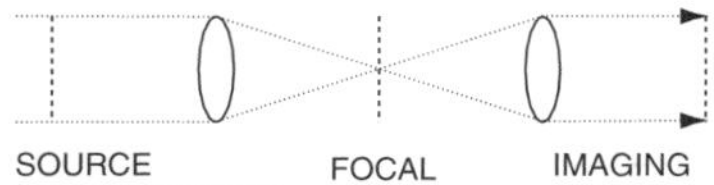

optical transfer function (OTF): Informally, the OTF is a measure of how well spatially varying patterns are observed by an optical system. More formally, in a 2D image, let $X(f_h, f_v)$ and $Y(f_h, f_v)$ be the Fourier transforms of the input $x(h, v)$ and output $y(h, v)$ images. Then, the OTF of a horizontal and vertical spatial frequency pair (f_h, f_v) is $H(f_h, f_v)/H(0, 0)$, where $H(f_h, f_v) = Y(f_h, f_v)/X(f_h, f_v)$. The optical transfer function is usually a complex number encoding both the reduction in signal strength at each spatial frequency and the phase shift.

optics: A general term for the manipulation and transformation of light and images using lenses and mirrors.

optimal basis encoding: A general technique for encoding image or other data by projecting onto some basis functions of a linear space and then using the projection coefficients instead of the original data. Optimal basis functions produce projection coefficients that allow the best discrimination between different classes of objects or members in a class (such as for face recognition).

optimization: A general term for finding the values of the parameters that maximize or minimize some quantity.

optimization parameter estimation: See optimization.

OR operator: A pixelwise logic operator defined on binary variables. It takes as input two binary images, I_1 and I_2, and returns an image I_3 in which the value of each pixel is 0 if both I_1 and I_2 are 0, and 1 otherwise. The rightmost image below shows the result of ORing the left and middle figures (note that the white pixels have value 1):

order statistic filter: A filter based on order statistics, a technique that sorts the pixels of a neighborhood by intensity value, and assigns a rank (the position in the sorted sequence) to each. An order statistics filter replaces the central value of the filtering neighborhood with the value at a given rank in the sorted list. A popular example is the median filter. As this filter is less sensitive to outliers, it is often used in robust statistics processes. See also rank order filter.

ordered texture: See macrotexture.

ordering: Sorting a collection of objects by a given property, for instance, intensity values in a order statistic filter.

orientation: The property of being directed towards or facing a particular region of space, or of a line; also, the pose or attitude of a body in space. For instance, the orientation of a vector (where the vector points to), specified by its unit vector; the orientation of an ellipsoid, specified by its principal directions; the orientation of a wire-frame model, specified by its own reference frame with respect to a world reference frame.

orientation error: The amount of error associated with an orientation value.

orientation representation: See pose representation.

oriented texture: A texture in which a preferential direction can be detected. For instance, the direction of the bricks in a regular brick wall. See also texture direction, texture orientation.

orthogonal image transform: Orthogonal Transform Coding is a well-known class of techniques for image compression. The key process is the projection of the image data onto a set of orthogonal basis functions. See, for instance, the discrete cosine, Fourier or Haar transforms. This is a special case of the linear integral transform.

orthogonal regression: Also known as total least squares. Traditionally seen as the generalization of linear regression to the case where both x and y are measured quantities and subject to error. Given samples x_i and y_i, the objective is to find estimates of the "true" points $(\tilde{x}_i, \tilde{y}_i)$, and line parameters (a, b, c) such that $a\tilde{x}_i + b\tilde{y}_i + c = 0, \forall i$, and such that the error

$\sum(x_i - \tilde{x}_i)^2 + (y_i - \tilde{y}_i)^2$ is minimized. This estimate is easily obtained as the line (or plane, etc., in higher dimensions) passing through the centroid of the data, in the direction of the eigenvector of the data scatter matrix that has smallest eigenvalue.

orthographic: The characteristic property of orthographic (or perpendicular) projection onto the image plane. See orthographic projection.

orthographic camera: A camera in which the image is formed according to a orthographic projection.

orthographic projection: Rendering of a 3D scene as a 2D image by a set of rays orthogonal to the image plane. The size of the objects imaged does not depend on their distance from the viewer. As a consequence, parallel lines in the scene remain parallel in the image. The equations of orthographic projections are

$$x = X \quad y = Y$$

where x, y are the image coordinates of an image point in the camera reference frame (that is, in millimeters, not pixels), and X, Y, Z are the coordinates of the corresponding scene point. An example is seen here:

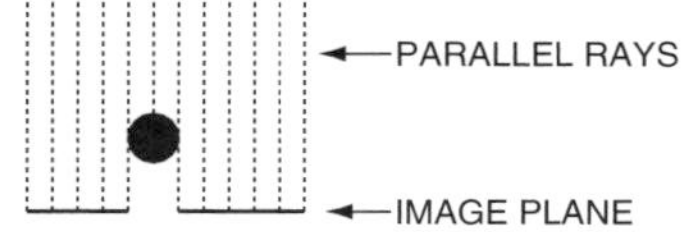

orthoimage: In photogrammetry, the warp of an aerial photograph to an approximation of the image that would have been taken had the camera pointed directly downwards. See also orthographic projection.

orthonormal: A property of a set of basis functions or vectors. If $<,>$ is the inner product function and a and b are any two different members of the set, then we have $<a, a> = <b, b> = 1$ and $<a, b> = 0$.

OTF: See optical transfer function.

outlier: If a set of data mostly conforms to some regular process or is well represented by a model, with the exception of a few data points, then these exception points are outliers. Classifying points as outliers depends on both the models used and the statistics of the data. This figure shows a line fit to some points and an outlying point.

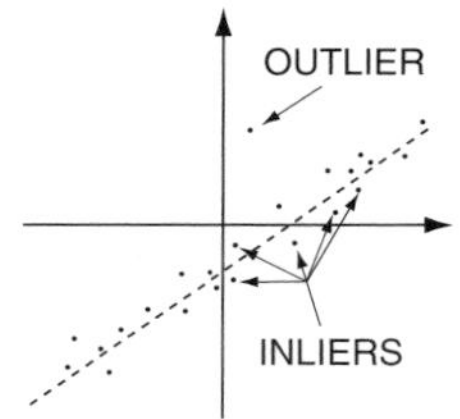

outlier rejection: Identifying outliers and removing them from the current process. Identification is often a difficult process.

over-segmented: Describing the output of a segmentation algorithm. Given an image where a desired segmentation result is known, the algorithm over-segments if the desired regions are represented by too many algorithmically output regions. This image should be segmented into three regions but it was oversegmented into five regions:

P

paired boundaries: See paired contours.

paired contours: A pair of contours occurring together in images and related by a spatial relationship, for instance the contours generated by river banks in aerial images, or the contours of a human limb (arm, leg). Co-occurrence can be exploited to make contour detection more robust. See also feature extraction. An example is seen here:

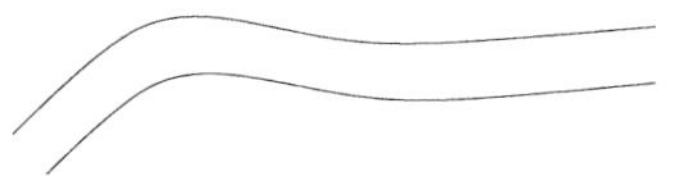

pairwise geometric histogram: A line- or edge- based shape representation used for object recognition, especially 2D. Histograms are built by computing, for each line segment, the relative angle and perpendicular distance to all other segments. The representation is invariant to rotation and translation. PGHs can be compared using the Bhattacharyya metric.

PAL camera: A camera conforming to the European PAL standard (Phase Alternation by Line). See also NTSC, RS-170, CCIR camera.

palette: The range of colors available.

pan: Rotation of a camera about a single axis through the camera center and (approximately) parallel to the image vertical:

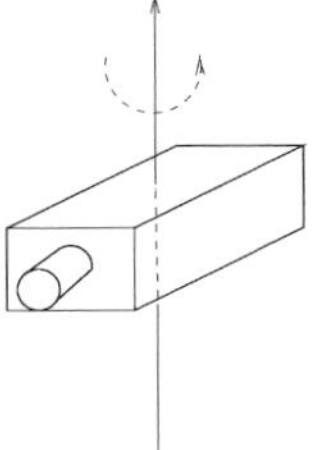

panchromatic: Sensitive to light of all visible wavelengths.

Dictionary of Computer Vision and Image Processing R.B. Fisher, K. Dawson-Howe, A. Fitzgibbon, C. Robertson and E. Trucco

Panchromatic images are gray scale images where each pixel averages light equally over the visible range.

panoramic: Associated with a wide field-of-view often created or observed by a panned camera.

panoramic image mosaic: A class of techniques for collating a set of partially overlapping images into a panoramic, single image. This is a mosaic build

from the frames of a hand-held camera sequence. Typically, the mosaic yields both very high resolution and large field of view, which cannot be simultaneously achieved by a physical camera. There are several ways to build panoramic mosaic, but, in general, there are three necessary steps: first, determining correspondences (see stereo correspondence problem) between adjacent images; second, using the correspondences to find a warping transformation between the two images (or between the current mosaic and a new image); third, blending the new image into the current mosaic.

panoramic image stereo: A stereo system working with a very large field of view, say 360° in azimuth and 120° in elevation. Disparity maps and depths are recovered for the whole field of view simultaneously. A normal stereo system would have to be moved and results registered to achieve the same result. See also binocular stereo, multi-view stereo, omnidirectional sensing.

Pantone matching system (PMS): A color matching system used by the printing industry to print spot colors. Colors are specified by the Pantone name or number. PMS works well for spot colors but not for process colors, usually specified by the CMYK color model.

Panum's fusional area: The region of space within which single vision is possible (that is, you do not perceive double images of objects) when the eyes fixate a given point.

parabolic point: A point on a smooth surface where the Gaussian curvature is positive. See also HK segmentation.

parallax: The angle between the two straight lines that join a point (possibly a moving one) to two viewpoints. In motion analysis, *motion parallax* occurs when two scene points that project to the same image point at one viewpoint later project to different points as the camera moves. The vector between the two new points is the parallax. See:

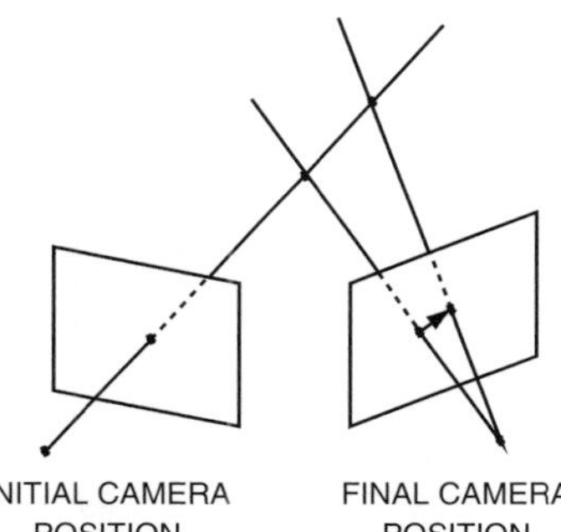

parallel processing: An algorithm is executed in parallel, or through parallel processing, when it can be divided into a number of computations that are performed simultaneously on separate hardware. See also single instruction multiple data, multiple instruction multiple data, pipeline parallelism, task parallelism.

parallel projection: A generalization of orthographic projection in which a scene is projected onto the image plane by a set of parallel rays not necessarily perpendicular to the image plane. This is a good approximation of perspective projection, up to a uniform scale factor, when the scene is small in comparison to its distance from the center of projection. Parallel projection is a subset of weak perspective viewing, where the weak perspective projection matrix is subject not only to orthogonality of the rows of the left 2×3 submatrix, but also to the constraint that the rows have equal norm. In orthographic projection, both rows have unit norm.

parameter estimation: A class of techniques aimed to estimate the parameters of a given parametric model. For instance, assuming that a set of image points lie on an ellipse, and considering the implicit ellipse model $ax^2 + bxy + cy^2 + dx + ey + f$, the parameter vector $[a, b, c, d, e, f]$ can be estimated, for instance, by least square surface fitting.

parametric edge detector: An edge detection technique that seeks to match image data using a parametric model of edge points and thus detects edges when the image data fits the edge model well. See Hueckel edge detector.

parametric mesh: A type of surface modeling primitive for 3D models in which the surface is defined by a mesh of points. A typical example is NURBS (non-uniform rational b-splines).

parametric model: A mathematical model expressed as function of a set of parameters, for instance, the parametric equation of a curve or surface (as opposed to its implicit form), or a parametric edge model (see parametric edge detector).

paraperspective: An approximation of perspective projection, whereby a scene is divided into parts that are imaged separately by parallel projection with different parameters.

part recognition: A class of techniques for recognizing assemblies or articulated objects from their subcomponents (parts), *e.g.*, a human body from head, trunk, and limbs. Parts have been represented by 3D models like generalized cones, superquadrics, and others. In industrial contexts, part recognition indicates the recognition of specific items (parts) in a production line, typically for classification and quality control.

part segmentation: A class of techniques for partitioning a set of data into components (parts) with an identity of their own, for instance a human body into limbs, head, and trunk. Part segmentation methods exist for both 2D and 3D data, that is, intensity images and range images, respectively. Various geometric models have been adopted for the parts, *e.g.*, generalized cylinders, superellipses, and superquadrics. See also articulated object segmentation.

partially constrained pose: A situation whereby an object is subject to a number of constraints restricting the number of admissible orientations or positions, but not fixing one univocally. For instance, cars on a road are constrained to rotate around an axis perpendicular to the road.

particle counting: An application of particle segmentation to counting the instances of small objects (particles) like pebbles, cells, or water droplets, in images or sequences, such as in this image:

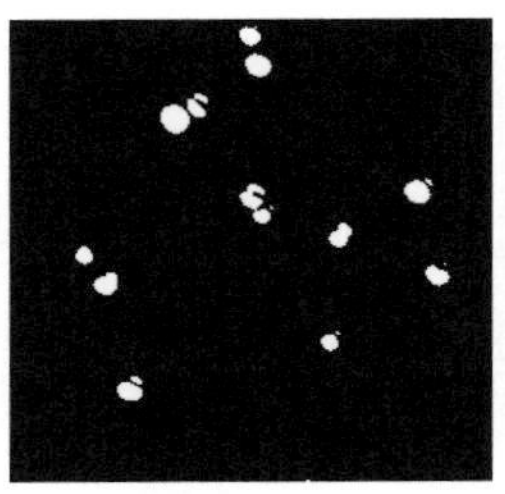

particle filter: A tracking strategy where the probability density of the model parameters is represented as a set of *particles*. A particle is a single sample of the model parameters, with an associated weight. The probability density represented by the particles is typically a set of delta functions or a set of Gaussians with means at the particle centers. At each tracking iteration, the current set of particles represents a prior on the model parameters, which is updated via a dynamical model and observation model to produce the new set representing the posterior distribution. See also condensation tracking.

particle segmentation: A class of techniques for detecting individual instances of small objects (particles) like pebbles, cells, or water droplets, in images or sequences. A typical problem is severe occlusion caused by overlapping particles. This problem has been approached successfully with the watershed transform.

particle tracking: See condensation tracking

Parzen: A Parzen window is a linearly increasing and decreasing weighting window (triangle-shaped) used to limit leakage to spurious frequencies when computing the power spectrum of a signal:

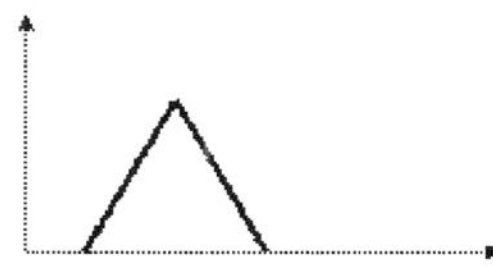

See also windowing, Fourier transform.

passive sensing: A sensing process that does not emit any stimulus or where the sensor does not move is passive. A normal stationary camera is passive. Structured light triangulation or a moving video camera are active.

passive stereo: A passive stereo algorithm uses only the information obtainable using a stationary set of cameras and ambient illumination. This contrasts with the active vision paradigm in stereo, where the camera(s) might move or some projected stimulus might be used to help solve the stereo correspondence problem.

patch classification: The problem of attributing a surface patch to a particular class in a shape catalogue, typically computed from dense range data using curvature estimates or shading. See also curvature sign patch classification, mean and Gaussian curvature shape classification.

path coherence: A property used in tracking objects in an image sequence. The assumption is that the object motion is mostly smooth in the scene and thus the observed motion in a projected image of the scene is also smooth.

path finding: The problem of determining a path with given properties in a graph, for example, the shortest path connecting two given nodes, or two nodes with given properties. A path is defined as a linear subgraph. Path finding is a characteristic problem of state-space methods, inherited from symbolic artificial intelligence. See also graph searching. This term is also used in the context of dynamic programming search, for instance applied to the stereo correspondence problem.

pattern grammar: See shape grammar.

pattern recognition: A large research area concerned with the recognition and classification of structures, relations or patterns in data. Classic techniques include syntactic, structural and statistical pattern recognition.

PCA: See principal component analysis.

PDM: See point distribution model.

peak: A general term for when a signal value is greater than the neighboring signal values. An example of a signal peak measured in one dimension is when crossing a bright line lying on a dark surface along a scanline. A cross-section along a scanline of an image of a light line on a dark background might observe the pixel values 7, 45, 105, 54, 7. The peak would be at 105. A two dimensional example is when observing the image of a bright spot on a darker background.

pedestrian surveillance: See person surveillance.

pel: See pixel.

pencil of lines: A bundle of lines passing through the same point. For example, if $\vec{p}$ is a generic bundle point and $\vec{p}_0$ the point through which all lines pass, the bundle is

$$\vec{p} = \vec{p}_0 + \lambda \vec{v}$$

where λ is a real number and $\vec{v}$ the direction of the individual line (both are parameters). An example is:

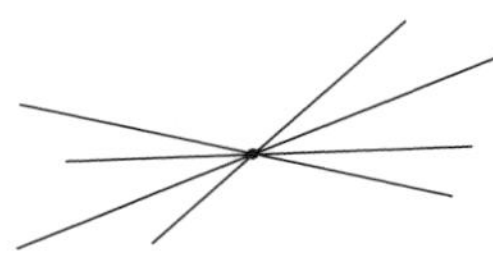

percentile method: A specialized thresholding technique used for selecting the threshold. The method assumes that the percentage of the scene that belongs to the desired object (*e.g.*, a darker object against a lighter background) is known. The threshold that selects that percentage of pixels is used.

perception: The process of understanding the world through the analysis of sensory input (such as images).

perceptron: A computational element $\phi(\vec{w} \cdot \vec{x})$ that acts on a data vector $\vec{x}$, where $\vec{w}$ is a vector of weights and $\phi()$ is the activation function. Perceptrons are often used for classifying data into one of two sets (*i.e.*, if $\phi(\vec{w} \cdot \vec{x}) \geq 0$ or $\phi(\vec{w} \cdot \vec{x}) < 0$). See also classification, supervised classification, pattern recognition.

perceptron network: A multi-layer arrangement of perceptrons, closely related to the well-known back-propagation networks.

perceptual grouping: See perceptual organization.

perceptual organization: A theory based on Gestalt psychology, centered on the tenet that certain organizations (or interpretations) of visual stimuli are preferred over others by the human visual system. A famous example is that a drawing of a wire-frame cube is immediately interpreted as a 3D object, instead of a 2D collection of lines. This concept has been used in several low-level vision systems, typically to find groups of low-level features most probably generated by interesting objects. See also grouping and Lowe's curve segmentation. A more complex example is below, where the line of feature endings suggests a virtual horizontal line.

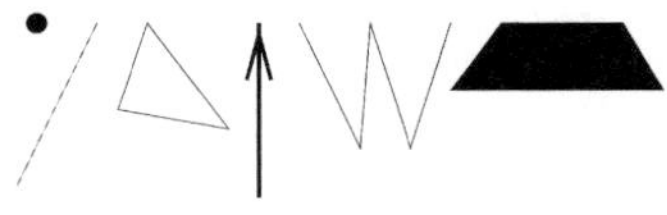

performance characterization: A class of techniques aimed to assess the performance of computer vision systems in terms of, for instance, accuracy, precision, robustness to noise, repeatability, and reliability.

perimeter: 1) The perimeter of a binary image is the set of foreground pixels that touch the background. 2) The length of the path through those pixels.

periodicity estimation: The problem of estimating the period of a periodic phenomenon, *e.g.*, given a texture created by the repetition of a fixed pattern, determine the pattern's size.

person surveillance: A class of techniques aimed at detecting, tracking, counting, and recognizing people or their behavior in CCTV videos, for security purposes. For examples, systems have been reported for the automated surveillance of car parks, banks, airports and the like. A typical system must detect the presence of a person, track the person's movement over time, possibly identify the person using a database of known faces, and classify the person's behavior according to a small class of pre-defined behaviors (*e.g.*, normal or anomalous). See also anomalous behavior detection, face recognition, and face tracking.

perspective: The rendering of a 3D scene as a 2D image according to perspective projection, the key characteristic of which is, intuitively, that the size of the imaged objects depend on their distance from the viewer. As a consequence, the image of a bundle of parallel lines is a bundle of lines converging into a point, the vanishing point. The geometry of perspective was formalized by the master painters of the Italian Quattrocento and Renaissance.

perspective camera: A camera in which the image is formed according to perspective projection. The corresponding mathematical model is commonly known as the pinhole camera model. An example of the projection in the perspective camera is:

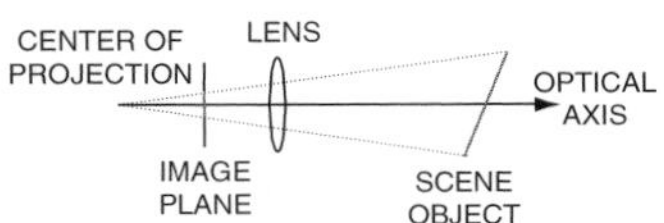

perspective distortion: A type of distortion in which lines that are parallel in the real world appear to converge in a perspective image. In the example notice how the train tracks appear to converge in the distance.

perspective inversion: The problem of determining the position of a 3D object from its image. *i.e.*, solving the perspective projection equations for the 3D coordinates. See also absolute orientation.

perspective projection: Imaging a scene with foreshortening. The projection equation of perspective is

$$x = f\frac{X}{Z} \quad y = f\frac{Y}{Z},$$

where x, y are the image coordinates of an image point in the camera reference frame (*e.g.*, in millimeters, not pixels), f is the focal length and X, Y, Z are the coordinates of the corresponding scene point.

PET: See positron emission tomography.

phase congruency: The property whereby components of the Fourier transform of an image are maximally in phase at feature points like step edges or lines. Phase congruency is invariant to image brightness and contrast and has been therefore used as an absolute measure of the significance of feature points. See also image feature.

phase correlation: A motion estimation method that uses the translation-phase duality property of the Fourier transform, that is, a shift in the spatial domain is equivalent to a phase shift in the frequency domain. When using log-polar coordinates, and the rotation and scale properties of the Fourier transform, spatial rotation and scale can be estimated from the frequency shift, independent of spatial translation. See also planar motion estimation.

phase matching stereo algorithm: An algorithm for solving the stereo correspondence problem by looking for similarity of the phase of the Fourier transform.

phase-retrieval problem: The problem of reconstructing a signal based on only the magnitude (not the phase) of the Fourier transform.

phase spectrum: The Fourier transform of an image can be decomposed into its phase spectrum and its power spectrum. The phase spectrum is the relative phase offset of the given spatial frequency.

phase unwrapping technique: The process of reconstructing the true phase shift from phase estimates "wrapped" into $[-\pi, \pi]$. The true phase shift values may not fall in this interval but instead be mapped into the interval by addition or subtraction of multiples of 2π. The technique maximizes the smoothness of the phase image by adding or subtracting multiples of 2π at various image locations. See also Fourier transform.

phi–s curve (ϕ–s): A technique for representing planar contours. Each point in the contour is represented by the angle ϕ formed by the line through P and the shape's center (*e.g.*, the barycentrum or center of mass) with a fixed direction, and the distance s from the center to P: See also shape representation.

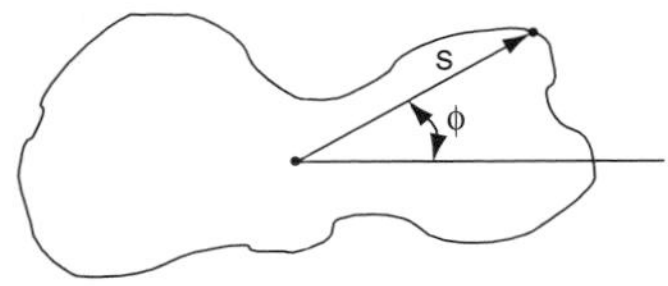

photo consistency: See shape from photo consistency.

photodiode: The basic element, or pixel, of a CCD or other solid state sensor, converting light to an electric signal.

photogrammetry: A research area concerned with obtaining reliable and accurate measurements from noncontact imaging, *e.g.*, a digital height map from a pair of overlapping satellite images. Consequently, accurate camera calibration is a primary concern. The techniques used overlap many typical of image processing and pattern recognition.

photometric invariant: A feature or characteristic of an image that is insensitive to changes in illumination. See also invariant.

photometric decalibration: The correction of intensities in an image so that the same surface (at the same orientation) will give the same response regardless of the position in which it appears in the image.

photometric stereo: A technique recovering surface shape (more precisely, the surface normal at each surface point) using multiple images acquired from a single viewpoint but under different illumination conditions. These lead to different reflectance maps, that together constrain the surface normal at each point.

photometry: A branch of optics concerned with the measurement of the amount or the spectrum of light. In computer vision, one frequently uses photometric models expressing the amount of light emerging from a surface, be it fictitious, or the surface of a radiating source, or from an illuminated object. A well-known photometric model is Lambert's law.

photon noise: Noise generated by the statistical fluctuations associated with photon counting over a finite time interval in the CCD or other solid state sensor of a digital camera. Photon noise is not independent of the signal, and is not additive. See also image noise, digital camera.

photopic response: The sensitivity-wavelength curve modeling the response of the human eye to normal lighting conditions. In such conditions, the cones are the photoreceptors on the retina that best respond to light. Their response curve peaks at 555 nm, indicating that the eye is maximally sensitive to green-yellow colors in normal lighting conditions. When light intensity is very low, the rods determine the eye's response, modeled by the *scotopic curve*, which peaks near to 510 nm.

photosensor spectral response: The spectral response of a photosensor characterizing the sensor's output as a function of the input light's spectral frequency. See also Fourier transform, frequency spectrum, spectral frequency.

physics based vision: An area of computer vision seeking to apply physics laws or methods (of optics, surfaces, illumination, etc.) to the analysis of images and videos. Examples include polarization based methods, in which physical properties of the scene surfaces are estimated via estimates of the state of polarization of the incoming light, and the use of detailed radiometric models of image formation.

picture element: A pixel. It is an indivisible image measurement. This is the smallest directly measured image feature.

picture tree: A recursive image and 2D shape representation in which a tree data structure is used. Each node in the tree represents a region that is then decomposed into subregions. These are represented by child nodes. The figure below shows a segmented image with four regions (left) and the corresponding picture tree.

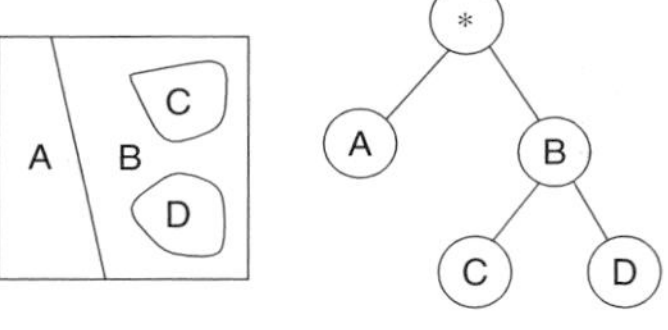

piecewise rigidity: The property of an object or scene that some of its parts, but not the object or scene as a whole, are rigid. Piecewise rigidity can be a convenient assumption, *e.g.*, in motion analysis.

pincushion distortion: A form of radial lens distortion where image points are displaced away from the center of distortion by an amount that increases with the distance to the center. A straight line that would have been parallel to an image side is bowed towards the center of the image. This is the opposite of barrel distortion.

pinhole camera model: The mathematical model for an ideal perspective camera formed by an image plane and a point aperture, through which all incoming rays must pass. For equations, see perspective projection. This is a good model for simple convex lens camera, where all rays pass through the virtual pinhole at the focal point.

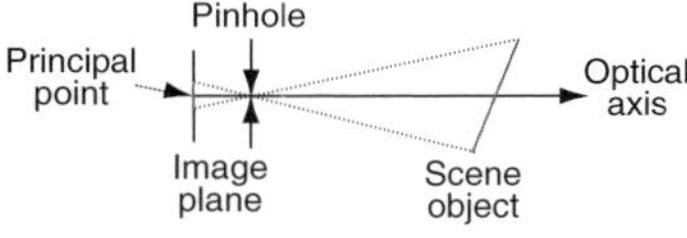

pink noise: Noise that is not white, *i.e.*, when there is a correlation between the noise at two pixels or at two times.

pipeline parallelism: Parallelism achieved with two or more, possibly dissimilar, computation devices. The nonparallel process comprises steps A and B, and will operate on a sequence of items $x_i, i > 0$, producing outputs y_i. The result of B depends on the result of A, so a sequential computer will compute $a_i = A(x_i); y_i = B(a_i)$; for each i. A parallel computer cannot compute a_i and y_i simultaneously as they are dependent, so the computation requires the following steps

$$
\begin{aligned}
a_1 &= A(x_1) \\
y_1 &= B(a_1) \\
a_2 &= A(x_2) \\
y_2 &= B(a_2) \\
a_3 &= A(x_3) \\
&\ldots \\
a_i &= A(x_i) \\
y_i &= B(a_i)\ldots
\end{aligned}
$$

However, notice that we compute y_i just after y_{i-1}, so the computation can be arranged as

$$
\begin{aligned}
&a_1 = A(x_1) \\
&a_2 = A(x_2) \qquad y_1 = B(a_1) \\
&a_3 = A(x_3) \qquad y_2 = B(a_2) \\
&\ldots \\
&a_{i+1} = A(x_{i+1}) \qquad y_i = B(a_i) \\
&\ldots
\end{aligned}
$$

where steps on the same line may be computed concurrently as they are independent. The output values y_i therefore arrive at a rate of one every cycle rather than one every two cycles without pipelining. The pipeline process can be visualized as:

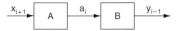

x_{i+1} → [A] → a_i → [B] → y_{i-1}

pit: 1) A general term for when a signal value is lower than the neighboring signal values. Unlike signal peaks, pits usually refer to two dimensional images. For example, a pit occurs when observing the image of a dark spot on a lighter background. 2) A local point-like concave shape defect in a surface.

pitch: A 3D rotation representation (along with yaw and roll) often used for cameras or moving observers. The pitch component specifies a rotation about a horizontal axis to give an up–down change in orientation. This figure shows the pitch rotation direction:

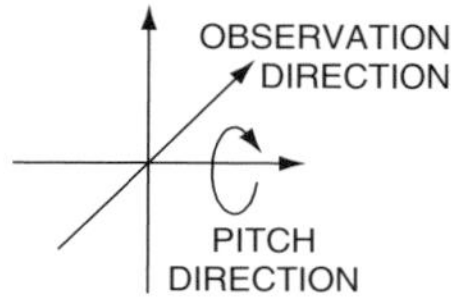

pixel: The intensity values of a digital image are specified at the locations of a discrete rectangular grid; each location is a pixel. A pixel is characterized by its coordinates (position in the image) and intensity value (see intensity and intensity image). Values can express physical quantities other than intensity for different kinds of images, as in, *e.g.*, infrared imaging. In physical terms, a pixel is the photosensitive cell on the CCD or other solid state sensor of a digital camera. The CCD pixe has a precise size, specified by the manufacturer and determining the CCD's aspect ratio. See also intensity sensor and photosensor spectral response.

pixel addition operator: A low-level image processing operator taking as input two gray scale images, I_1 and I_2, and returning an image I_3 in which the value of each pixel is $I_3 = I_1 + I_2$. This figure shows at the right the sum of the two images at the left (the sum divided by 2 to rescale to the original intensity level):

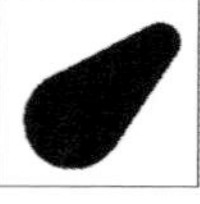 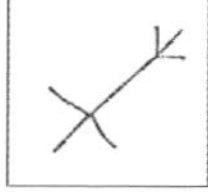 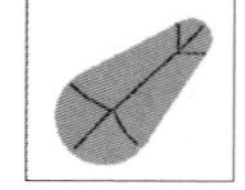

pixel classification: The problem of assigning the pixels of an image to certain classes. See also image segmentation, supervised classification, and clustering. This image shows the pixels of the left image classified into four classes denoted by the four different shades of grey:

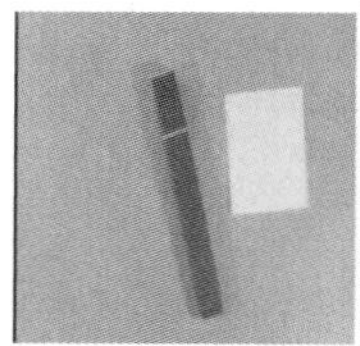 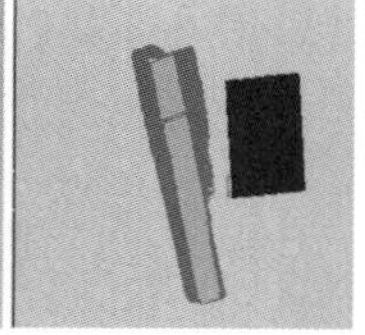

pixel connectivity: The pattern specifying which pixels are considered neighbors of a given one (X) for the purposes of computation. Common connectivity schemes are 4 connectedness and 8 connectedness, as seen in the left and right images here:

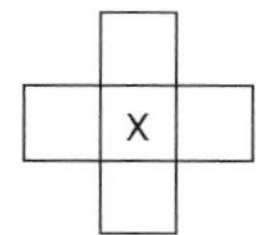

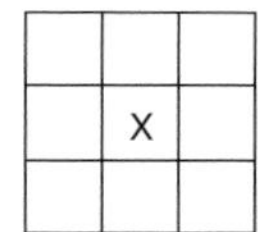

pixel coordinates: The coordinates of a pixel in an image. Normally these are the row and column position.

pixel coordinate transformation: The mathematical transformation linking two image reference frames, specifying how the coordinates of a pixel in one reference frame are obtained from the coordinate of that pixel in the other reference frame. One linear transformation can be specified by

$$i_1 = ai_2 + bj_2 + e$$

$$j_1 = ci_2 + dj_2 + f$$

where the coordinates of $\vec{p}_2 = (i_2, j_2)$ are transformed into $\vec{p}_1 = (i_1, j_1)$. In matrix form, $\vec{p}_1 = \mathbf{A}\vec{p}_2 + \vec{t}$, with $\mathbf{A} = \begin{pmatrix} a & b \\ c & d \end{pmatrix}$ a rotation matrix and $\vec{t} = \begin{pmatrix} e \\ f \end{pmatrix}$ a translation vector. See also Euclidean, affine and holography transforms.

pixel counting: A simple algorithm to determine the area of an image region by counting the numbers of pixels composing the region. See also region.

pixel division operator: An operator taking as input two gray scale images, I_1 and I_2, and returning an image I_3 in which the value of each pixel is $I_3 = I_1 / I_2$.

pixel exponential operator: A low-level image processing operator taking as input one gray scale image, I_1, and returning an image I_2 in which the value of each pixel is $I_2 = cb^{I_1}$. This operator is used to change the dynamic range of an image. The value of the basis b depends on the desired degree of compression of the dynamic range. c is a scaling factor. See also logarithmic transformation, pixel logarithm operator. The right image is 1.005 raised to the pixel values of the left image:

pixel gray scale resolution: The number of different gray levels that can be represented in a pixel, depending on the number of bits associated with each pixel. For instance, an 8-bit pixel (or image) can represent $2^8 = 256$ different intensity values. See also intensity, intensity image, and intensity sensor.

pixel interpolation: See image interpolation.

pixel jitter: A frame grabber must estimate the pixel sampling clock of a digital camera, *i.e.*, the clock used to read out the pixel values, which is not included in the output signal of the camera. Pixel jitter is a form of image noise generated by time variations in the frame grabber's estimate of the camera's clock.

pixel logarithm operator: An image processing operator taking as input one gray scale image, I_1, and returning an image I_2 in which the value of each pixel is $I_2 = c \log_b (| I_1 + 1 |)$. This operator is used to change the dynamic range of an image (see also contrast enhancement), such as for the enhancement of the magnitude of the Fourier transform. The base b of the logarithm function is often e, but it does not actually matter because the relationship between logarithms of any two bases is only one of scaling. See also pixel exponential operator. The right image is the scaled logarithm of the pixel values of the left image:

pixel multiplication operator: An image processing operator taking as input two gray scale images, I_1 and I_2, and returning an image I_3 in which the value of each pixel is $I_3 = I_1 * I_2$. The right image is the product of the left and middle images (scaled by 255 for contrast here):

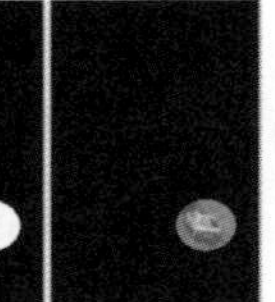

pixel subsampling: The process of producing a smaller image from a given one by including only one pixel out of every N. Subsampling is rarely applied this literally, however, as severe aliasing is introduced; scale space filtering is applied instead.

pixel subtraction operator: A low-level image processing operator taking as input two gray scale images, I_1 and I_2, and returning an image I_3 in which the value of each pixel is $I_3 = I_1 - I_2$. This operator implements the simplest possible change detection algorithm. The right image (with 128 added) is the middle image subtracted from the left image:

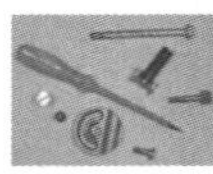

planar facet model: See surface mesh.

planar mosaic: A panoramic image mosaic of a planar scene. If the scene is planar, the transformation linking different views is a homography.

planar motion estimation: A class of techniques aiming to estimate the motion parameters of bodies moving on a planes in space. See also motion estimation.

planar patch extraction: The problem of finding planar regions, or patches, most commonly in range images. Plane extraction can be useful, for instance, in 3D pose estimation, as several model-based matching techniques yield higher accuracy with planar than non-planar surfaces.

planar patches: See surface triangulation.

planar projective transformation: See homography.

planar rectification: A class of rectification algorithms projecting the original images onto a plane parallel to the baseline of the cameras. See also stereo and stereo vision.

planar scene: 1) When the depth of a scene is small with respect to its distance from the camera, the scene can be considered planar, and useful approximations can be adopted; for instance, the transformation between two views taken by a perspective camera is a homography. See also planar mosaic. 2) When all of the surfaces in a scene are planar, *e.g.*, a blocks world scene.

plane: The locus of all points $\vec{x}$ such that the surface normal $\vec{n}$ of the plane and a point in the plane $\vec{p}$ satisfy the relation $(\vec{x}-\vec{p})\cdot\vec{n}=0$. In 3D space, for instance, a plane is defined by two vectors and a point lying on the plane, so that the plane's parametric equation is

$$\vec{p}=a\vec{u}+b\vec{v}+\vec{p}_0$$

where $\vec{p}$ is the generic plane point, $\vec{u},\vec{v},\vec{p}_0$ are the two vectors and the point defining the plane, respectively. The implicit equation of a plane is $ax+by+cz+d=0$, where $[x,y,z]$ are the coordinates of the generic plane point. In vector form, $\vec{p}\cdot\vec{n}=d$, where $\vec{p}=[x,y,z]$, $\vec{n}=[a,b,c]$ is a vector perpendicular to the plane, and $\frac{d}{\sqrt{\|\vec{n}\|}}$ is the distance of the plane from the origin. All of these definitions are equivalent.

plane conic: Any of the curves defined by the intersection of a plane with a 3D double cone, namely ellipse, hyperbola and parabola. Two intersecting lines and a single point represent degenerate conics, defined by special configurations of the cone and plane. The implicit equation of a conic is $ax^2+bxy+cy^2+dx+ey+f=0$. See also conic fitting. This figure shows an ellipse formed by intersection:

plane projective transfer: An algorithm based on projective invariants that, given two images of a planar object, I_1 and I_2, and four feature correspondences, determines the position of any other point of I_1 in I_2. Interestingly, no knowledge of the scene or of the imaging system's parameters is necessary.

plane projective transformation: The linear transformation between the coordinates of two projective planes, also known as homography. See also projective geometry, projective plane, and projective transformation.

plenoptic function representation: A parameterized function for describing everything that is visible from a given point in space, a fundamental representation in image based rendering.

Plessey corner finder: A well-known corner detector also known as Harris corner detector, based on the local autocorrelation of first-order image derivatives. See also feature extraction.

Plücker line coordinates: A representation of lines in projective 3D space. A line is represented by six numbers $(l_{12}, l_{13}, l_{14}, l_{23}, l_{24}, l_{34})$ that must satisfy the constraint that $l_{12}l_{34} + l_{13}l_{24} + l_{14}l_{23} = 0$. The numbers are the entries of the *Plücker matrix*, L, for the line. For any two points A, B on the line, L is given by $l_{ij} = A_iB_j - B_iA_j$. The pencil of planes containing the line are the nullspace of L. The six numbers may also be seen as a pair of 3-vectors, one a point $\vec{a}$ on the line, one the direction $\vec{n}$ with $\vec{a} \cdot \vec{n} = 0$.

PMS: See Pantone matching system.

point: A primitive concept of Euclidean geometry, representing an infinitely small entity. In computer vision, pixels are regarded as image points, and one speaks of "points in the scene" as positions in the 3D space observed by the cameras.

point distribution model (PDM): A shape representation for flexible 2D contours. It is a type of deformable template model and its parameters can be learned by supervised learning. It is suitable for 2D shapes that undergo general but correlated deformations or variations, such as component motion or shape variation. For instance, fronto-parallel images of leaves, fish or human hands, resistors on a board, people walking in surveillance videos, and the like. The shape variations of the contour in a series of examples are captured by principal component analysis.

point feature: An image feature that occupies a very small portion of an image, ideally one pixel, and is therefore local in nature. Examples are corners (see corner detection) or edge pixels. Notice that, although point features occupy only one pixel, they require a neighborhood to be defined; for instance, an edge pixel is characterized by a sharp variation of image values in a small neighborhood of the pixel.

point invariant: A property that 1) can be measured at a point in an image and 2) is invariant to some transformation. For instance, the ratio of a pixel's observed intensity to that of its brightest neighbor is invariant to changes in illumination. Another example: the magnitude of the gradient of intensity at a point is invariant to translation and rotation. (Both of these examples assume ideal images and observation.)

point light source: A point-like light source, typically radiating energy radially, whose intensity decreases as $\frac{1}{r^2}$, where r is the distance to the source.

point matching: A class of algorithms solving the matching or correspondence problem for point features.

point of extreme curvature: A point where the curvature achieves an extremum, that is, a maximum or a minimum. This figure shows one of each type circled:

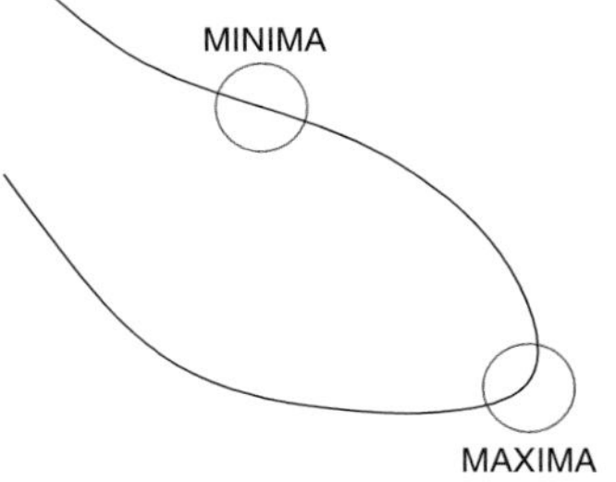

point sampling: Selection of discrete points of data from a continuous signal. For example a digital camera samples a continuous image function into a digital image.

point similarity measure: A function measuring the similarity of image points (actually small neighborhoods to include sufficient information to characterize the image location), for instance cross correlation, SAD (sum of absolute differences), or SSD (sum of squared differences).

point source: A point light source. An ideal illumination source in which all light comes from a single spatial point. The alternative is an extended light source. The assumption of being a point source allows easier interpretation of shading and shadows, etc.

point spread function: The response of a 2D system or filter to an input Dirac impulse. The response is typically spread over a region surrounding the point of application of the impulse, hence the name. Analogous to the impulse response of a 1D system. See also filter, linear filter.

polar coordinates: A system of coordinates specifying the position of a point P in terms of the direction of the line through P and the origin, and the distance from P to the origin along that line. For example, the transformation between polar (r, θ) and Cartesian coordinates (x, y) in the plane is given by $x = r\cos\theta$ and $y = r\sin\theta$, or $r = \sqrt{x^2 + y^2}$ and $\theta = atan(\frac{y}{x})$.

polar rectification: A rectification algorithm designed to cope with any camera geometry in the context of uncalibrated vision, reparameterizing the images in polar coordinates around the epipoles.

polarization: The characterizing property of polarized light.

polarized light: Unpolarized light results from the nondeterministic superposition of the x and y components of the electric field. Otherwise, the light is said to be polarized, and the tip of the electric field evolves on an ellipse (elliptically polarized light). Light is often partially polarized, that is, it can be regarded as the sum of completely polarized and completely unpolarized light. In computer vision, polarization analysis is an area of physics based vision, and has been used for metal–dielectric discrimination, surface reconstruction, fish classification, defect detection, and in structured light triangulation.

polarizer: A device changing the state of polarization of light to a specific polarized state, for example, producing linearly polarized light in a given plane.

polycurve: A simple curve C that is smooth everywhere but at a finite set of points, and such that, given any point P on C, the tangent to C converges to a limit approaching P from each direction. Computer vision shape models often describe boundary shapes using polycurve models consisting of a sequence of curved or straight segments, such as in this example using four circular arcs. See also polyline.

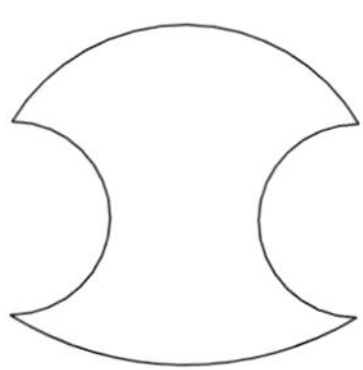

polygon: A closed, piecewise linear, 2D contour. Squares, rectangles and pentagons are examples of regular polygons, where all sides have equal length and all angles formed by contiguous sides are equal. This does not hold for a general polygon.

polygon matching: A class of techniques for matching polygonal shapes. See polygon.

polygonal approximation: A polyline approximating a curve. This circular arc is (badly) approximated by the polyline:

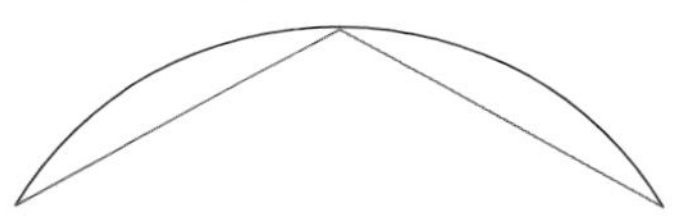

polyhedron: A 3D object with planar faces, a "3D polygon". A subset of 3 whose boundary is a subset of finitely many planes. The basic primitive of many 3D modeling schemes, as many hardware accelerators process polygons particularly quickly. A tetrahedron is the simplest polyhedron:

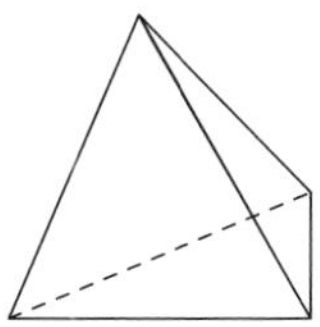

polyline: A piecewise linear contour. If closed, it becomes a polygon. See also polycurve, contour analysis and contour representation.

pose: The location and orientation of an object in a given reference frame, especially a world or camera reference frame. A classic problem of computer vision is pose estimation.

pose clustering: A class of algorithms solving the pose estimation problem using clustering techniques (see clustering/cluster analysis). See also pose, k-means clustering.

pose consistency: An algorithm seeking to establish whether two shapes are equivalent. Given two sets of points G_1 and G_2, for example, the algorithm finds a sufficient number of point correspondences to determine a transformation T between the two sets, then applies T to all other points of G_1. If the transformed points are close to points in G_2, consistency is satisfied. Also known as *viewpoint consistency*. See also feature point correspondence.

pose determination: See pose estimation.

pose estimation: The problem of determining the orientation and translation of an object, especially a 3D one, from one or more images thereof. Often the term means finding the transformation that aligns a geometric model with the image data. Several techniques exist for this purpose. See also alignment, model registration, orientation estimation, and rotation representation.

pose representation: The problem of representing the angular position, or pose, of an object (especially 3D) in a given reference frame.

A common representation is the rotation matrix, which can be parameterized in different ways, *e.g.*, Euler angles, pitch-, yaw-, roll-angles, rotation angles around the coordinate axes, axis-angle, and quaternions. See also orientation estimation and rotation representation.

position: Location in space (either 2D or 3D).

position dependent brightness correction: A technique seeking to counteract the brightness variation caused by a real imaging system, typically the fact that brightness decreases as one moves away from the optical axis in a lens system with finite aperture. This effect may be noticeable only in the periphery of the image. See also lens.

position invariant: Any property that does not vary with position. For instance, the length of a 3D line segment is invariant to the line's position in 3D space, but the length of the line's projection on the image plane is not. See also invariant.

positron emission tomography (PET): A medical imaging method that can measure the concentration and movement of a positron-emitting isotope in living tissue.

postal code analysis: A set of image analysis techniques concerned with understanding written or printed postal codes. See handwritten and optical character recognition.

posture analysis: A class of techniques aiming to estimate the posture of an articulated body, for instance a human body (*e.g.*, pointing, sitting, standing, crouching, etc.).

potential field: A mathematical function that assigns some (usually scalar) value at every point in some space. In computer vision and robotics, this is usually a measure of some scalar property at each point of a 2D or 3D space or image, such as the distance from a structure. The representation is used in path planning, such that the potential at every point indicates, for example, the ease/difficulty of getting to some destination.

power spectrum: In the context of computer vision, normally the amount of energy at each spatial frequency. The term could also refer to the amount of energy at each light frequency. Also called the power spectrum density function or spectral density function.

precision: 1) The repeatability of the accuracy of a vision system (in general, of an instrument) over many measures carried out in the same conditions. Typically measured by the standard deviation of a target error measure. For instance, the precision of a vision system measuring linear size would be assessed by taking thousands of measurements of a perfectly known object and computing the standard deviation of the measurements. See also accuracy. 2) The number of significant bits in a floating point or double precision number that lie to the right of the decimal point.

predictive compression method: A class of image compression algorithms using redundancy information, mostly correlation, to build an estimate of a pixel value from values of neighboring pixels.

pre-processing: Operations on an image that, for example, suppress some distortion(s) or enhance some feature(s). Examples include geometric transformations, edge detection, image restoration, etc. There is no clear distinction between image pre-processing and image processing.

Prewitt gradient operator: An edge detection operator based on template matching. It applies a set of convolution masks, or kernels (see Prewitt kernel), implementing matched filters for edges at various (generally eight) orientations. The magnitude (or strength) of the edge at a given pixel is the maximum of the responses to the masks. Alternatively, some implementations use the sum of the absolute value of the responses from the horizontal and vertical masks.

Prewitt kernel: The mask used by the Prewitt gradient operator. The horizontal and vertical masks are:

−1	0	+1
−1	0	+1
−1	0	+1

Gx

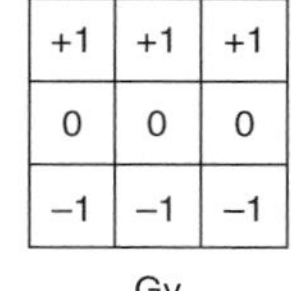

+1	+1	+1
0	0	0
−1	−1	−1

Gy

primal sketch: A representation for early vision introduced by Marr, focusing on low-level features like edges. The full primal sketch groups the information computed in the raw primal sketch (consisting largely of edge, bar, end and blob feature information extracted from the images), for instance by forming subjective contours. See also Marr–Hildreth edge detection and raw primal sketch.

primary color: A color coding scheme whereby a range of perceivable colors can be made by a weighted combination of primary colors. For example, color television and computer screens use red, green and blue light-emitting chemicals to produce these three primary colors. The ability to use only three colors to generate all others arises from the trichromacy of the human eye, which has cones that respond to three different color spectral ranges. See also additive and subtractive color.

principal component analysis (PCA): A statistical technique useful for reducing the dimensionality of data, at the basis of many computer vision techniques (*e.g.*, point distribution models and eigenspace based recognition). In essence, the deviation of a random vector, $\vec{x}$, from the population mean, μ, can be expressed as the product of A, the matrix of eigenvectors of the covariance matrix of the population, and a vector y of projection weights:

$$\vec{y} = A(\vec{x} - \vec{\mu})$$

so that

$$\vec{x} = A^{-1}\vec{y} + \vec{\mu}$$

Usually only a subset of the components of $\vec{y}$ is sufficient to approximate $\vec{x}$. The elements of this subset correspond to the largest eigenvalues of the covariance matrix. See also Karhunen–Loève transformation.

principal component basis space: In principal component analysis, the space generated by the basis formed by the eigenvectors, or eigendirections, of the covariance matrix.

principal component representation: See principal component analysis.

principal curvature: The maximum or minimum normal curvature at a surface point, achieved along a principal direction. The two principal curvatures and directions, together completely specify the local surface shape. The principal curvatures in the two directions at the point X on the cylinder of radius r below are 0 (along axis) and $\frac{1}{r}$ (across axis).

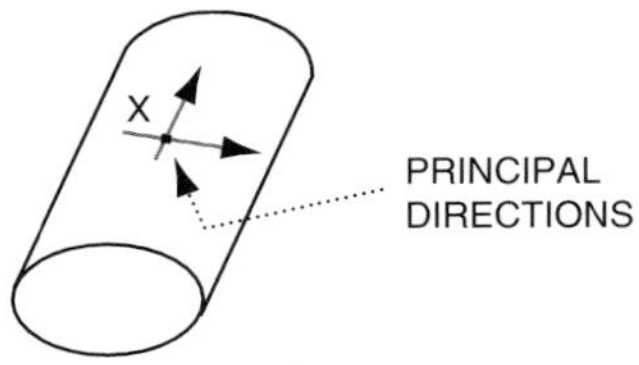

principal curvature sign class: See mean and Gaussian curvature shape classification.

principal direction: The direction in which the normal curvature achieves an extremum, that is, a principal curvature. The two principal curvatures and directions, together, specify completely the local surface shape. The principal directions at the point X on the cylinder below are parallel to the axis and around the cylinder.

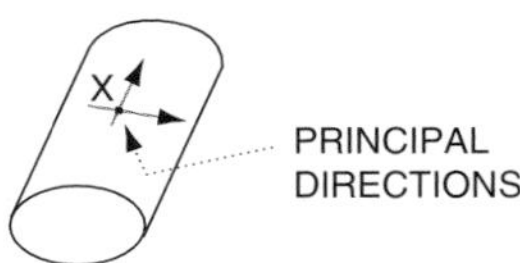

principal point: The point at which the optical axis of a pinhole camera model intersects the image plane, as in:

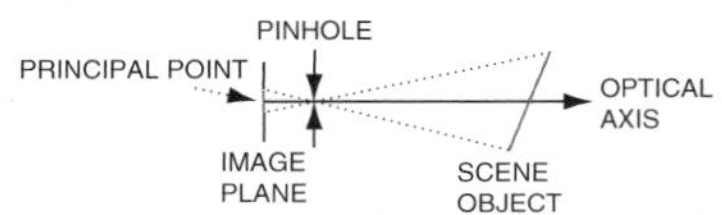

principal texture direction: An algorithm identifying the direction of a texture. A directional or oriented texture in a small image patch generates a peak in the Fourier transform. To determine the direction, the Fourier amplitude plot is regarded as a distribution of physical mass, and the minimum-inertia axis identified.

privileged viewpoint: A viewpoint where small motions cause image features to appear or disappear. This contrasts with a generic viewpoint.

probabilistic causal model: A representation used in artificial intelligence for causal models. The simplest causal model is a *causal graph*, in essence an acyclic graph in which nodes represents variables and directed arcs represent cause and effect. A probabilistic causal model is a causal graph with the probability distribution of each variable conditional to its causes.

probabilistic Hough transform: The probabilistic Hough transform computes an approximation to the Hough transform by using only a percentage of the image data. The goal is to reduce the computational cost of the standard Hough transform. A threshold effect has been observed so that if the percentage sampled is above the threshold level then few false positives are detected.

probabilistic model learning: A class of Bayesian learning algorithms based on probabilistic networks, that allow you to input information at any node (unlike neural networks), and associate uncertainty coefficients to classification answers. See also Bayes' rule, Bayesian model, Bayesian network.

probabilistic principal component analysis: A technique defining a probability model for principal component analysis (PCA). The model can be extended to mixture models, trained using the expectation maximization (EM) algorithm. The original data is modeled as being generated by the reduced-dimensionality subset typical of PCA plus Gaussian noise (called a *latent variable model*).

probabilistic relaxation: A method of data interpretation in which local inconsistencies act as inhibitors and local consistencies act as excitors. The hope is that the combination of these two influences constrains the probabilities.

probabilistic relaxation labeling: An extension of relaxation labeling in which each entity to be labeled, for instance each image feature, is not simply assigned to a label, but to a set of probabilities, each giving the likelihood that the feature could be assigned a specific label.

probability: A measure of the confidence one may have in the occurrence of an event, on a scale from 0 (impossible) to 1 (certain), and defined as the proportion of favorable outcomes to the total number of possibilities. For instance, the probability of getting any number from a dice in a single throw is $\frac{1}{6}$. Probability theory, an important part of statistics, is the basis of several vision techniques.

probability density estimation: A class of techniques for estimating the density function or its parameters given a sample from a population. A related problem is testing whether a particular sample has been generated by a process characterized by a particular probability distribution. Two common tests are the *goodness-of-fit* and the *Kolmogorov–Smirnov* tests. The former is a parametric test best used with large samples; the latter gives good results with smaller samples, but is a non-parametric test and, as such, does not produce estimates of the population parameters. See also non-parametric method.

procedural representation: A class of representations used in artificial intelligence that are used to encode how to perform a task (procedural knowledge). A classic example is the production system. In contrast, *declarative representations* encode how an entity is structured.

Procrustes analysis: A method for comparing two data sets through the minimization of squared errors, by translation, rotation and scaling.

production system: 1) An approach to computerized logical reasoning, whereby the logic is represented as a set of "production rules". A rule is of the form "LHS→RHS". This states that if the pattern or set of conditions encoded in the left-hand side (LHS) are true or hold, then do the actions specified in the right-hand side (RHS), which may simply be the assertion of some conclusion. A sample rule might be "If the number of detected edge fragments is less than 10, then decrease the threshold by 10%". 2) An industrial system that manufactures some product. 3) A system that is to be actually used, as compared to a demonstration system.

profiles: A shape signature for image regions, specifying the number of pixels in each column (vertical profile) or row (horizontal profile). Used in pattern recognition. See also shape, shape representation.

progressive image transmission: A method of transmitting an image in which a low-resolution version is first transmitted, followed by details that allow progressively higher

FIRST IMAGE

BETTER IMAGE

BEST IMAGE

resolution versions to be recreated.

progressive scan camera: A camera that transfers an entire image in the order of left-to-right, top-to-bottom, without the alternate line interlacing used in television standards. This is much more convenient for machine vision and other computer-based applications.

projection: 1) The transformation of a geometric structure from one space to another, *e.g.*, the projection of a 3D point onto the nearest point in a given plane. The projection may be specified by a linear function, *i.e.*, for all points $\vec{p}$ in the initial structure, the points $\vec{p}'$ in the projected structure are given by $\vec{p}' = \mathbf{M}\vec{p}$ for some matrix **M**. Alternatively, the projection need not be linear, *e.g.*, $\vec{p}' = \vec{f}(\vec{p}')$. 2) The specific case of projection of a scene that creates an image on a plane by use of, for example, a perspective camera, according to the rules of perspective.

projection matrix: The matrix transforming the homogeneous projective coordinates of a 3D scene point $(x, y, z, 1)$ into the pixel coordinates $(u, v, 1)$ of the point's image in a pinhole camera. It can be factored as the product of the two matrices of the intrinsic camera parameters and extrinsic camera parameters. See also camera coordinates, image coordinates, scene coordinates.

projective geometry: A field of geometry dealing with projective spaces and their properties. A projective geometry is one where only properties preserved by projective transformations are defined. Projective geometry provides a convenient and elegant theory to model the geometry of the common perspective camera. Most notably, the perspective projection equations become linear.

projective invariant: A property, say I, that is not affected by a projective transformation. More specifically, assume an invariant, $I(\vec{P})$, of a geometric structure described by a parameter vector $\vec{P}$. When the structure is subject to a projective transformation (**M**) this gives a structure with parameter vector $\vec{p}$, and $I(\vec{P}) = I(\vec{p})$. The most fundamental projective invariant is the cross ratio. In some applications, invariants of weight w occur, which transform as $I(\vec{p}) = I(\vec{P})(\det \mathbf{M})^{w}$.

projective plane: A plane, usually denoted by P^2, on which a projective geometry is defined.

projective reconstruction: The problem of reconstructing the geometry of a scene from a set or sequence of images in a projective space. The transformation from projective to Euclidean coordinates is easy if the Euclidean coordinates of the five points in a projective basis are known. See also projective geometry and projective stereo vision.

projective space: A space of $(n+1)$-dimensional vectors, usually denoted by P^n, on which a projective geometry is defined.

projective stereo vision: A class of stereo algorithms based on projective geometry. Key concepts expressed elegantly by the projective framework are epipolar geometry, fundamental matrix, and projective reconstruction.

projective stratum: A layer in the stratification of 3D geometries. Moving from the simplest to the most complex, we have the projective, affine, metric and Euclidean strata. See also projective geometry, projective reconstruction.

projective transformation: Also known as *projectivity*, from one projective plane to another. It can be represented by a non-singular 3×3 matrix acting on homogeneous coordinates. The transformation has eight degrees of freedom, as only the ratio of projective coordinates is significant.

property based matching: The process of comparing two entities (*e.g.*, image features or patterns) using their properties, *e.g.*, the moments of a region. See also classification, boundary property, metric property.

property learning: A class of algorithms aiming at learning and characterizing attributes of spatio-temporal patterns. For example, learning the color and texture distributions that differentiate beween normal and cancerous cells. See also boundary property, metric property, unsupervised learning and supervised learning.

prototype: An object or model serving as representative example for a class, capturing the defining characteristics of the class.

proximity matrix: A matrix $\mathbf{M}$ occurring in cluster analysis. $\mathbf{M}(i, j)$ denotes the distance (*e.g.*, the Hamming distance) between clusters i and j.

pseudocolor: A way of assigning a color to pixels that is based on an interpretation of the data rather than the original scene color. The usual purpose of pseudocoloring is to label image pixels in a useful manner. For example, one common pseudocoloring assigns different colors according to the local surface shape class. A pseudocoloring scheme for aerial or satellite images of the earth assigns colors according to the land type, such as water, forest, wheat field, etc.

PSF: See point spread function.

purposive vision: An area of computer vision linking perception with purposive action; that is, modifying the position or parameters of an imaging system purposively, so that a visual task is facilitated or made possible. Examples include changing the lens parameters so to obtain information about depth, as in depth from defocus, or moving around an object to achieve full shape information.

pyramid: A representation of an image including information at several spatial scales. The pyramid is constructed by the original image (maximum resolution) and a scale operator that reduces the content of the image (*e.g.*, a Gaussian filter) by discarding details at coarser scales:

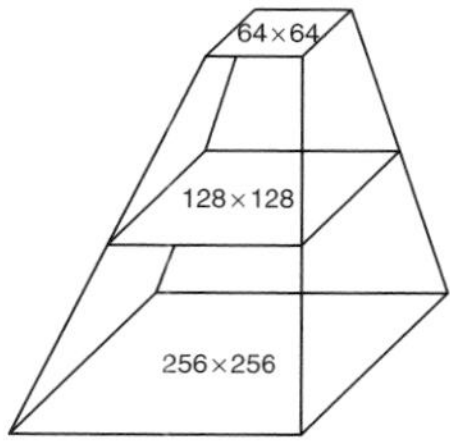

Applying the operator and subsampling the resulting image leads to the next (lower-resolution)level of the pyramid. See also scale space, image pyramid, Gaussian pyramid, Laplacian pyramid, pyramid transform.

pyramid architecture: A computer architecture supporting pyramid-based processing, typically occurring in the context of multi-scale processing. See also scale space, pyramid, image pyramid, Laplacian pyramid, Gaussian pyramid.

pyramid transform: An operator for building a pyramid from an image. See pyramid, image pyramid, Laplacian pyramid, Gaussian pyramid.

Q

QBIC: See query by image content.

quadratic variation: 1) Any function (here, expressing a variation of some variables) that can be modeled by a quadratic polynomial. 2) The specific measure of surface shape deformation $f_{xx}^2 + 2f_{xy}^2 + f_{yy}^2$ of a surface $f(x, y)$. This measure has been used to constrain the smoothness of reconstructed surfaces.

quadrature mirror filter: A class of filters occurring in wavelet and image compression filtering theory. The filter splits a signal into a high pass component and a low pass component, with the low pass component's transfer function a mirror image of that of the high pass component.

quadric: A surface defined by a second-order polynomial. See also conic.

quadric patch: A quadric surface defined over a finite region of the independent variables or parameters; for instance, in range image analysis, a part of a range surface that is well approximated by a quadric (*e.g.*, an elliptical patch).

quadric patch extraction: A class of algorithms aiming to identify the portions of a surface that are well approximated by quadric patches. Techniques are similar to those applied for conic fitting. See also surface fitting, least square surface fitting.

quadrifocal tensor: An algebraic constraint imposed on quadruples of corresponding points by the geometry of four simultaneous views, analogous to the epipolar constraint for the two-camera case and to the trifocal tensor for the three-camera case. See also stereo correspondence, epipolar geometry.

quadrilinear constraint: The geometric constraint on four views of a point (*i.e.*, the intersection of four epipolar lines). See also epipolar constraint and trilinear constraint.

quadtree: A hierarchical structure representing 2D image regions, in which each node represents a region, and the whole image is the root of the tree. Each non-leaf node, representing a region *R*, has four children, that represent the four subregions into which *R* is divided:, as illustrated below.

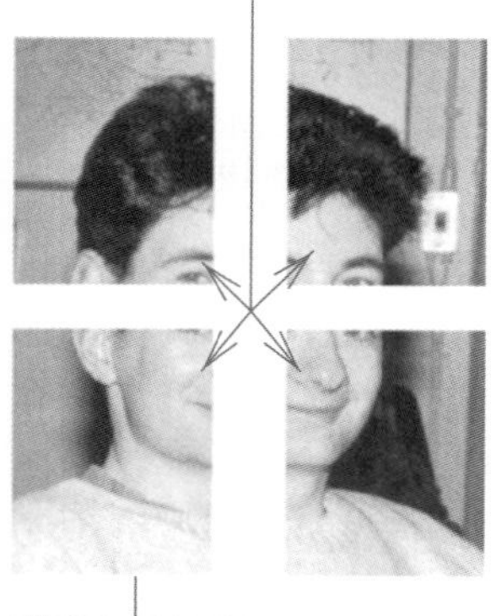

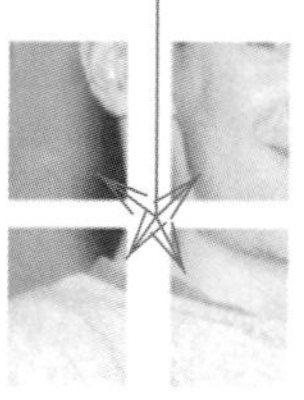

Hierarchical subdivision continues until the remaining regions have constant properties. Quadtrees can be used to create a compressed image structure. The 3D extension of a quadtree is the octree.

qualitative vision: A paradigm based on the idea that many perceptual tasks could be better accomplished by computing only qualitative descriptions of objects and scenes from images, as opposed to quantitative information like accurate measurements. Suggested in the framework of computational theories of human vision.

quantization: See spatial quantization.

quantization error: The approximation error created by the quantization of a continuous variable, typically using a regularly spaced scale of values. This figure

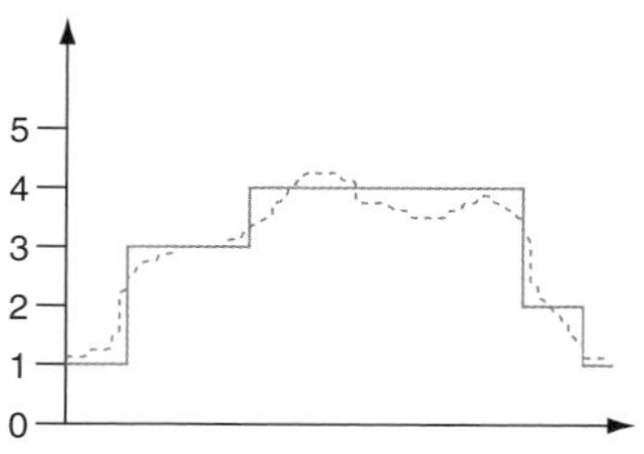

shows a continuous function (dashed) and its quantized version (solid line) using six values only. The quantization error is the vertical distance between the two curves. For instance, the intensity values in a digital image can only take on a certain number (often 256) of discrete values. See also sampling theorem and Nyquist sampling rate.

quantization noise: See quantization error.

quasi-invariant: An approximation of an invariant. For instance, quasi-invariant parameterizations of image curves have been built by approximating the invariant arc length with lower spatial derivatives.

quaternion: A forerunner of the modern vector concept, invented by Hamilton, used in vision to represent rotations. Any rotation matrix, **R**, can be parameterized by a vector of four numbers, $\vec{q} = (q_0, q_1, q_2, q_3)$, such that $\sum_{k=0}^{3} q_k^2 = 1$, that define uniquely the rotation. A rotation has two representations, $\vec{q}$ and $-\vec{q}$. See rotation matrix for alternative representations of rotations.

query by image content (QBIC): A class of techniques for selecting members from a database of images by using examples of the desired image content (as opposed to textual search). Examples of contents include color, shape, and texture. See also image database indexing.

R

R–S curve: A contour representation giving the distance, r, of each point of the contour from an origin chosen arbitrarily, as a function of the arc length, s. Allows rotation-invariant comparison of contour. See also contour, shape representation.

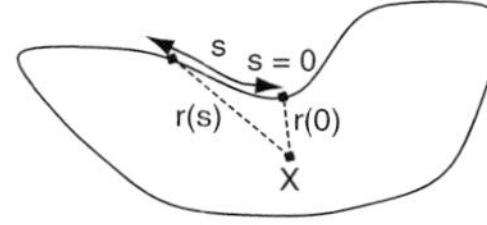

radar: An active sensor detecting the presence of distant objects. A narrow beam of very high-frequency radio pulses is transmitted and reflected by a target back to the transmitter. The direction of the reflected beam and the time of flight of the pulse determine the target's position. See also time-of-flight range sensor.

radial lens distortion: A type of geometric distortion introduced by a real lens. The effect is to shift the position of each image point, p, away from its true position, along the line through the image center and p. See also lens, lens distortion, barrel distortion, tangential distortion, pin cushion distortion, distortion coefficient. This figure shows the typical deformations of a square (exaggerated):

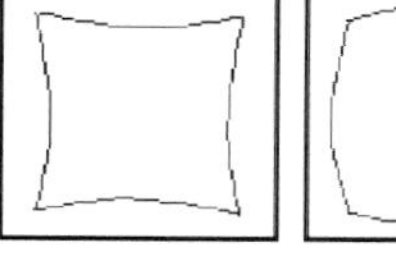
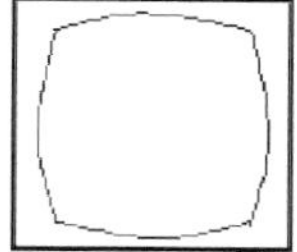

radiance: The amount of light (radiating energy) leaving a surface. The light can be generated by the surface itself, as in a light source, or reflected by it. The surface can be real (*e.g.*, a wall) or imaginary (*e.g.*, an infinite plane). See also irradiance, radiometry.

radiance map: A map of radiance for a scene. Sometimes

Dictionary of Computer Vision and Image Processing R.B. Fisher, K. Dawson-Howe, A. Fitzgibbon, C. Robertson and E. Trucco

used to refer to a high dynamic range image.

radiant flux: The radiant energy per time unit, that is, the amount of energy transmitted or absorbed per time unit. See also radiance, irradiance, radiometry.

radiant intensity: See radiant flux.

radiometric calibration: A process seeking to estimate radiance from pixel values. The rationale for radiometric calibration is that the light entering a real camera (the radiance) is, in general, altered by the camera itself. A simple calibration model is $E(i,j) = g(i,j)I + o(i,j)$, where, for each pixel (i,j), E is the radiance to estimate, I the measured intensity, and g and o a pixel-specific gain and offset to be calibrated. Ground truth values for E can be measured using *photometers*.

radiometry: The measurement of optical radiation, *i.e.*, electromagnetic radiation between 3×10^{11} and 3×10^{16} Hz (wavelengths between 0.01 and 1000 μm). This includes ultraviolet, visible and infrared. Common units encountered are $\frac{watts}{m^2}$ and $\frac{photons}{sec-steradian}$. Compare with photometry, which is the measurement of visible light.

radius vector function: A contour or boundary representation based about a point $\vec{c}$ in the center of the figure (usually the center of gravity or a physically meaningful point). The representation then records the distance $r(\theta)$ from $\vec{c}$ to points on the boundary, as a function of θ, which is the angle between the direction and some reference direction. The representation has problems when the vector at angle θ intersects the boundary more than one time. See:

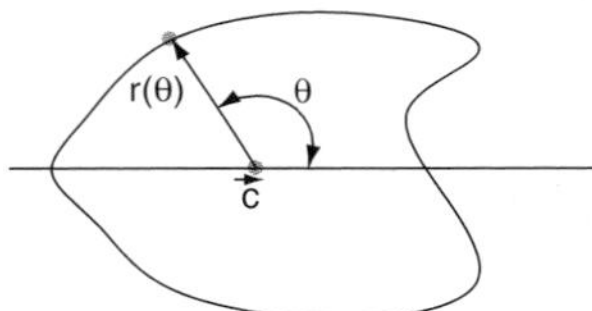

radon transform: A transformation mapping an image into a parameter space highlighting the presence of lines. It can be regarded as an extension of the Hough transform. One definition is

$$g(\rho, \theta) = \iint I(x, y) \times \delta(\rho - x \cos\theta - y \sin\theta)\ dx dy$$

where $I(x, y)$ is the image (gray values) and $\rho = x\cos\theta + y\sin\theta$ is a parametric line in the image. Lines are identified by peaks in the ρ, θ space. See also Hough transform line finder.

RAG: See region adjacency graph.

random access camera: A random access camera is characterized by the possibility of accessing any image location directly. The name was introduced to distinguish such cameras from sequential scan cameras, where image values are transmitted in a standard order.

random dot stereogram: A stereo pair formed by one random dot image (that is, binary images in which each pixel is assigned to black or white at random), and a second image that is derived from the first. This figure

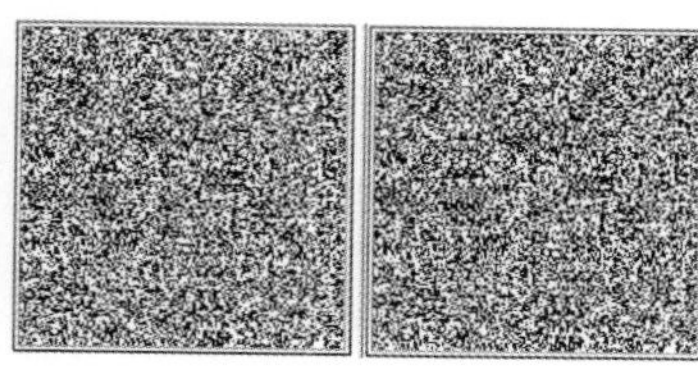

shows an example, in which a central square is shifted horizontally. Looking cross-eyed at close distance, you should perceive a strong 3D effect. See also stereo and stereo vision.

random sample consensus: See RANSAC.

random variable: A scalar or a vector variable that takes on a random value. The set of possible values may be describable by a standard distribution, such as the Gaussian, mixture of Gaussians, uniform, or Poisson distributions.

randomized Hough transform: A variation of the standard Hough transform designed to produce higher accuracy with less computational effort. The line-finding variant of the algorithm selects pairs of image edge points randomly and increments the accumulator cell corresponding to the line through these two points. The selection process is repeated a fixed number of times.

range compression: Reducing the dynamic range of an image to enhance the appearance of the image. This is often needed for images resulting from the magnitude of the Fourier transform which might have pixels with both large and very low values. Without range compression it will be hard to see the structure in the pixels with the low values. The left image shows the magnitude of a 2D Fourier transform with a single bright spot in the middle. The right image shows the logarithm of the left image, revealing more details.

range data: A representation of the spatial distribution of a set of 3D points. The data is often acquired by stereo vision or by a range sensor. In computer vision, range data are often represented as *cloud of points*, *i.e.*, a set of triplets representing the X, Y, Z coordinate of each point, or as range images, also known as *Moirè patch*. The figure below shows a range image of an industrial part, where brighter pixels are closer:

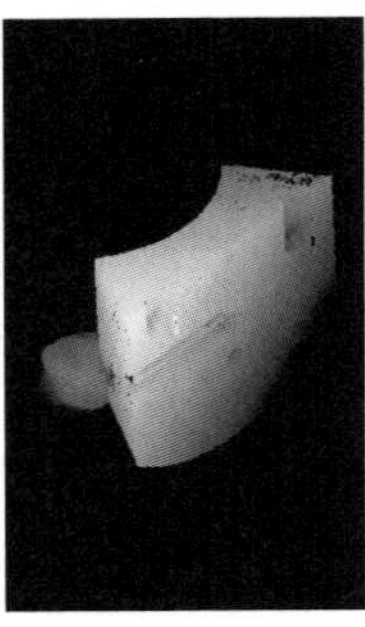

range data fusion: The merging of multiple sets of range data, especially for the purpose of 1) extending the portion of an object's surface described by the range data, or 2) increasing the accuracy of measurements by exploiting the redundancy of multiple measures available for each point of surface area. See also information fusion, fusion, sensor fusion.

range data integration: See range data fusion.

range data registration: See registration.

range data segmentation: A class of techniques partitioning range data into a set of regions. For instance, a well-known method for segmenting range images is HK segmentation, which produces a set of surface patches covering the initial surface. The right image shows the plane, cylinder and spherical patches extracted from the left range image. See also surface segmentation.

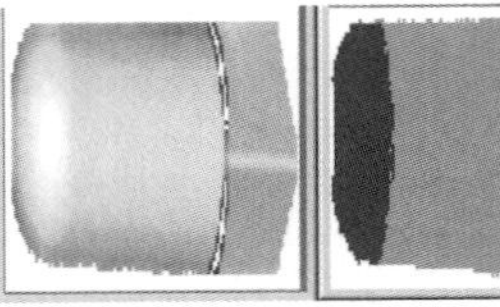

range edge: See surface shape discontinuity

range flow: A class of algorithms for the measurement of motion in time-varying range data, made possible by the evolution of fast range sensors. See also optical flow.

range image: A representation of range data as an image. The pixel coordinates are related to the spatial position of each point on the range surface, and the pixel value represents the distance of the surface point from the sensor (or from an arbitrary, fixed background). The figure below shows a range image of a face, where darker pixels are closer:

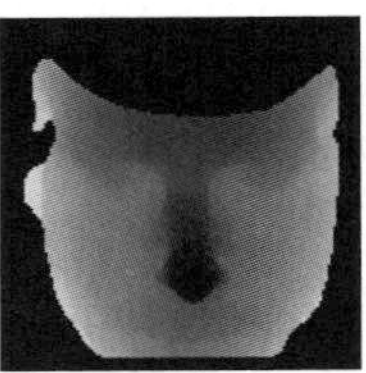

range image edge detector: An edge detector working on range images. Typically, edges occur where depths or surface normal directions (fold edge) change rapidly. See also edge detection, range images. The right image shows the depth and fold edges extracted from the left range image:

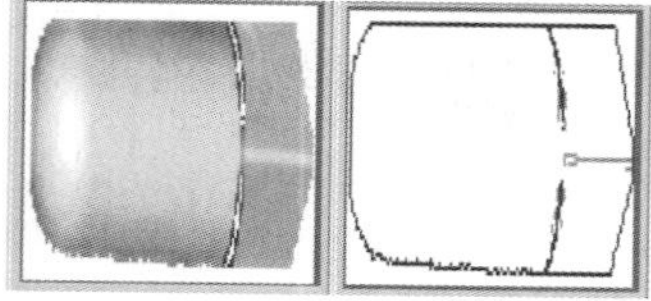

range sensor: Any sensor acquiring range data. The most popular range sensors in computer vision are based on optical and acoustic technologies. A laser range sensor often uses structured light triangulation. A time-of-flight range sensor measures the round-trip time of an acoustic or optical pulse. See also depth estimation. An example of a triangulation range sensor is

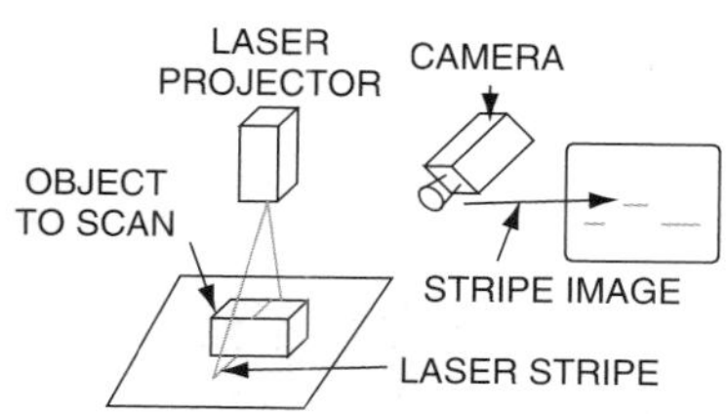

rank order filtering: A class of filters the output of which depends on an ordering (ranking) of the pixels within the region of support. The classic example is the median filter which selects the middle value of the set of input values. More generally, the filter selects the kth largest value in the input set.

RANSAC: Acronym for *random sample consensus*, a robust estimator seeking to counter the effect of outliers in data used, for example, in a least square estimation problem. In essence, RANSAC considers a number of data subsets of the minimum size necessary to solve the problem (say a parametric surface fit), then looks for statistical agreement of the results. See also least median square estimation, M-estimation, outlier rejection.

raster scan: “Raster” refers to the region of a monitor, *e.g.*, a cathode ray tube (CRT) or a liquid crystal display (LCD) capable of rendering images. In a CRT, the raster is a sequence of horizontal lines that are scanned rapidly with an electron beam from left to right and top to bottom, largely in the same way as a TV picture tube is scanned. In an LCD, the raster (usually called a “grid”) covers the whole device area and is scanned differently, in that image elements are displayed individually.

rate-distortion: A statistical method useful in analog-to-digital conversion. It determines the minimum number of bits required to encode data while tolerating a given level of distortion, or *vice versa*.

rate-distortion function: The number of bits per sample (the rate R_d) to encode an analog image (or other signal) value given the allowable distortion D (or mean square of the error). Also needed is the variance σ^2 of the input value (assuming it is a Gaussian random variable). Then $R_d = max(0, \frac{1}{2}log_2(\frac{\sigma^2}{D}))$.

raw primal sketch: The first representation built in the perception process according to Marr's theory of vision, heavily based on detection of local edge features. It represents the location, orientation, contrast and scale of center–surround, edge, bar and truncated bar features. See also primal sketch.

RBC: See recognition by components.

real time processing: Any computation performed within the time limits imposed by a given process. For example, in visual servoing a tracking system feeds positional data to a control algorithm generating control signals; if the control signals are generated too slowly, the whole system may become unstable. Different processes can impose very different constraints for real time processing. When processing video-stream data, real time means complete processing of one frame of data in the time before the next frame is acquired (possibly with several frames lag time as in a pipeline parallel process).

receiver operating curves and performance analysis for vision: A receiver operating curve (ROC) is a diagram showing the performance of a classifier. It plots the number or percentage of true positives against the number or percentage of true negatives. Performance analysis is a substantial topic in computer vision and the object of an ongoing debate. See also performance characterization, test, classification.

receptive field: 1) The retinal area generating the response to a photostimulus. The main cells responsible for visual perception in the retina are the *rods* and the *cones*, active in high- and low-intensity situations respectively. See also photopic response. 2) The region of visual space giving rise to that response. 3) The region of an image that is input to the calculation of each output value. (See region of support.)

recognition: See identification.

recognition by components (RBC): 1) A theory of human image understanding devised by Biederman. The foundation is a set of 3D shape primitives called geons, reminiscent of Marr's generalized cones. Different combinations of geons yield a large variety of 3D shapes, including articulated objects. 2) The recognition of a complex object by recognizing subcomponents and then combining these to recognize more complex objects. See also hierarchical matching, shape representation, model based recognition, object recognition.

recognition by parts: See recognition by components.

structural decomposition-recognition by structural decomposition: See recognition by components.

reconstruction: The problem of computing the shape of a 3D object or surface from one or more intensity or range images. Typical techniques include model acquisition and the many shape from X methods reported (see shape from contour and following entries).

reconstruction error: Inaccuracies in a model when compared to reality. These can be caused by inaccurate sensing or compression. (See lossy compression.)

rectification: A technique warping two images into some form of geometric alignment, *e.g.*, so that the vertical pixel coordinates of corresponding points are equal. See also stereo image rectification. This figure shows a stereo pair (top row) and its rectified version (bottom row), highlighting some of the corresponding scanlines, where corresponding image features lie:

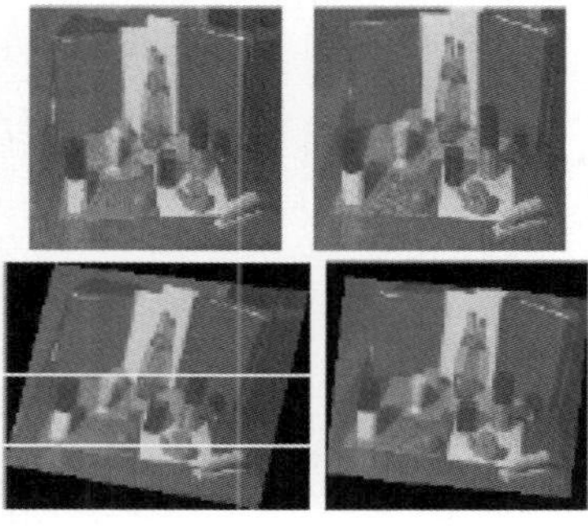

recursive region growing: A class of recursive algorithms for region growing. An initial pixel is chosen. Given an adjacency rule to determine the neighbors of a pixel, (*e.g.*, 8-adjacency), the neighboring pixels are explored. If any meets the criteria for addition to the region, the growing procedure is called recursively on that pixel. The process continues until all connected image pixels have been examined. See also adjacent, image connectedness, neighborhood recursive splitting.

recursive splitting: A class of recursive algorithms for region segmentation, dividing an image into a region set. The region set is initialized to the whole image. A homogeneity criterion is then applied; if not satisfied, the image is split according to a given scheme (*e.g.*, into four sub-images, as in a quadtree), leading to a new region set. The procedure is applied recursively to all regions in the new region set, until all remaining regions are homogeneous. See also region segmentation, region based segmentation, recursive region growing.

reference frame transformation: See coordinate system transformation.

reference image: An image of a known scene or of a scene at a particular time used for comparison with a current image. See, for example, change detection.

reference views: In iconic recognition, the views chosen as most representative for a 3D object. See also eigenspace based recognition, characteristic view.

reference white: A sample image value which corresponds to a known white object. The knowledge of such a value facilitates white balance corrections.

reflectance: The ratio of reflected to incident flux, in other words the ratio of reflected to incident (light) power. See also bidirectional reflectance distribution function.

reflectance estimation: A class of technique for estimating the bidirectional reflectance distribution function (BDRF). Used notably within the techniques for shape from shading and image based rendering, which seeks to render arbitrary images of scenes from video material only. All information about geometry and photometry (*e.g.*, the BDRF) is derived from video. See also physics based vision.

reflectance map: The reflectance map expresses the reflectance of a material in terms of a viewer-centered representation of local surface orientation. The most commonly used is the Lambertian reflectance map, based on Lambert's law. See also shape from shading, photometric stereo.

reflectance ratio: A photometric invariant used for segmentation and recognition. It is based on the observation that the illumination on both sides of a reflectance or color edge is nearly the same. So, although we cannot factor out the reflectance and illumination from only the observed lightness, the ratio of the lightnesses on both sides of the edge equals the ratio of the reflectances, independent of illumination. Thus the ratio is invariant to illumination and local surface geometry for a significant class of reflectance maps. See also invariant, physics based vision.

reflection: 1) A mathematical transformation where the output image is the input image flipped over about a given transformation line in the image plane. See reflection operator. 2) An optics phenomenon whereby all incident light incident on a surface is deflected away, without absorption, diffusion or scattering. An ideal mirror is the perfect reflecting surface. Given a single ray of light incident on a reflecting surface, the angle of incidence equals the angle of reflection, as shown below. See also specular reflection.

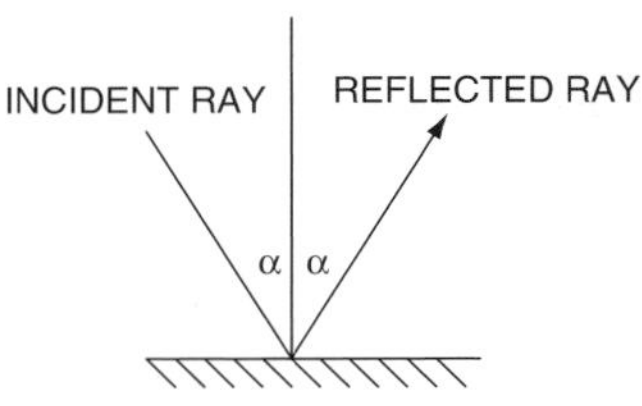

reflection operator: A linear transformation intuitively changing each vector or point of a given space to its mirror image, as shown below. The transformation corresponding matrix, say **H**, has the property **HH** = **I**, *i.e.*, $\mathbf{H}^{-1} = \mathbf{H}$: a reflection matrix is its own inverse. See also rotation.

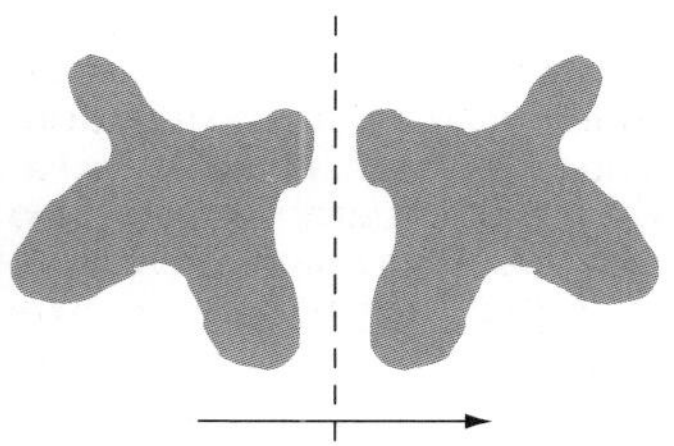

refraction: An optical phenomenon whereby a ray of light is deflected while passing through different optic mediums, *e.g.*, from air to water. The amount of deflection is governed by the difference between the refraction indices of the two mediums, according to *Snell's law*:

$$\frac{n_1}{sin(\alpha_1)} = \frac{n_2}{sin(\alpha_2)}$$

where n_1 and n_2 are the refraction indices of the two media, and α_1, α_2 the respective refraction angles:

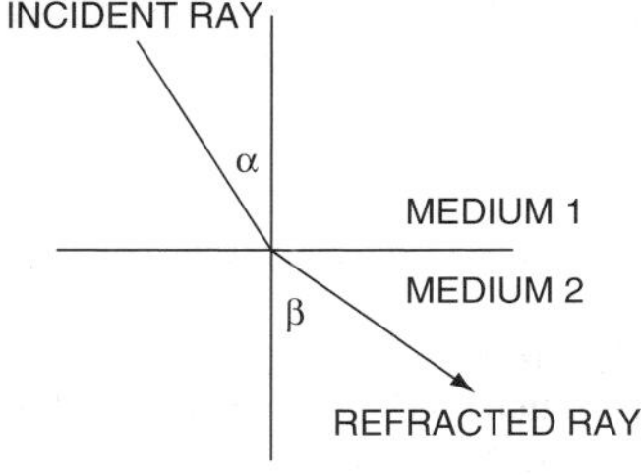

region: A connected part of an image, usually homogeneous with respect to a given criterion.

region adjacency graph (RAG): A graph expressing the adjacency relations among image regions, for instance generated by a segmentation algorithm. See also region segmentation and region based segmentation. The adjacency relations of the regions in the left figure are encoded in the RAG at the right:

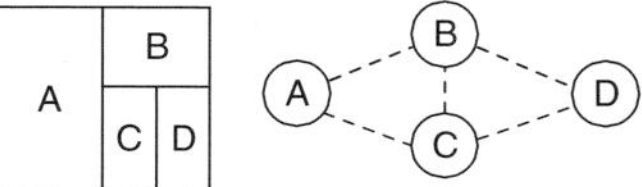

region based segmentation: A class of segmentation techniques producing a number of image regions, typically on the basis of a given homogeneity criterion. For instance, intensity image regions can be homogeneous by color (see color image segmentation) or texture properties (see texture field segmentation); range image regions can be homogeneous by shape or curvature properties (see HK segmentation).

region boundary extraction: The problem of computing the boundary of a region, for example, the contour of a region in an intensity image after color based segmentation.

region decomposition: A class of algorithms aiming to partition an image or region thereof into regions. See also region based segmentation.

region descriptor: 1) One or more properties of a region, such as compactness or moments. 2) The data structure containing all data pertaining to a region. For instance, for image regions this could include the region's position in the image (*e.g.*, the coordinates of the center of mass), the region's contour (*e.g.*, a list of 2D coordinates), some indicator of the region shape (*e.g.*, compactness or perimeter squared over area), and the value of the region's homogeneity index.

region detection: A vast class of algorithms seeking to partition an image into regions with particular properties. See for details region identification, region labeling, region matching, region based segmentation.

region filling: A class of algorithms assigning a given value to all the pixels in the interior of a closed contour identifying a region. For instance, one may want to fill the interior of a closed contour in a binary image with zeros or ones. See also morphology, mathematical morphology, binary mathematical morphology.

region growing: A class of algorithms that construct a connected region by incrementally expanding the region, usually at the boundary. New data are merged into the region when the data are consistent with the previous region. The region is often redescribed after each new set of data is added to it. Many region growing algorithms have the form: 1) Describe the region based on the current pixels that belong to the region (*e.g.*, fit a linear model to the intensity distribution). 2) Find all pixels adjacent to the current region. 3) Add an adjacent pixel to the region if the region description also describes this pixel (*e.g.*, it has a similar intensity). 4) Return to step 1 as long as new pixels continue to be added. A similar algorithm exists for region growing with 3D points, giving a surface fitting. The data points could come from a regular grid (pixel or voxel) or from an unstructured list. In the latter case, it is harder to determine adjacency.

region identification: A class of algorithms seeking to identify regions with special properties, for instance, a human figure in a surveillance video, or road vehicles in an aerial sequence. Region identification covers a very wide area of techniques spanning many applications, including remote sensing, visual surveillance, surveillance, and agricultural and forestry surveying. See also target recognition, automatic target recognition (ATR), binary object recognition, object recognition, pattern recognition.

region invariant: 1) A property of a region that is invariant (does not change) after some transformation is applied to the region, such as translation, rotation or perspective projection. 2) A property or function which is invariant over a region.

region labeling: A class of algorithms which are used to assign a label or meaning to each image region in a given image segmentation to achieve an appropriate image interpretation. Representative techniques are relaxation labeling, probabilistic relaxation labeling, and interpretation trees (see interpretation tree search). See also labeling problem.

region matching: 1) Establishing the correspondences between matching members of two sets of regions. 2) The degree of similarity between two regions, *i.e.*, solving the matching problem for regions. See, for instance, template matching, color matching, color histogram matching.

region merging: A class of algorithms fusing two image regions into one if a given homogeneity criterion is satisfied. See also region, region based segmentation, region splitting.

region of interest: A subregion of an image where processing is to occur. Regions of interest may be used to: 1) reduce the amount of computation that is required or 2) to focus processing so that image data outside the region do not distract from or distort results. As an example, when tracking a target through an image sequence, most algorithms for locating the target in the next video frame only consider image data from a region of interest surrounding the predicted target position. The figure shows a boxed region of interest:

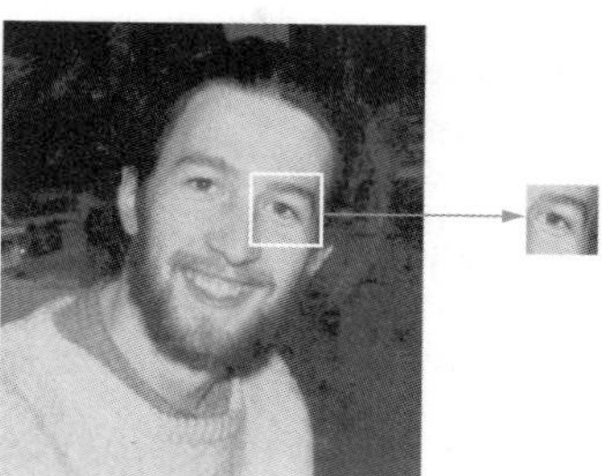

region of support: The subregion of an image that is used in a particular computation. For example, an edge detector usually only uses a subregion of pixels neighboring the pixel currently being considered for being an edge.

region neighborhood graph: See region adjacency graph.

region propagation: The problem of tracking moving image regions.

region representation: A class of methods to represent the defining characteristics of an image region. For encoding the shapes, see axial representation, convex hull, graph model, quadtree, run-length coding, skeletonization. For encoding a region by its properties, see moments, curvature scale space, Fourier shape descriptor, wavelet descriptor, shape representation.

region segmentation: See region based segmentation.

region snake: A snake representing the boundary of some region. The operation of computing of the snake may be used as a region segmentation technique.

region splitting: A class of algorithms dividing an image, or a region thereof, into parts (sub regions) if a given homogeneity criterion is not satisfied over the region. See also region, region based segmentation, region merging.

registration: A class of techniques aiming to align, superimpose, or match two objects of the same kind (*e.g.*, images, curves, models); more specifically, to compute the geometric transformation superimposing one to the other. For instance, image registration determines the region common to two images, thereby finding the planar transformation (rotation and translation) aligning them; similarly, curve registration determines the transformation aligning the similar (or same) part of two curves. This figure shows the registration (right) of the solid (left) and dashed (middle) curves. The transformation needs not be rigid; non-rigid registration is common in medical imaging, for instance in digital subtraction angiography. Notice also that most often there is no exact solution, as the two objects are not exactly the same, and the best approximate solution must be found by least squares or more complex methods. See also Euclidean transformation, medical image registration, model registration, multi-image registration.

regression: 1) In statistics, the relationship between one variable and another, as in linear regression. A particular case of curve and surface fitting. 2) Regression testing verifies that changes to the implementation of a system have not caused a loss of functionality, or *regression* to the state where that functionality did not exist.

regularization: A class of mathematical techniques to solve an ill-posed problem. In essence, to determine a single solution, one introduces the constraint that the solution must be smooth, in the intuitive sense that similar inputs must correspond to similar outputs. The problem is then cast as a variational problem, in which the variational integral depends both on the data and on the smoothness constraint. For instance, a regularization approach to the problem of estimating a function f from a set of values $y_1, y_2, \ldots, y_n$ at the data point $\vec{x}_1, \ldots, \vec{x}_n$, leads to the minimization of the functional

$$H(f) = \sum_{i=1}^{N} (f(\vec{x}_i) - y_i)^2 + \lambda\Phi(f)$$

where $\Phi(f)$ is the *smoothness functional*, and λ a positive parameter called the *regularization number*.

relational graph: A graph in which the arcs express relations between the properties of image entities (*e.g.*, regions or other features) which are the nodes in the graph. For regions, for instance, commonly used properties are adjacency, inclusion, connectedness, and relative area size. See also region adjacency graph (RAG), shape representation. The adjacency relations of the regions in the left figure are encoded in the RAG at the right:

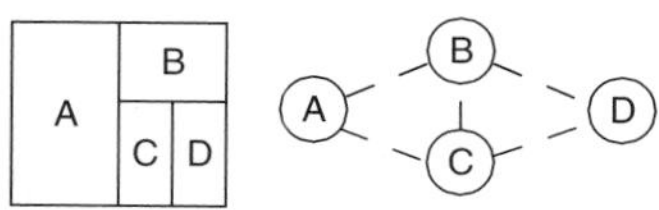

relational matching: A class of matching algorithms based on relational descriptors. See also relational graph.

relational model: See relational graph.

relational shape description: A class of shape representation techniques based on relations between the properties of image entities (*e.g.*, regions or other features). For regions, for instance, commonly used properties are adjacency, inclusion, connectedness, and relative area size. See also relational graph, region adjacency graph.

relative depth: The difference in depth values (distance from some observer) for two points. In certain situations while it may not be possible to compute actual or absolute depth, it may be possible to compute relative depth.

relative motion: The motion of an object with respect to some other, possibly also moving, frame of reference (typically the observer's).

relative orientation: The problem of computing the orientation of an object with respect to another coordinate system, such as that of the sensor. More specifically, the rotation matrix aligning the reference frames attached to the object and second object. See also pose and pose estimation.

relaxation: A technique for assigning values from a continuous or discrete set to the node of a network or graph by propagating the effects of local constraints. The network can be an image grid, in which case the pixels are nodes, or features, for instance edges or regions. At each iteration, each node interacts with its neighbors, altering its value according to the local constraints. As the number of iterations increases, the effect of local constraints are propagated to farther and farther parts of the network. Convergence is achieved when no more changes occur, or changes become insignificant. See also discrete relaxation, relaxation labeling, probabilistic relaxation labeling.

relaxation labeling: A relaxation technique for assigning a label from a discrete set to each node of a network or graph. A well-known example, a classic in artificial intelligence, is Waltz's line labeling algorithm (see also line drawing analysis).

relaxation matching: A relaxation labeling technique for model matching, the purpose of which is to label (match) each model primitive with a scene primitive. Starting from an initial labeling, the algorithm harmonizes iteratively neighboring labels using a coherence measure for the set of matches. See also discrete relaxation, relaxation labeling, probabilistic relaxation labeling.

relaxation segmentation: A class of segmentation techniques based on relaxation. See also image segmentation.

remote sensing: The acquisition, analysis and understanding of imagery, mainly of the Earth's surface, acquired by airplanes or satellites. Used frequently in agriculture, forestry, meteorological and military applications. See also multi-spectral analysis, multi-spectral image, geographic information system (GIS).

representation: A description or model specifying the properties defining an object or class of objects. A classic example is shape representation, a group of techniques for describing the geometric shape of 2D and 3D objects. See also Koenderink's surface shape classification. Representations can be symbolic or non-symbolic (see symbolic object representation and non-symbolic representation), a distinction inherited from artificial intelligence.

resection: The computation of the position of a camera given the images of some known 3D points. Also known as camera calibration, or pose estimation.

resolution: The number of pixels per unit area, length, visual angle, etc.

restoration: Given a noisy sample of some true data, the goal of restoration is to recover the best possible estimate of the original true data, using only the noisy sample.

reticle: The network of fine wires or receptors placed in the focal plane of an optical instrument for measuring the size or position of the objects under observation.

retinal image: The image which is formed on the retina of the human eye.

retinex: An image enhancement algorithm based on *retinex theory*, aimed to compute an illuminant- independent quantity called lightness at each image pixel. The key observation is that normal illumination on a surface changes slowly, leading to slow changes in the observed brightness of a surface. This contrasts with strong changes in brightness at reflectance and fold edges. The retinex algorithm removes the slowly varying components by exploiting the fact that the observed brightness $B = L \times I$ is product of the surface lightness (or reflectance) L and the illumination I. By taking the logarithm of B at each pixel, the product of L and I become a sum of logarithms. Slow changes can be detected by differentiation and then removed by thresholding. Re-integration of the result produces the lightness image (up to an arbitrary scale factor).

reverse engineering: The problem of generating a model of a 3D object from a set of views, for instance a VRML or a triangulated model. The model can be purely geometric, that is, describing just the object's shape, or combine shape and textural properties. Techniques exists for reverse engineering from both range images and intensity images. See also geometric model, model acquisition.

RGB: A format for color images, encoding the Red, Green, and Blue component of each pixel in separate channels. See also YUV, color image.

ribbon: A shape representation for pipe-like planar objects whose contours are approximately parallel, *e.g.*, roads in aerial imagery. See also generalized cones, shape representation.

ridge: A particular type of discontinuity of the intensity function, giving rise to thick edges and lines. This figure shows a characteristic dark-to-light-to-dark intensity ridge profile

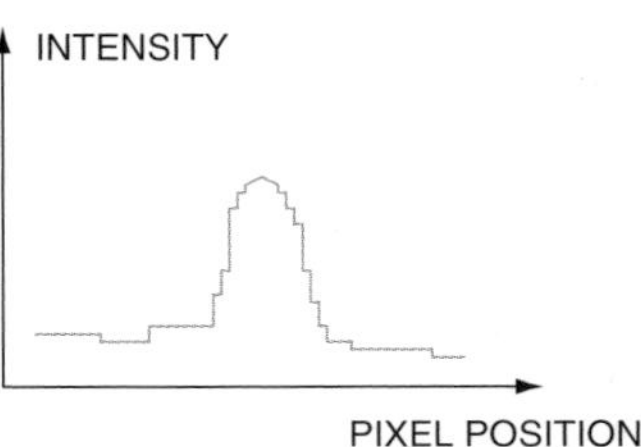

along a scanline. See also step edge, roof edge, edge detection.

ridge detection: A class of algorithms, especially edge and line detectors, for detecting ridges in images.

right-handed coordinate system: A 3D coordinate system with the XYZ axes arranged as follows. The alternative is a left-handed coordinate system.

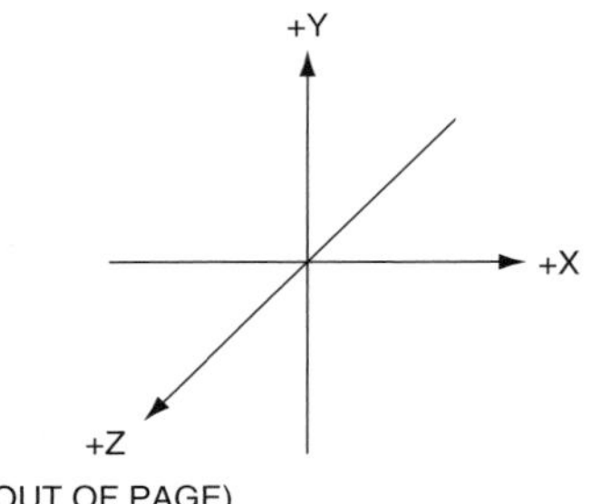

rigid body segmentation: The problem of partitioning automatically the image of an articulated or deformable body into a number of rigid subcomponents. See also part segmentation, recognition by components (RBC).

rigid motion estimation: A class of techniques aiming to estimate the 3D motion of a rigid body or scene in space from a sequence of images by assuming that there are no changes in shape. Rigidity simplifies the problem significantly so that changes in appearance arise solely from changes in relative position and projection. Techniques exist for using known 3D models, or estimating the motion of a general cloud of 3D points, or from image feature points or estimating motion from optical flow. See also motion estimation, egomotion.

rigid registration: Registration where neither the model nor data is allowed to deform. This reduces registration to estimating the Euclidean transformation that aligns the model with the data. See also non-rigid registration.

rigidity constraint: The assumption that a scene or object under analysis is rigid, implying that all 3D points remain in the same relative position in space. This constraint can simplify significantly many algorithms, for instance shape reconstruction (see shape and following "shape from" entries) and motion estimation.

road structure analysis: A class of techniques which are used to derive information about roads from images. These can be close-up images (*e.g.*, images of the tarmac as acquired from a moving vehicles, to map defects automatically over extended distances) or remotely sensed images (*e.g.*, to analyze the geographical structure of road networks).

Roberts cross gradient operator: An operator used for edge detection, computing an estimate of perpendicular components of the image gradient at each pixel. The image is convolved with the two Roberts kernels, yielding two components, G_x and G_y, for each pixel. The gradient magnitude $\sqrt{G_x^2 + G_y^2}$ and orientation $\arctan \frac{G_y}{G_x}$ can then be estimated as for any 2D vector. See also edge detection, Canny edge detector, Sobel gradient operator, Sobel kernel, Deriche edge detector, Hueckel edge detector, Kirsch edge detector, Marr–Hildreth edge detector, O'Gorman edge detector, Robinson edge detector.

Roberts kernel: A pair of kernels, or masks, used to estimate perpendicular components of the image gradient within the Roberts cross gradient operator:

0	1
−1	0

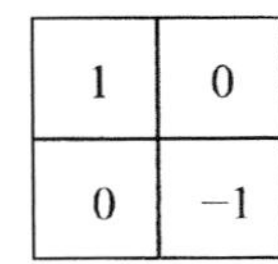

1	0
0	−1

The masks respond maximally to edge oriented to plus or minus 45° from the vertical axis of the image.

Robinson edge detector: An operator for edge detection, computing an estimate of the directional first derivatives of the image in eight directions. The image is convolved with the eight kernels, three of which as shown here

1	1	1
1	−2	1
−1	−1	−1

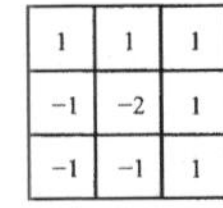

1	1	1
−1	−2	1
−1	−1	1

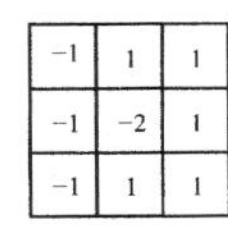

−1	1	1
−1	−2	1
−1	1	1

Two of these, typically those responding maximally to differences along the coordinate axes, can be taken as estimates of the two components of the gradient, G_x and G_y. The gradient magnitude $\sqrt{G_x^2 + G_y^2}$ and orientation $\arctan \frac{G_y}{G_x}$ can then be estimated as for any 2D vector. See also edge detection, Roberts cross gradient operator, Sobel gradient operator, Sobel kernel, Canny edge detector, Deriche edge detector, Hueckel edge detector, Kirsch edge detector, Marr–Hildreth edge detector, O'Gorman edge detector.

robust: A general term referring to a technique which is insensitive to noise or other perturbations.

robust estimator: A statistical estimator which, unlike normal least squares estimators, is not distracted by even significant percentages of outliers in the data. Popular robust estimators in computer vision include RANSAC, least median of squares, and M-estimators. See also outlier rejection.

robust regression: A form of regression that does not use outlier values in computing the fitting parameters. For example, if doing a least square straight line fit to a set of data, normal regression methods use all data points, which can give distorted results if even one point is very far away from the "true" line. Robust processes either eliminate these outlying points or reduce their contribution to the results. The figure below shows a rejected outlying point:

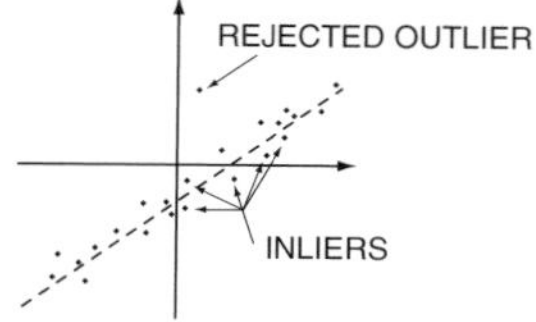

robust statistics: A general term describing statistical methods which are not significantly influenced by outliers.

robust technique: See robust estimator.

ROC: See receiver operating characteristic.

roll: A 3D rotation representation component (along with pitch and yaw) often used for cameras or moving observers. The roll component specifies a rotation about the optical axis or line of sight. This figure shows the roll rotation direction:

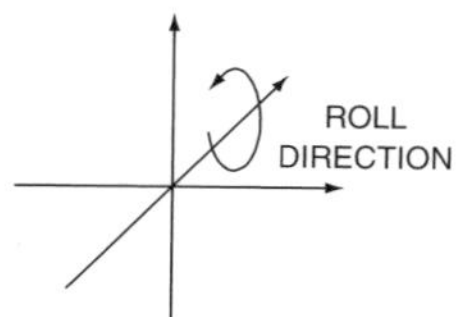

roof edge: 1) An image edge where the values increase continuously to a maximum and then decrease continuously, such as the brightness values on a Lambertian cylinder when lit by a point light source, or an orientation discontinuity (or fold edge) in a range image. 2) A scene edge where an orientation discontinuity occurs. The figure shows a horizontal roof edge in a range image:

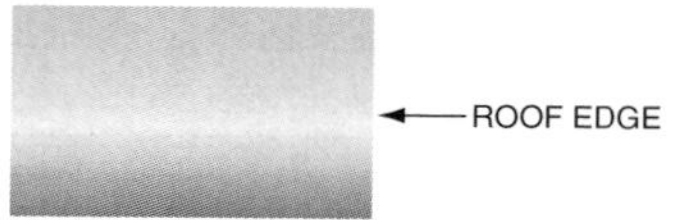

rotating mask: A mask which is considered in a number of orientations relative to some pixel. See, for example, the masks used in the Robinson edge detector. Most commonly used as a type of average smoothing in which the most homogeneous mask is used to compute the smoothed value for every pixel. In the example, notice how although image detail has been reduced the major boundaries have not been smoothed.

rotation: A circular motion of a set of points or object around a given point (2D) or line (3D, called the axis of rotation).

rotation estimation: The problem of estimating rotation from raw or processed image, video or range data, typically from two sets of corresponding points (or lines, planes, etc.) taken from rotated versions of a pattern. The problem usually appears in one of three forms: 1) estimating the 3D rotation from 3D data (three points are needed), 2) estimating the 3D rotation from 2D data (three points are needed but lead to multiple solutions), or 3) estimating the 2D rotation from 2D data (two points are needed). A second issue to consider is the effect of noise: typically more than the minimum number of points are needed to counteract the effects of noise, which leads to least square algorithms.

rotation invariant: A property that keeps the same value even if the data values, the camera, the image or the scene from which the data comes is rotated. One needs to distinguish between 2D (*i.e.*, in the image) and 3D (*i.e.*, in the scene) rotation invariance. For example, the angle between two image lines is invariant to image rotation, but not to rotation of the lines in the scene.

rotation matrix: A linear operator rotating a vector in a given space. The inverse of a rotation matrix equals its transpose. A rotation matrix has only three degrees of freedom in 3D and one in 2D. In 3D space, there are three eigenvalues, namely -1, $\cos\theta + i\sin\theta$, $\cos\theta - i\sin\theta$, where i is the imaginary unit. A rotation matrix in 3D has nine entries but only three degrees of freedom, as it must satisfy six orthogonality constraints. It can be parameterized in various ways, usually through Euler angles, yaw, pitch, roll, rotation angles around the coordinate axes, and axis-angle, etc. See also orientation estimation, rotation representation, quaternions.

rotation operator: A linear operator expressed by a rotation matrix.

rotation representation: A formalism describing rotations and their algebra. The most frequent is definitely the rotation matrix, but quaternions, Euler angles, yaw, pitch, roll, rotation angles around the coordinate axes, and axis-angle, etc. have also been used.

rotational symmetry: The property of a set of point or object to remain unchanged after a given rotation. For instance, a cube has several rotational symmetries, with respect to any 90° rotation around any axis passing through the centers of opposite faces. See also rotation, rotation matrix.

RS-170: The standard black-and-white video format in the

United States. The EIA (Electronic Industry Association) is the standards body that originally defined the 525-line, 30 frame per second TV standard for North America, Japan, and a few other parts of the world. The EIA standard, also defined under US standard RS-170A, defines only the monochrome picture component but is mainly used with the NTSC color encoding standard. A version exists for PAL cameras.

rubber sheet model: See membrane model.

rule-based classification: A method of object recognition drawn from artificial intelligence in which logical rules are used to infer object type.

run code: See run length coding.

run length coding: A lossless compression technique used to reduce the size of a repeating string of characters, called a "run", also applicable to images. The algorithm encodes a run of symbols into two bytes, a count and a symbol. For instance, the 6-byte string "xxxxxx" would become "6x" occupying 2 bytes only. It can compress any type of information content, but the content itself affects, obviously, the compression ratio. Compression ratios are not high compared to other methods, but the algorithm is easy to implement and quick to execute. Run-length coding is supported by bitmap file formats such as TIFF, BMP and PCX. See also image compression, video compression, JPEG.

run length compression: See run length coding.

S

saccade: A movement of the eye or camera, changing the direction of fixation sharply.

saliency map: A representation encoding the saliency of given image elements, typically features or groups thereof. See also salient feature, Gestalt, perceptual grouping, perceptual organization.

salient feature: A feature associated with a high value of a saliency measure, quantifying feature suggestiveness for perception (from the Latin *salire*, to leap). For instance, inflection points have been indicated as salient features for representing contours. Saliency is a concept originated from Gestalt psychology. See also perceptual grouping, perceptual organization.

salt-and-pepper noise: A type of impulsive noise. Let $x, y \in [0, 1]$ be two uniform random variables, I the true image value at a given pixel, and I_n the corrupted (noisy) version of I. We can define the effect of salt-and-pepper noise as $I_n = i_{min} + y(i_{max} - i_{min})$ iff $x \geq l$, where l is a parameter controlling how much of the image is corrupted, and i_{min}, i_{max} the range of the noise. See also image noise, Gaussian noise. This image was corrupted with 1% noise:

sampling: The transformation of a continuous signal into a discrete one by recording its values at discrete instants or locations. Most digital images are sampled in space, time and intensity, as intensity values are defined only on a regular

Dictionary of Computer Vision and Image Processing R.B. Fisher, K. Dawson-Howe, A. Fitzgibbon, C. Robertson and E. Trucco

spatial grid, and can only take integer values. This shows an example of a continuous signal and its samples:

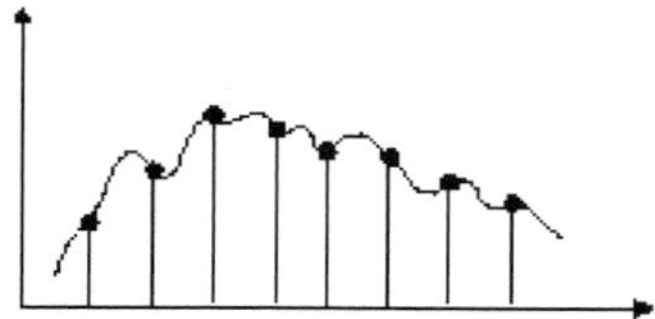

sampling density: The density of a sampling grid, that is, the number of samples collected per unit interval. See also sampling.

sampling theorem: If an image is sampled at a rate higher than its Nyquist frequency then an analog image could be reconstructed from the sampled image whose mean square error with the original image converges to zero as the number of samples goes to infinity.

Sampson approximation: An approximation to the geometric distance in the fitting of implicit curves or surfaces that are defined by a parameterized function of the form $f(\vec{a};\vec{x}) = 0$ for $\vec{x}$ on the surface $S(\vec{a})$ defined by parameter vector $\vec{a}$. Fitting the surface to the set of points $\{\vec{x}_1, \ldots, \vec{x}_n\}$ consists in minimizing a function of the form $e(\vec{a}) = \sum_{i=1}^{n} d(\vec{x}_i, S(\vec{a}))$. Simple solutions are often available if the distance function $d(\vec{x}, S(\vec{a}))$ is the algebraic distance $d(\vec{x}, S(\vec{a})) = f(\vec{a};\vec{x})^2$, but under certain common assumptions, the optimal solution arises when d is the more complicated geometric distance $d(\vec{x}, S(\vec{a})) = \min_{\vec{y}\in S} \|\vec{x} - \vec{y}\|^2$. The Sampson approximation defines

$$d(\vec{x}, S(\vec{a})) = \frac{f(\vec{a};\vec{x})^2}{\|\nabla f(\vec{a};\vec{x})\|^2}$$

which is a first-order approximation to the geometric distance. If an efficient algorithm for minimizing weighted algebraic distance is available, then the *Sampson iterations* are a further approximation, where the kth iterate $\vec{a}_k$ is the solution to

$$\vec{a}_k = \operatorname*{argmin}_{\vec{a}} \sum_{i=1}^{n} w_i\, f(\vec{a};\vec{x}_i)^2$$

with weights computed using the previous estimate so $w_i = 1/\|\nabla f(\vec{a}_{k-1};\vec{x}_i)\|^2$.

SAR: see synthetic aperture radar.

SAT: See symmetric axis transform.

satellite image: An image of a section of the Earth acquired using a camera mounted on an orbiting satellite.

saturation: Reaching the upper limit of a dynamic range. For instance, intensity saturation occurs for a 8-bit monochromatic image when intensities greater than 255 are recorded: any such value is encoded as 255, the largest possible value in the range.

Savitzky–Golay filtering: A class of filters achieving least squares fitting of a polynomial to a moving window of a signal. Used for fitting and data smoothing. See also linear filter, curve fitting.

scalar: A one dimensional entity; a real number.

scale: 1) The ratio between the size of an object, image, or feature and that of a reference or model. 2) The property that some image features are apparent only when viewed at a given size, such as a line being enlarged so much that it appears as a pair of parallel edge features. 3) A measure of the degree to which fine features have been removed or reduced in an image. One can analyze images at multiple spatial scales, whereby only features in certain size ranges appear at each scale (see scale space and pyramid).

scale invariant: A property that keeps the same value even if the data, the image or the scene from which the data comes is shrunk or enlarged. The ratio $\frac{perimeter^2}{area}$ is invariant to image scaling.

scale operator: An operator suppressing details (high-frequency contents) in an image, *e.g.*, Gaussian smoothing. Details at small scales are discarded. The resulting content can be represented in a smaller-size image. See also scale space, image pyramid, Gaussian pyramid, Laplacian pyramid, pyramid transform.

scale reduction: The result of the application of a scale operator.

scale space: A theory for early vision developed to account properly for the multi-scale nature of images. The rationale is that, in the absence of *a priori* information on the optimal spatial scale at which a specific problem should be treated (*e.g.*, edge detection), images should be analyzed at all possible scales, the coarser ones representing simplifications of the finer ones. The finest scale is the input image itself. See scale space representation for details.

scale space filtering: The filtering operation that transforms one resolution level into another in a scale space, for instance Gaussian filtering.

scale space matching: A class of matching techniques that compare shape at various scales. See also scale space and image matching.

scale space representation: A representation of an image, and more generally of a signal, making explicit the information contained at multiple spatial scales, and establishing a causal relationship between adjacent scale levels. The scale level is identified by a scalar parameter, called *scale parameter*. A crucial requirement is that coarser levels, obtained by successive applications of a scale operator, should constitute simplifications of previous (finer) levels, *i.e.*, introduce no spurious details. A popular scale space representation is the Gaussian scale space, in which the next coarser image is obtained by convolving the current image with a Gaussian kernel. The variance of this kernel is the scale parameter. See also scale space, image pyramid, Gaussian smoothing.

scaling: 1) The process of zooming or shrinking an image. 2) Enlarging or shrinking a model to fit a set of data. 3) The process of transforming a set of values so that they lie inside a standard range (*e.g.*, $[-1, 1]$), often to improve numerical stability.

scanline: A single (horizontal) line of an image. Originally this term was used for cameras in which the image is acquired line by line by a sensing element that generally scans each pixel on a line and then moves onto the next line.

scanline slice: The cross section of a structure along an image scanline. For instance, the scanline slice of a convex polygon in a binary image is:

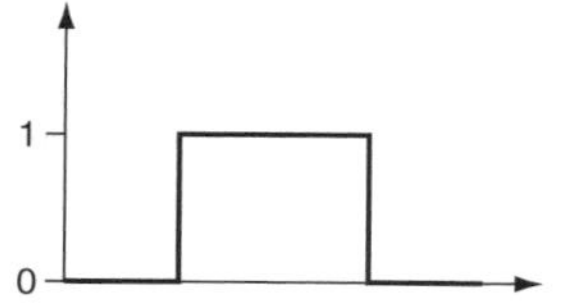

scanline stereo matching: The stereo matching problem with rectified images, whereby corresponding points lie on scanlines with the same index. See also rectification, stereo correspondence.

scanning electron microscope (SEM): A scientific microscope introduced in 1942. It uses a beam of highly energetic electrons to examine objects on a very fine scale. The imaging process is essentially the same as for a light microscope apart from the type of radiation used. Magnification is much higher than what can be achieved with light. The images are rendered in gray shades. This technique is particularly useful for investigating microscopic details of surfaces.

scatter matrix: For a set of d dimensional points represented as column vectors $\{\vec{x}_1, \ldots, \vec{x}_n\}$, with mean $\vec{\mu} = \frac{1}{n}\sum_{i=1}^{n} \vec{x}_i$, the scatter matrix is the $d \times d$ matrix

$$\mathbf{S} = \sum_{i=1} (\vec{x}_i - \vec{\mu})(\vec{x}_i - \vec{\mu})^\top$$

It is $(n-1)$ times the sample covariance matrix.

scattergram: See scatterplot.

scatterplot: A data display technique in which each data item is plotted as a single point in an appropriate coordinate system, that might help a person to better understand the data. For example, if a set of estimated surface normals is plotted in a 3D scatterplot, then planar surfaces should produce tight clusters of points. The figure shows a set of data points plotted according to their values of features 1 and 2:

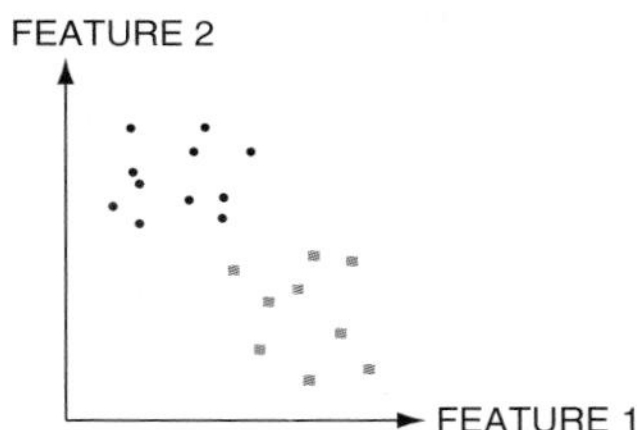

scene: The part of 3D space captured by an imaging sensor, and every visible object therein.

scene analysis: The process of examining an image or video, for the purpose of inferring information about the scene in view, such as the shape of the visible surfaces, the identity of the objects in the scene, and their spatial or dynamic relationships. See also shape from contour and the following "shape from" entries, object recognition, and symbolic object representation.

scene constraint: Any constraint imposed on the image data by the nature of the scene, for instance, rigid motion, or the orthogonality of walls and floors, etc.

scene coordinates: A 3D coordinate system that describes the position of scene objects relative to a given coordinate system origin. Alternative coordinate systems are camera coordinates, viewer centered coordinates or object centered coordinates.

scene labeling: The problem of identifying scene elements from image data, associating them to labels representing their nature and roles. See also labeling problem, region labeling, relaxation labeling, image interpretation, scene understanding.

scene reconstruction: The problem of estimating the 3D geometry of a scene, for example the shape of visible surfaces or contours, from image data. See also reconstruction, shape from contour and the following "shape from" entries or architectural model, volumetric, surface and slice based reconstruction.

scene understanding: The problem of constructing a semantic interpretation of a scene from image data, that is, describing the scene in terms of object identities and relationships among objects. See also image interpretation, object recognition, symbolic object representation, semantic net, graph model, relational graph.

SCERPO: Spatial Correspondence, Evidential Reasoning and Perceptual Organization. A well known vision system developed by David Lowe that demonstrated recognition of complex polyhedral objects (*e.g.*, razors) in a complex scene.

screw motion: A 3D transformation comprising a rotation about an axis $\vec{a}$ and translation along $\vec{a}$. The general Euclidean transformation $\vec{x} \mapsto \mathbf{R}\vec{x} + \vec{t}$ is a screw transformation if $\mathbf{R}\vec{t} = \vec{t}$.

search tree: A data structure that records the choices that could be made in a problem-solving activity, while searching through a space of alternative choices for the next action or decision. The tree could be explicitly created or be implicit in the sequence of actions. For example, a tree that records alternative model-to-data feature matching is a specialized search tree called an interpretation tree. If each non-leaf node has two children, we have a binary search tree. See also decision tree, tree classifier.

SECAM: SECAM (*Sequential Couleur avec Mémoire*) is the television broadcast standard in France, the Middle East, and most of Eastern Europe. SECAM broadcasts 819 lines per second. It is one of three main television standards throughout the world, the other two being PAL (see PAL camera) and NTSC.

second derivative operator: A linear filter estimating the second derivative from an image at a given point and in a given direction. Numerically, a simple approximation of the second derivative of a 1D function f is the central (finite) difference, derived from the Taylor approximation of f:

$$f_i'' = \frac{f_{i+1} - 2f_i + f_{i-1}}{h^2} + O(h)$$

where h is the sampling step (assumed constant), and $O(h)$ indicates that the truncation error vanishes as h. A similar but more complicated approximation exists for estimating the second derivative in a given direction in an image. See also first derivative filter.

second fundamental form: See surface curvature.

seed region: The initial region used in a region growing process such as surface fitting in range data or intensity region finding in an intensity image. The patch on the surface here is a potential seed region for region growing the full cylindrical patch:

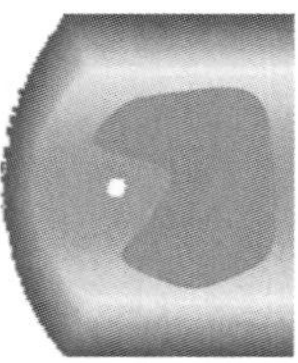

segmentation: The problem of dividing a data set into parts according to a given set of rules. The assumption is that the different segments correspond to different structures in the original input domain observed in the image. See for instance image segmentation, color image segmentation, curve segmentation, motion segmentation, part segmentation, range data segmentation, texture segmentation.

self-calibration: The problem of estimating the calibration parameters using only information extracted from a sequence or set of images (typically feature point correspondences in subsequent frames of a sequence or in several simultaneous views), as opposed to traditional calibration in photogrammetry, that adopt specially built calibration objects. Self calibration is intimately related with basic concepts of multi-view geometry. See also camera calibration, autocalibration, stratification, projective geometry.

self-localization: The problem of estimating the sensor's position within an environment from image or video data. The problem can be cast as geometric model matching if models of sufficiently complex objects are available, *i.e.*, containing enough points to allow a full solution of the pose estimation problem. In some situations it is possible to identify a sufficient number of landmark points (see landmark detection). If no information at all is available about the scene, one can still apply tracking or optical flow techniques to get corresponding points over time, or stereo correspondences in multiple simultaneous frames. See also motion estimation, egomotion.

self-occlusion: Occlusion in which part of an object is occluded by another part of the same object. In the following example the left leg of the person is occluding their right leg.

SEM: See scanning electron microscope.

semantic net: A graph representation in which nodes represent the objects of a given domain, and arcs properties and relations between objects. See also symbolic object representation, graph model, relational graph. A simple example: an arch and its semantic net representation:

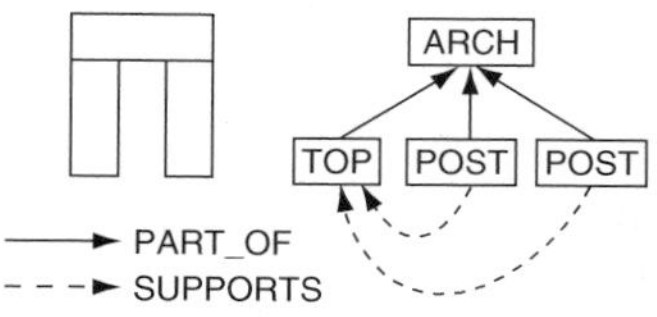

semantic region growing: A region merging scheme incorporating *a priori* knowledge about adjacent regions; for instance, in aerial imagery of countryside areas, the fact that roads are usually surrounded by fields. Constraint propagation can then be applied to achieve a globally optimal region segmentation. See also constraint satisfaction, relaxation labeling, region segmentation, region based segmentation, recursive region growing.

sensor: A general word for a mechanism that records information from the "outside world", generally for processing by a computer. The sensor might obtain raw measurements, *e.g.*, a video camera, or partially processed information, *e.g.*, depth from a stereo triangulation process.

sensor fusion: A vast class of techniques aiming to combine the different information contained in data from different sensors, in order to achieve a richer or more accurate description of a scene or action. Among the many paradigms for fusing sensory information are the Kalman filter, Bayesian models, fuzzy logic, Dempster–Shafer evidential reasoning, production systems and neural networks.

sensor motion compensation: A class of techniques aiming to suppress the motion of a sensor (or its effects) in a video sequence, or in data extracted from the sequence. A typical example is image sequence stabilization, in which a target moving across the image in the original sequence appears stationary in the output sequence. Another example is keeping a robot stationary in front of a target using only visual data (*station keeping*). Suppression of jitter in handheld video recorders is now commercially available. Basic ingredients are tracking and motion estimation. See also egomotion.

sensor motion estimation: See egomotion.

sensor path planning: See sensor planning.

sensor placement determination: See camera calibration and sensor planning.

sensor planning: A class of techniques aimed to determine optimal sensing strategies for a reconfigurable sensor system, normally given a task and a geometric model of the target object (that may be partially acquired in previous views). For example, given a geometric feature on an object for which a CAD-like model is known, and the task to verify the feature's size, a sensor planning system would determine the best position and orientation of, say, a single camera and associated illumination for estimating the size of each feature. The two basic approaches have been *generate-and-test*, in which sensor configurations are generated and then evaluated with respect to the task constraints, and *synthetic methods*, in which task constraints are characterized analytically and the resulting equations solved to yield the optimal sensor configuration. See also active vision, purposive vision.

sensor position estimation: See pose estimation.

sensor response: The output of a sensor, or a characterization of some key output quantities, given a set of inputs. Typically expressed in the frequency domain, as a function linking the magnitude and phase of the Fourier transform of the output signal with the known frequency of the input. See also phase spectrum, power spectrum, spectral response.

sensor sensitivity: In general, the weakest input signal that a sensor can detect. It can be inferred from the sensor response curve. For the common CCD sensor of video cameras, sensitivity depends on various parameters, mainly the *fill factor* (the percentage of the sensor's area actually sensitive to light) and *well capacity* (the amount of charge that a photosensitive element can hold). The larger the values of the above parameters, the more sensitive the camera. See also sensor spectral sensitivity.

sensor spectral sensitivity: A characterization of a sensor's response in frequency. For example,

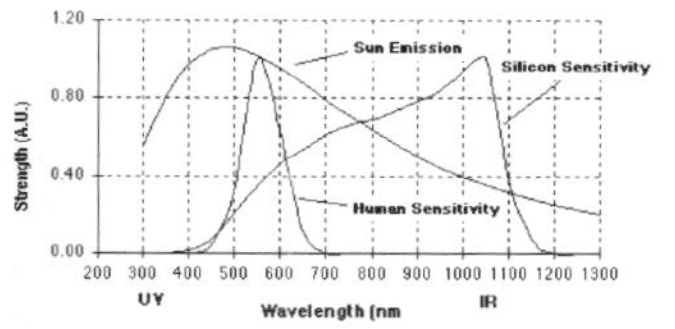

shows the spectral sensitivity of a typical CCD sensor (actually its spectral response, from which the spectral sensitivity can be inferred). Notice that the high sensitivity of silicon in the infrared means that IR blocking filters should be considered for fine measurements depending on camera intensities. We also notice that a CCD camera makes a very good sensor for the near-infrared range (750–1000 nm).

separability: A term used in classification problems referring to whether the data is capable of being split into distinct subclasses by some automatic decision process. If property values of two classes overlap, then the classes are not separable. The circle class is *linearly* separable in the figure below, but the × and box classes are not:

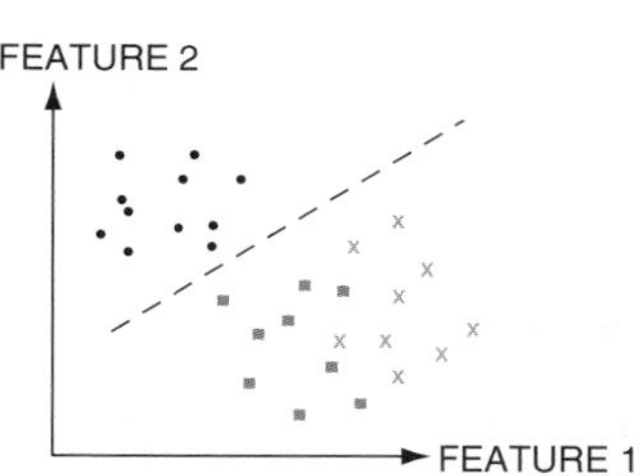

separable filter: A 2D (in image processing) filter that can be expressed as the product of two filters, each of which acts independently on rows and columns. The classic example is the linear Gaussian filter (see Gaussian convolution).

Separability implies a significant reduction in computational complexity, typically reducing processing costs from $O(N^2)$ to $O(2N)$, where N is the filter size. See also linear filter, separable template.

separable template: A template or structuring element in a filter, for instance a morphological filter (see morphology), that can be decomposed into a sequence of smaller templates, similarly to separable kernels for linear filters. The main advantage is a reduction in the computational complexity of the associated filter. See also separable filter.

set theoretic modeling: See constructive solid geometry.

shading: The pattern formed by the graded areas of an intensity image, suggesting light and dark. Variations in the lightness of surfaces in the scene may be due to vartiations in illumination, surface orientation and surface reflectance. See also illumination, shadow.

shading correction: A class of techniques for changing undesirable shading effects, for instance strongly uneven brightness distribution caused by nonuniform illumination. All techniques assume a shading model, *i.e.*, a photometric model of image formation, formalizing the dependency of the measured image brightness on camera parameters (typically gain and offset), illumination and object reflectance. See also shadow, photometry.

shading from shape: A technique recovering the reflectance of isolated objects given a single image and a geometric model, but not exactly the inverse of the classic shape from shading problem. See also photometric stereo.

shadow: A part of a scene that direct illumination does not reach because of self-occlusion (attached shadow or self-shadow) or occlusion caused by other objects (cast shadow). Therefore, this region appears darker than its surroundings. See also shape from shading, shading from shape, photometric stereo. See:

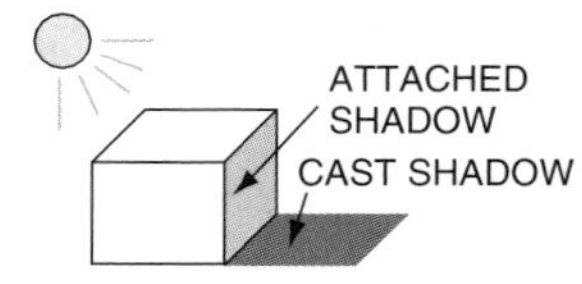

shadow, attached: A shadow caused by an object on itself by self-occlusion. See also shadow, cast.

shadow, cast: A shadow thrown by an object on another object. See also shadow, attached.

shadow detection: The problem of identifying image regions corresponding to shadows in the scene, using photometric properties. Useful for true color estimation and region analysis. See also color, color image segmentation, color matching, photometry, region segmentation.

shadow type labeling: A problem similar to shadow detection, but requiring classification of different types of shadows.

shadow understanding: Estimating various properties of a 3D scene based on the appearance or size of shadows, *e.g.*, building height. See also shadow type labeling.

shape: Informally, the form of an image or scene object. Typically described in computer vision through geometric representations (see shape representation), *e.g.*, modeling image contours with polynomials or b-spline, or range data patches with quadric surfaces. More formally, definitions are: **1.** (adj) The quality of an object that is invariant to changes of the coordinate system in which it is expressed. If the coordinate system is Euclidean, this corresponds to the conventional idea of shape. In an affine coordinate system, the change of coordinates may be affine, so that, for example, an ellipse and a circle have the same *shape*. **2.** (n) A family of point sets, any pair being related by a coordinate system transformation. **3.** (n) A specific set of n-dimensional points, *e.g.*, the set of squares. For example a curve in $\mathbb{R}^2$ defined parametrically as $\vec{c}(t) = (x(t), y(t))$ comprises the point set or *shape* $\{\vec{c}(t) \mid -\infty < t < \infty\}$. The volume inside the unit sphere in 3D is the shape $\{\vec{x} \mid \|\vec{x}\| < 1, \vec{x} \in \mathbb{R}^3\}$.

shape class: One in a set of classes representing different types of shape in a given classification, for instance, "locally convex" or "hyperbolic" in HK segmentation of a range image.

shape decomposition: See segmentation and hierarchical modeling.

shape from contours: A class of algorithms for estimating the shape of a 3D object from the contour it generates in an image. A well-known technique, shape from silhouettes, consists in extracting the object's silhouette from a number of views, and intersecting the 3D cones generated by the silhouettes' contours and the centers of projections. The intersection volume is known as the visual hull. Work also exists on understanding shape from the differential properties of apparent contours.

shape from defocus: A class of algorithms for estimating scene depth at each image pixel, and therefore surface shape, from multiple images acquired at different, controlled focus settings. A closed-form model of the relation between depth and image focus is assumed, containing a number of parameters (*e.g.*, the optics parameters) that must be calibrated. Depth is estimated using this model once image readings (pixel values) are available. Notice that the camera uses a large aperture, so that the points in the scene are in focus over the smallest possible depth interval. See also shape from focus.

shape from focus: A class of algorithms for estimating scene depth at each image pixel, and therefore surface shape, by varying the focus setting of a camera until the image achieves optimal focus (minimum blur) in a neighborhood of the pixel under examination. Obviously, pixels corresponding to different depths would achieve optimal focus for different settings. A model of the relation between depth and image focus is assumed, containing a number of parameters (*e.g.*, the optics parameters) that must be calibrated. Notice that the camera uses a large aperture, so that the smallest possible depth interval generates in-focus image points. See also shape from defocus.

shape from line drawings: A class of symbolic algorithms inferring 3D properties of scene objects (as opposed to exact shape measurements, as in other "shape from" methods) from line drawings. First, assumptions are made about the type of line drawings admissible, *e.g.*, polyhedral objects only, no surface markings or shadows, maximum three lines forming an image junction. Then, a dictionary of line junctions is formed, assigning a symbolic label to every possible appearance of the line junctions in space under the given assumptions. This figure shows part of a simple dictionary of junctions and a labeled shape:

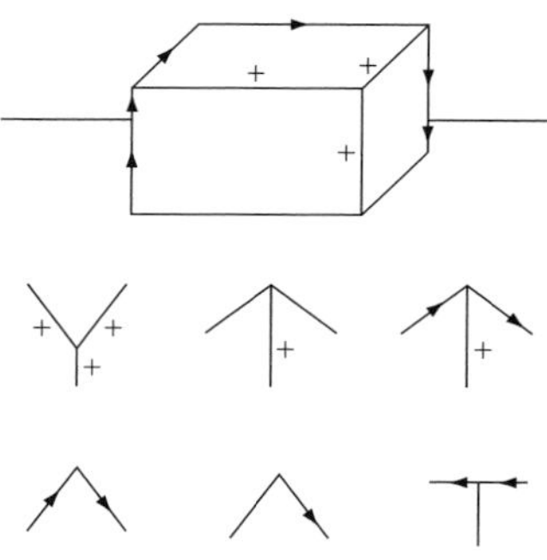

where + means planes intersecting in a convex shape, − in a concave shape, and the arrows a discontinuity (occlusion) between surfaces. Each image junction is then assigned the set of all possible labels that its shape admits locally (*e.g.*, all possible two-line junction labels for a two-line junction). Finally, a constraint satisfaction algorithm is used to prune labels inconsistent with the context. See also Waltz's line labeling, relaxation labeling.

shape from monocular depth cues: A class of algorithms estimating shape from information related to depth detected in a single image, *i.e.*, from monocular cues. See shape from contours, shape from line drawings, shape from perspective, shape from shading, shape from specularity, shape from structured light, shape from texture.

shape from motion: A vast class of algorithms for estimating 3D shape (structure), and often depth, from the motion information contained in an image sequence. Methods exist that rely on tracking sparse sets of image features (for instance, the Tomasi–Kanade factorization) as well as dense motion fields, *i.e.*, optical flow, seeking to reconstruct dense surfaces. See also motion factorization.

shape from multiple sensors: A class of algorithms recovering shape from information collected from a number of sensors of the same type, or of different types. For the former class, see multi-view stereo, For the second class, see sensor fusion.

shape from optical flow: See optical flow.

shape from orthogonal views: See shape from contours.

shape from perspective: A class of techniques estimating depth for various features from perspective cues, for instance the fact that a translation along the optical axis of a perspective camera changes the size of the imaged objects. See also pinhole camera model.

shape from photo consistency: A technique based on space carving for recovering shape from multiple views (photos). The basic constraint is that the underlying shape must be "photo-consistent" with all the input photos, *i.e.*, roughly speaking, give rise to compatible intensity values in all cameras.

shape from photometric stereo: See photometric stereo.

shape from polarization: A technique recovering local shape from the polarization properties of a surface under observation. The basic idea is to illuminate a surface with known polarized light, estimate the polarization state of the reflected light, then use this estimate in a closed-form model linking the surface normals with the measured polarization parameters. In practice, polarization estimates can be noisy. This method can be useful wherever intensity images do not provide information, *e.g.*, featureless specular surfaces. See also polarization based methods.

shape from shading: The problem of estimating shape, here in the sense of a field of normals from which a surface can be recovered up to a scale factor, from the shading pattern (light and shadows) of an image. The key idea is that, assuming a reflectance map for the scene (typically Lambertian), an image irradiance equation can be written linking the surface normals to the illumination direction and the image intensity. The constraint can be used to recover the normals assuming local surface smoothness.

shape from shadows: A technique for recovering geometry from a number of images of an outdoor scene acquired at different times, *i.e.*, with the sun at different angles. Geometric information can be recovered under various assumptions and knowledge of the sun's position. Also called "shape from darkness". See also shape from shading and photometric stereo.

shape from silhouettes: See shape from contours.

shape from specularity: A class of algorithms for estimating local shape from surface specularities. A specularity constrains the surface normal as the incident and reflection angles must coincide. The detection of specularities in images is, in itself, a non-trivial problem.

shape from structured light: See structured light triangulation.

shape from texture: The problem of estimating shape, here in the sense of a field of normals from which a surface can be recovered up to a scale factor, from the image texture. The deformation of a planar texture recorded in an image (the texture gradient) depends on the shape of the surface to which the texture is applied. Techniques exist for shape estimation from statistical texture and regular texture patterns.

shape from X: A generic term for a method that generates 3D shape or position estimates from one of a variety of possible techniques, such as stereo, shading, focus, etc.

shape from zoom: The problem of computing shape (in the sense of the distance of each scene point from the sensor) from two or more images acquired at different zoom settings, achieved through a zoom lens. The basic idea is to differentiate the projection equations with respect to the focal length, f, achieving an expression linking the variations of f and pixel displacement with depth.

shape grammar: A grammar specifying a class of shapes, whose rules specify patterns for combining more primitive shapes. Rules are composed of two parts, 1) describing a specific shape and 2) how to replace or transform it. Used also in design, CAD, and architecture. See also production system, expert system.

shape index: A measure, usually indicated by S, of the type of shape of a surface patch in terms of its principal curvature. Formally

$$S = -\frac{2}{\pi} \arctan \frac{\kappa_M + \kappa_m}{\kappa_M - \kappa_m}$$

where κ_m and κ_M are the principal curvatures. S is undetermined for planar patches. A related parameter, R, called *curvedness*, measures the amount of curvedness of the patch:

$$\sqrt{(\kappa_M^2 + \kappa_m^2)/2}$$

All curvature-based shape classes map to the unit circle in the R–S plane, with planar patches at the origin. See also mean and Gaussian curvature shape classification, shape representation.

shape magnitude class: Part of a local surface curvature representation scheme in which each point has a curvature class, and a magnitude of curvature (shape magnitude). This representation is an alternative to the more common shape classification based on either the two principal curvatures or the mean and Gaussian curvature.

shape representation: A large class of techniques seeking to capture the salient properties of shapes, both 2D and 3D, for analysis and comparison purposes. Many representations have been proposed in the literature, including skeletons for 2D and 3D shapes (see medial axis skeletonization and distance transform), curvature-based representations (for instance, the curvature primal sketch, the curvature scale space, the extended Gaussian image), generalized cones for articulated objects, invariants, and flexible objects models (for instance snakes, deformable superquadrics, and deformable template model).

shape texture: The texture of a surface from the point of view of the variation in the shape, as contrasted to the variation in the reflectance patterns on the surface. See also surface roughness characterization.

sharp–unsharp masking: A form of image enhancement that makes the edges of image structures crisper. The operator can either add a weighted amount of a gradient or high-pass filter of the image or subtract a weighted amount of a smoothing or low pass filter of the image. The image on the right is an unsharp masked version of the one on the left:

shear transformation: An affine image transformation changing one coordinate only. The corresponding transformation matrix, S, is equal to the identity apart from $s_{12} = s_x$, which changes the first image coordinate. Shear on the second image coordinate is obtained similarly by $s_{21} = s_y$.

An example of the result of a shear transformation is:

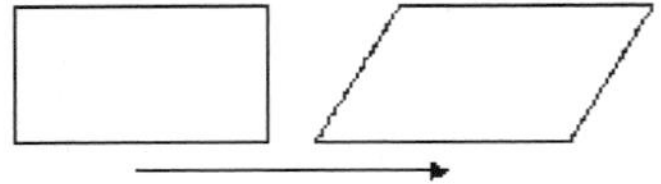

shock tree: A 2D shape representation technique based on the singularities (see singularity event) of the radius function along the medial axis (MA). The MA is represented by a tree with the same structure, and is divided into continuous segments of uniform behavior (local maximum, local minimum, constant, monotonic). See also medial axis skeletonization, distance transform.

short baseline stereo: See narrow baseline stereo.

shot noise: See impulse noise and salt-and-pepper noise.

shutter: A device allowing the light into a camera for enough time to form an image on a photosensitive film or chip. Shutters can be mechanical, as in traditional photographic cameras, or electronic, as in a digital camera. In the former case, a window-like mechanism is opened to allow the light to be recorded by a photosensitive film. In the latter case, a CCD or other type of sensor is triggered electronically to record the amount of incident light at each pixel.

shutter control: The device controlling the length of time that the shutter is open.

side looking radar: A radar projecting a fan-shaped beam illuminating a strip of the scene at the side of the instrument, typically used for mapping a large area. The map is produced as the instrument is carried along by a vehicle sweeping the surface to the side. See also sonar.

signal coding system: A system for encoding a signal into another, typically for compression or security purposes. See image compression, digital watermarking.

signal processing: The collection of mathematical and computational tools for the analysis of typically 1D (but also 2D, 3D, etc.) signals such as audio recordings or other intensity *versus* time or position measurements. Digital signal processing is the subset of signal processing which pertains to signals that are represented as streams of binary digits.

signal-to-noise ratio (SNR): A measure of the relative strength of the interesting and uninteresting (noise) part of a signal. In signal processing, SNR is usually expressed in decibels as the ratio of the power of signal and noise, *i.e.*, $10 \log_{10} \frac{P_s}{P_n}$. With statistical noise, the SNR can be defined as 10 times the log of the ratio of the standard deviations of signal and noise.

signature identification: A class of techniques for verifying a written signature. Also known as Dynamic Signature Verification. An area of biometrics. See also handwriting verification, handwritten character recognition, fingerprint identification, face identification.

signature verification: The problem of authenticating a signature automatically with image processing techniques; in practice, deciding whether a signature matches a specimen sufficiently well. See also handwriting verification and handwritten character recognition.

silhouette: See object contour.

SIMD: See single instruction multiple data.

similarity: The property that makes two entities (images, models, objects, features, shape, intensity values, etc.) or sets thereof similar, that is, resembling each other. A similarity transformation creates perfectly similar structures and a similarity metric quantifies the degree of similarity of two possibly non-identical structures. Examples of similar structures are 1) two polygons identical except for a change in size, and 2) two image neighborhoods whose intensity values are identical except for scaling by a multiplicative factor. The concept of similarity lies at the heart of several classic vision problems, including stereo correspondence, image matching, and geometric model matching.

similarity metric: A metric quantifying the similarity of two entities. For instance, cross correlation is a common similarity metric for image regions. For similarity metrics on specific objects encountered in vision, see feature similarity, graph similarity, gray scale similarity. See also point similarity measure, matching.

similarity transformation: A transformation changing an object into a similar-looking one; formally, a conformal mapping preserving the ratio of distances (the magnification ratio). The transformation matrix, $\mathbf{T}$, can be written as $\mathbf{T} = \mathbf{B}^{-1}\mathbf{A}\mathbf{B}$, where $\mathbf{A}$ and $\mathbf{B}$ are similar matrices, that is, representing the same transformation after a change of basis. Examples include rotation, translation, expansion and contraction (scaling).

simple lens: A lens composed by a single piece of refracting material, shaped in such a way to achieve the desired lens behavior. For example, a convex focusing lens.

simulated annealing: A coarse-to-fine, iterative optimization algorithm. At each iteration, a smoothed version of the energy landscape is searched and a global minimum located by a statistical (*e.g.*, random) process. The search is then performed at a finer level of smoothing, and so on. The idea is to locate the basin of the absolute minimum at coarse scales, so that fine-resolution search starts from an approximate solution close enough to the absolute minimum to avoid falling into surrounding local minima. The name derives from the homonymous procedure for tempering metal, in which temperature is lowered in stages, each time allowing the material to reach thermal equilibrium. See also coarse-to-fine processing.

single instruction multiple data (SIMD): A computer architecture allowing the same instruction to be simultaneously executed on multiple processors and thus different portions of the data set (*e.g.*, different pixels or image neighborhoods). Useful for a variety of low-level image processing operations. See also MIMD, pipeline parallelism, data parallelism, parallel processing.

single photon emission computed tomography (SPECT): A medical imaging technique that involves the rotation of a photon detector array around the body in order to detect photons emitted by the decay of previously injected radionuclides. This technique is particularly useful for creating a volumetric image showing metabolic activity. Resolution is lower than PET but imaging is cheaper and some SPECT radiopharmaceuticals may be used where PET nuclides cannot.

singular value decomposition (SVD): A factorization of any $m \times n$ matrix $\mathbf{A}$ into $\mathbf{A} = \mathbf{UDV}^T$. The columns of the $m \times m$ matrix $\mathbf{U}$ are mutually orthogonal unit vectors, as are the columns of the $n \times n$ matrix $\mathbf{V}$. The $m \times n$ matrix $\mathbf{D}$ is diagonal, and its nonzero elements, the *singular values* σ_i, satisfy $\sigma_1 \geq \sigma_2 \geq \ldots \geq \sigma_n \geq 0$. The SVD has extremely useful properties. For example:

- $\mathbf{A}$ is nonsingular if and only if all its singular values are nonzero, and the number of nonzero singular values gives the rank of $\mathbf{A}$;
- the columns of $\mathbf{U}$ corresponding to the nonzero singular values span the range of $\mathbf{A}$; the columns of $\mathbf{V}$ corresponding to the nonzero singular values span the null space of $\mathbf{A}$;
- the squares of the nonzero singular values are the nonzero eigenvalues of both $\mathbf{AA}^T$ and $\mathbf{A}^T\mathbf{A}$, and the columns of $\mathbf{U}$ are eigenvectors of $\mathbf{AA}^T$, those of $\mathbf{V}$ of $\mathbf{A}^T\mathbf{A}$.

Moreover, the pseudoinverse of a matrix, occurring in the solution of rectangular linear systems, can be easily computed from the SVD definition.

singularity event: A point in the domain of the map of a geometric curve or surface where the first derivatives vanish.

sinusoidal projection: A family of linear image transforms, C, the rows of which are the eigenvalues of a special symmetric tridiagonal matrix. This includes the discrete cosine transform (DCT).

skeleton: A curve, or tree-like set of curves, capturing the basic structure of an object. This figure shows an example of a linear skeleton for a puppet-like 2D shape:

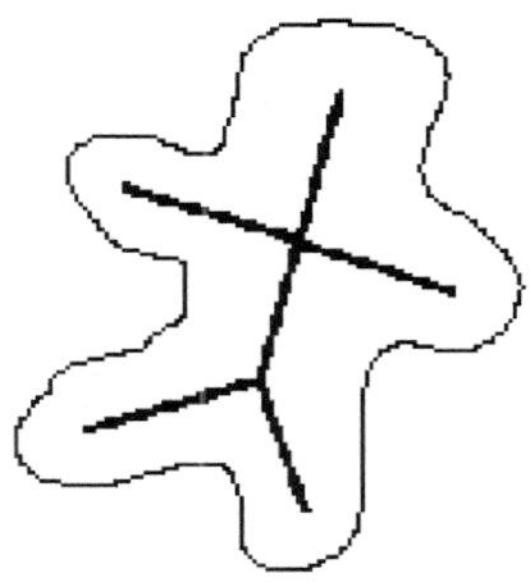

The curves forming the skeleton are typically central to the shape. Several algorithms exist for computing skeletons, for instance, the medial axis transform (see medial axis skeletonization) and the distance transform, for which the grassfire algorithm can be applied.

skeleton by influence zones (SKIZ): Commonly known as the Voronoi diagram.

skeletonization: A class of techniques that try to reduce a 2D (or 3D) binary image to a "skeleton" form in which every remaining pixel is a skeleton pixel, but the essential shape of the input image is captured. Definitions of the skeleton include the set of centers of circles bitangent to the object boundary and smoothed local symmetries.

skew: An error introduced in the imaging geometry by a non-orthogonal pixel grid, in which rows and columns of pixels do not form an angle of exactly 90°. This is usually considered only in high-accuracy photogrammetry applications.

skew correction: A transformation compensating for the skew error.

skew symmetry: A *skew symmetric contour* is a planar contour such that every straight line oriented at an angle ϕ with respect to a particular axis, called the *skew symmetry axis* of the contour, intersects the contour at two points equidistant from the axis. An example:

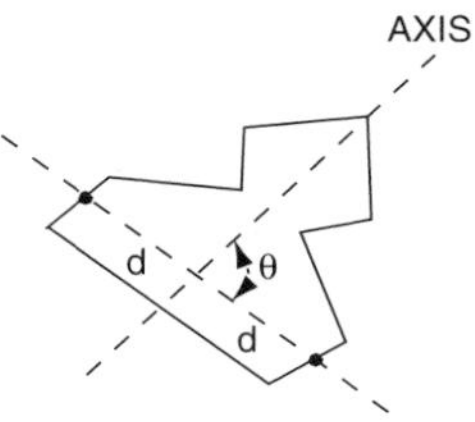

skin color analysis: A set of techniques for color analysis applied to images containing skin, for instance for retrieving images from a database (see color based image retrieval). See also color, color image, color image segmentation, color matching, and colorimetry.

SKIZ: See skeleton by influence zones.

slant: The angle between a surface normal in the scene and the viewing direction:

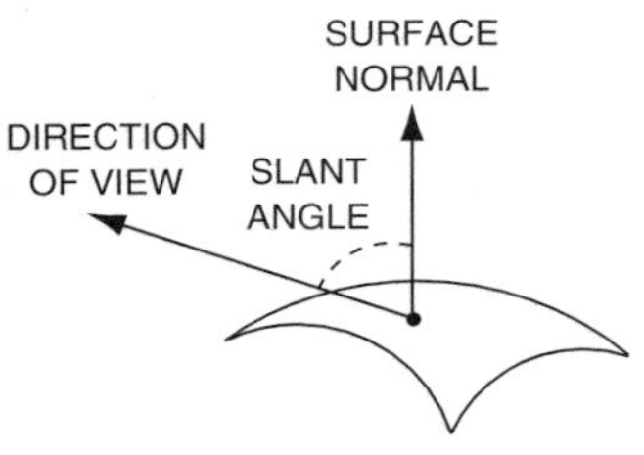

See also tilt, shape from texture.

slant normalization: A class of algorithms used in handwritten character recognition, transforming slanted cursive character into vertical ones. See handwritten character recognition, optical character recognition.

slice based reconstruction: The reconstruction of a 3D object from a number of planar slices, or sections taken across the object. The slice plane is typically advanced at regular spatial intervals to sweep the working volume. See also tomography, computerized tomography, single photon emission, computed tomography, and nuclear magneticresonance.

slope density function: This is the histogram of the tangential orientations (slopes) of a curve or region boundary. It can be used to represent the curve shape in a manner invariant to translation and rotation (up to a shift of the density function).

small motion model: A class of mathematical models representing very small (ideally, infinitesimal) camera-scene motion between frames. Used typically in shape from motion. See also optical flow.

smart camera: A hardware device incorporating a camera and an on-board computer in a single, small container, thus achieving a programmable vision system within the size of a normal video camera.

smooth motion curve: The curve defined by a motion that can be expressed by smooth (that is, differentiable: derivatives of all orders exist) parametric functions of the image coordinates. Notice that "smooth" is often used in an intuitive sense, not in the strict mathematical sense above (clearly, an exacting constraint), as, for example, in image smoothing. See also motion, motion analysis.

smoothed local symmetries: A class of skeletonization algorithms, associated with Asada and Brady. Given a 2D curve that bounds a closed region in the plane, the skeleton as computed by smoothed local symmetries is the locus of chord midpoints of bitangent circles. Compare the symmetric axis transform. Two skeleton points as defined by smoothed local symmetries are shown:

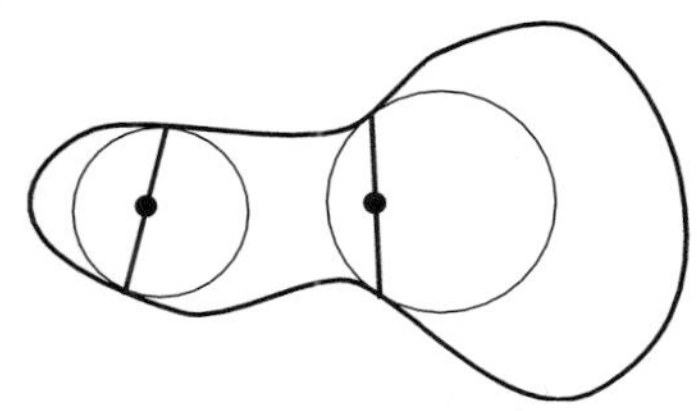

smoothing: Generally, any modification of a signal intended to remove the effects of noise. Often used to mean the attenuation of high spatial frequency components of a signal. As many models of noise have a flat power spectral density (PSD), while natural images have a PSD that decays toward zero at high spatial frequencies, suppressing the high frequencies increases the overall signal-to-noise ratio of the image. See also discontinuity preserving smoothing, anisotropic diffusion and adaptive smoothing.

smoothing filter: Smoothing is often achieved by convolution of the image with a smoothing filter to reduce noise or high spatial frequency detail. Such filters include discrete approximations to the symmetric probability densities such as the Gaussian, binomial and uniform distributions. For example, in 1D, the discrete signal $x_1 \ldots x_n$ is convolved with the kernel $[\frac{1}{6}\frac{4}{6}\frac{1}{6}]$ to produce the smoothed signal $y_1 \ldots y_{n+2}$ in which $y_i = \frac{1}{6}x_{i-1} + \frac{4}{6}x_i + \frac{1}{6}x_{i+1}$.

smoothness constraint: An additional constraint used in data interpretation problems. The general principle is that results derived from nearby data must themselves have similar values. Traditional examples of where the smoothness constraint can be applied are in shape from shading and optical flow. The underlying observation that supports this computational constraint is that the observed real world surfaces and motions are smooth almost everywhere.

snake: A snake is the combination of a deformable model and an algorithm for fitting that model to image data. In one common embodiment, the model is a parameterized 2D curve, for example a b-spline parameterized by its control points. Image data, which might be a gradient image or 2D points, induces forces on points on the snake that are translated to forces on the control points or parameters. An iterative algorithm adjusts the control points according to these forces and recomputes the forces. Stopping criteria, step lengths, and other issues of optimization are all issues that must be dealt with in an effective snake.

SNR: See signal-to- noise ratio.

Sobel edge detector: An edge detector based on the Sobel kernels. The edge magnitude image E is the square root of the sum of squares of the convolution of the image with horizontal and vertical Sobel kernels, given by $E = \sqrt{(K_x * I)^2 + (K_y * I)^2}$. The Sobel operator applied to the left image gives the right image:

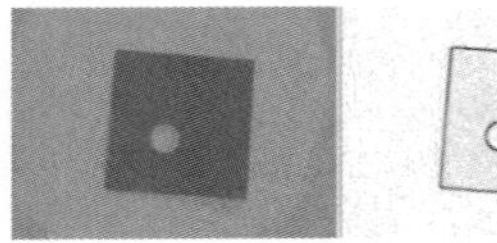

Sobel gradient operator: See Sobel kernel.

Sobel kernel: A gradient estimation kernel used for edge detection. The horizontal kernel is the convolution of a smoothing filter, $s = [1, 2, 1]$ in the horizontal direction and a gradient operator $d = [-1, 0, 1]$ in the vertical direction. The kernel

$$K_y = s * d^\top = \begin{pmatrix} -1 & -2 & -1 \\ 0 & 0 & 0 \\ 1 & 2 & 1 \end{pmatrix}$$

highlights horizontal edges. The vertical kernel K_x is the transpose of K_y.

soft mathematical morphology: An extension of gray scale morphology in which the min/max operations are replaced by other rank operations *e.g.*, replace each pixel in an image by the 90th percentile value in a 5×5 window centered at the pixel. Weighted ranks may be computed. See also fuzzy morphology.

soft morphology: See soft mathematical morphology.

soft vertex: A point on a polyline whose connecting line segments are almost collinear. Soft vertices may arise from segmentation of a smooth curve into line segments. They are called 'soft' because they may be removed if the segments of the polyline are replaced by curve segments.

solid angle: Solid angle is a property of a 3D *object*: the amount of the unit sphere's surface that the object's projection onto the unit sphere occupies. The unit sphere's surface area is 4π, so the maximum value of a solid angle is 4π steradians:

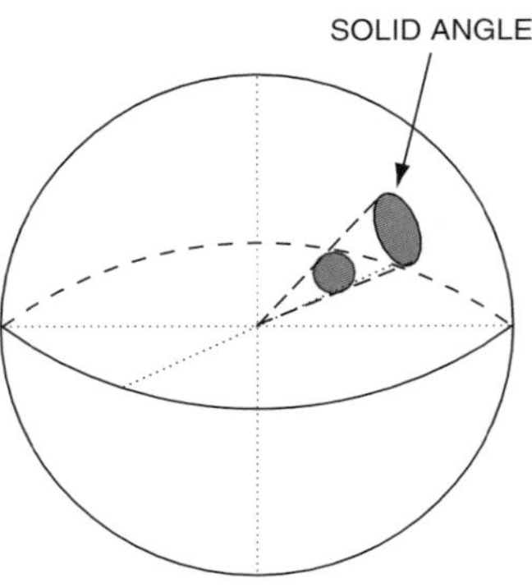

source: An emitter of energy that illuminate the vision system's sensors.

source geometry: See light source geometry.

source image: The image on which an image processing or an image analysis operation is based.

source placement: See light source placement.

space carving: A method for creating a 3D volumetric model from 2D images. Starting from a voxel representation in which a 3D cube is marked "occupied", voxels are removed if they fail to be photo-consistent in the set of 2D images in which they appear. The *order* in which the voxels are processed is a key aspect of space carving, as it allows otherwise intractable visibility computations to be avoided.

space curve: A curve that may follow a path in 3D space (*i.e.*, it is not restricted to lying in a plane).

space variant sensor: A sensor in which the pixels are not uniformly sampling the projected image data. For example, a log-polar sensor has rings of pixels of exponentially increasing size as one moves radially from the central point.

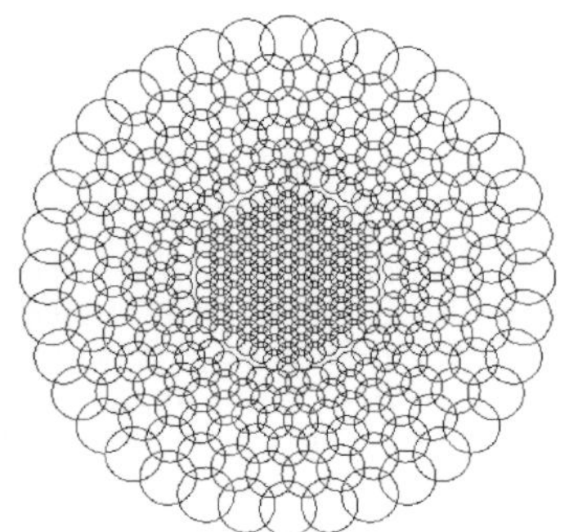

spatial angle: The area on a unit sphere that is bounded by a cone with its apex in the center of the sphere. Measured in steradians. This is frequently used when analyzing luminance.

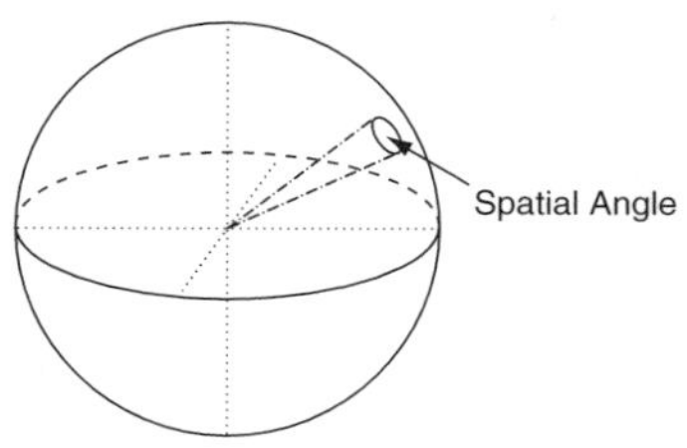

spatial averaging: The pixels in the output image are weighted averages of their neighboring pixels in the input image. Mean and Gaussian smoothing are examples of spatial averaging.

spatial domain smoothing: An implementation of smoothing in which each pixel is replaced by a value that is directly computed from other pixels in the image. In contrast, frequency domain smoothing first processes all pixels to create a linear transformation of the image, such as a Fourier transform and expresses the smoothing operation in terms of the transformed image.

spatial frequency: The rate of repetition of intensities *across* an image. In a 2D image the space to which *spatial* refers is the image's X–Y plane.

This image has significant repetition at a spatial frequency of $\frac{1}{10}$ pixel^{-1} in the horizontal direction. The 2D Fourier transform represents spatial frequency contributions in all directions, at all frequencies. A discrete approximation is efficiently computed using the fast Fourier transform (FFT).

spatial hashing: See spatial indexing.

spatial indexing: 1) Conversion of a *shape* to a number, so that it may be quickly compared to other shapes. Intimately linked with the computation of *invariants* to spatial transformations and imaging distortions of the shape. For example, a shape represented as a collection of 2D boundary points might be indexed by its compactness. 2) The design of efficient data structures for search and storage of geometric quantities. For example closest-point queries are made more efficient by the computation of spatial indices such as the Voronoi diagram, distance transform, k-D trees, or Binary Space Partitioning (BSP) trees.

spatial matched filter: See matched filter.

spatial occupancy: A form of object or scene representation in which a 3D space is divided into a grid of voxels. Voxels containing a part of the object are marked as being occupied and other voxels are marked as free space. This representation is particularly useful for tasks where properties of the object are less important than simply the presence and position of the object, as in robot navigation.

spatial proximity: The distance between two structures in real space (as contrasted with proximity in a feature or property space).

spatial quantization: The conversion of a signal defined on an infinite domain to a finite set of limited-precision samples. For example the function $f(x,y)$: $\mathbb{R}^2 \mapsto \mathbb{R}$ might be quantized to the image g, of width w and height h defined as $g(i,j)$: $\{1..w\} \times \{1..h\} \mapsto \mathbb{R}$. The value of a particular sample $g(i,j)$ is determined by the point-spread function $p(x,y)$, and is given by $g(i,j) = \int p(x-i, y-j) f(x,y) \mathrm{d}x \mathrm{d}y$.

spatial reasoning: Inference from geometric rather than

symbolic or linguistic information. See also geometric reasoning.

spatial relation: An association of two or more spatial entities, expressing the way in which such entities are connected or related. Examples include perpendicularity or parallelism of lines or planes, and inclusion of one image region in another.

spatial resolution: The smallest separation between distinct signal features that can be measured by a sensor. For a CCD camera, this is dictated by the distance between adjacent pixel centers. It is often specified as an angle: the angle between the 3D rays corresponding to adjacent pixels. The inverse of the highest spatial frequency that a sensor can represent without aliasing.

spatio-temporal analysis: The analysis of moving images by processing that operates on the 3D volume formed by the stack of 2D images in a sequence. Examples include kinetic occlusion, the epipolar plane image (EPI) and spatio-temporal autoregressive models (STAR).

special case motion: A subproblem of the general structure from motion problem, where the camera motion is known to be constrained *a priori*. Examples include planar motion, turntable motion or single-axis rotation, and pure translation. In each case, the constrained motion simplifies the general problem, yielding one or more of: closed-form solutions, greater efficiency, increased accuracy. Similar benefits can be obtained from approximations such as the affine camera and weak perspective.

speckle: A pattern of light and dark spots superimposed on the image of a scene that is illuminated by coherent light such as from a laser. Rough surfaces in the scene change the path lengths and thus the interference effects of different rays, so a fixed scene, laser and imager configuration results in a fixed speckle pattern on the imaging surface.

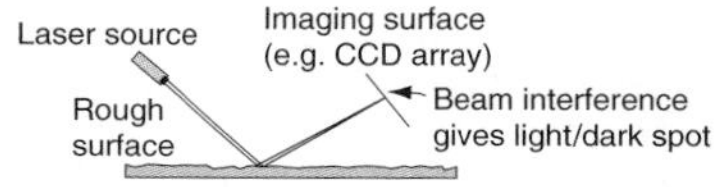

speckle reduction: Restoration of images corrupted with speckle noise, such as laser or ultrasound images.

SPECT: See single-photon emission computed tomography.

spectral analysis: 1) Analysis performed in either the spatial, temporal or electromagnetic frequency domain. 2) Generally, any analysis that involves the examination of eigenvalues. This is a nebulous concept, and consequently the number of "spectral techniques" is large. Often equivalent to PCA.

spectral decomposition method: See spectral analysis.

spectral density function: See power spectrum.

spectral distribution: The spatial power spectrum or electromagnetic spectrum distribution.

spectral factorization: A method for designing linear filters based on difference equations that have a given spectral density function when applied to white noise.

spectral filtering: Modifying the light before it enters the sensor by using a filter tuned to different spectral frequencies. A common use is with laser sensing, in which the filter is chosen to pass only light at the laser's frequency. Another usage is to eliminate ambient infrared light in order to increase the sharpness of an image (as most silicon- based sensors are also sensitive to infrared light).

spectral frequency: Electromagnetic or spatial frequency.

spectral reflectance: See reflectance.

spectral response: The response R of an imaging sensor illuminated by monochromatic light of wavelength λ is the product of the input light intensity I and the spectral response at that wavelength $s(\lambda)$, so $R = Is(\lambda)$.

spectrum: A range of values such as the electromagnetic spectrum.

specular reflection: Mirror-like reflection or highlight. Formed when a light source at 3D location L, surface point P, surface normal N at that point and camera center C are all coplanar, and the angles LPN and NPC are equal.

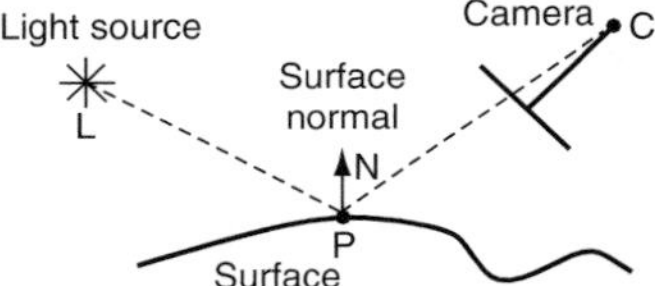

specularity: See specular reflection.

sphere: 1. A surface in any dimension defined by the $\vec{x}$ such that $\|\vec{x} - \vec{c}\| = r$ for a center $\vec{c}$ and radius r. 2. The volume of space bounded by the above, or $\vec{x}$ such that $\|\vec{x} - \vec{c}\| \leq r$.

spherical: Having the shape of, characteristics of, or associations with, a sphere.

spherical harmonic: A function defined on the unit sphere of the form

$$Y_l^m(\theta, \phi) = \eta_{lm} P_l^m(\cos\theta) e^{im\phi}$$

is a spherical harmonic, where η_{lm} is a normalizing factor, and P_l^m is a Legendre polynomial. Any real function defined on the sphere $f(\theta, \phi)$ has an expansion in terms of the spherical harmonics of the form

$$f(\theta, \phi) = \sum_{l=0}^{\infty} \sum_{m=-l}^{l} \alpha_l^m Y_l^m(\theta, \phi)$$

that is analogous to the Fourier expansion of a function defined on the plane, with the α_l^m analogous to the Fourier coefficients. Polar plots of the first ten spherical harmonics, for $m = 0 \ldots 2, l = 0 \ldots m$. The plots show $r = 1 + Y_l^m(\theta, \phi)$ in polar coordinates:

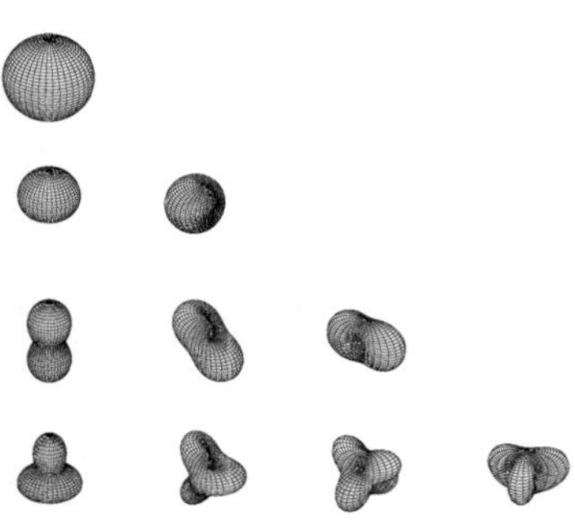

spherical mirror: Sometimes used in catadioptric cameras. A mirror whose shape is a portion of a sphere.

spin image: A local surface representation of Johnson and Hebert. At selected points $\vec{p}$ with surface normal $\vec{n}$, all other surface points $\vec{x}$ can be represented in a 2D basis as $(\alpha, \beta) = (\sqrt{\|\vec{x} - \vec{p}\|^2 - (\vec{n} \cdot (\vec{x} - \vec{p}))^2}, \quad \vec{n} \cdot (\vec{x} - \vec{p}))$. The spin image is the histogram of all of the (α, β) values for the surface. Each selected points $\vec{p}$ leads to a different spin image. Matching points compares their spin images by correlation. Key advantages of the representation are 1) it is independent of pose and 2) it avoids ambiguities of representation that can occur with nearly flat surfaces.

splash: An invariant representation of the region about a 3D point. It gives a local shape representation useful for position invariant object recognition.

spline: 1) A curve $\vec{c}(t)$ defined as a weighted sum of *control points*: $\vec{c}(t) = \sum_{i=0}^{n} w_i(t)\vec{p}_i$, where the control points are $\vec{p}_{1 \ldots n}$ and one weighting (or "blending") function w_i is defined for each control point. The curve may interpolate the control points or approximate them. The construction of the spline offers guarantees of continuity and smoothness. With *uniform* splines the weighting functions for each point are translated copies of each other, so $w_i(t) = w_0(t - i)$. The form of w_0 determines the type of spline: for B-splines and Bezier curves, $w_0(t)$ is a polynomial (typically cubic) in t. *Nonuniform* splines reparameterize the t axis, $\vec{c}(t) = \vec{c}(u(t))$ where $u(t)$ maps the integers $k = 0..n$ to knot points $t_{0..n}$ with linear interpolation for non-integer values of t. *Rational* splines with n-D control points are perspective projections of normal splines with $(n + 1)$-D control points. 2) *Tensor-product* splines define a 3D surface $\vec{x}(u, v)$ as a product of splines in u and v.

spline smoothing: Smoothing of a discretely sampled signal $x(t)$ by replacing the value at t_i

by the value predicted at that point by a spline $\hat{x}(t)$ fitted to neighboring values.

split and merge: A two-stage procedure for segmentation or clustering. The data is divided into subsets, with the initial division being a single set containing all the data. In the *split* stage, subsets are repeatedly subdivided depending on the extent to which they fail to satisfy a coherence criterion (for example, similarity of pixel colors). In the *merge* stage, pairs of adjacent sets are found that, when merged, will again satisfy a coherence criterion. Even if the coherence criteria are the same for both stages, the merge stage may still find subsets to merge.

SPOT: *Systeme Probatoire de l'Observation de la Terre*. A series of satellites launched by France that are a common source of satellite images of the earth. SPOT-5 for example was launched in May 2002 and provides complete coverage of the earth every 26 days.

spot detection: An image processing operation for locating small bright or dark locations against contrasting backgrounds. The issues here are what size of spot and amount of contrast.

spur: A short segment attached to a more significant line or edge. Spurs often arise when linear structures are tracked through noisy data, such as by an edge detector. This figure shows some spurs:

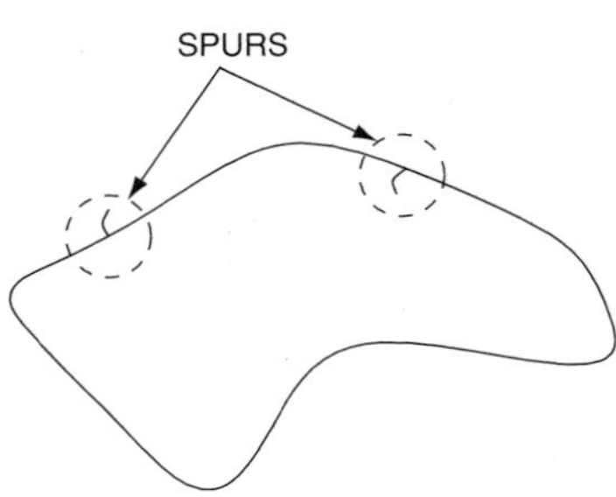

squared error clustering: A class of clustering algorithms that attempt to find cluster centers $\vec{c}_1 \ldots \vec{c}_n$ that minimize the squared error $\sum_{\vec{x} \in \mathcal{X}} \min_{i \in \{1 \ldots n\}} (\vec{x} - \vec{c}_i)^2$ where $\mathcal{X}$ is the set of points to be clustered.

stadimetry: The computation of distance to an object of known size based on its apparent size in the camera's field of view.

stationary camera: A camera whose optical center does not move. The camera may pan, tilt and rotate about its optical center, but not translate. Images taken by a stationary camera are always related by a planar homography. Also known as a *rotating camera* or *non-translating* camera. The term may also refer to a camera that does not move at all.

statistical classifier: A function mapping from a space of input data to a set of *labels*. Input data are points $\vec{x} \in \mathbb{R}^n$ and labels are scalars. The classifier $c(\vec{x}) = l$ assigns the label l to point $\vec{x}$. The classifier is typically a parameterized function, such as a neural network (with weights as parameters) or a support vector machine. The classifier parameters could be set by optimizing performance on a training set of known $(\vec{x}, l)$ pairs or by a self-organizing learning algorithm.

statistical pattern recognition: Pattern recognition that depends on classification rules learned from examples rather than constructed by designers. Compare structural pattern recognition.

statistical shape model: A parameterized shape model where the parameters are assumed to be random variables drawn from a known probability distribution. The distribution is learned from training examples. Examples include point distribution models.

statistical texture: A texture whose description is in terms of the statistics of image neighborhoods. General examples are co-occurrence statistics of pairs of neighboring pixels, Fourier texture descriptors, autocorrelation and autoregressive models. A specific example is the statistics of the distribution of entries in 5×5 neighborhoods. These statistics may be learned from a set of training images or automatically discovered via clustering.

steerable filter: A filter applied to a 2D image, whose response is dependent on an scalar "orientation" parameter θ, but for which the response at any arbitrary value of θ may be computed as a function of a small number of *basis* responses, thus saving computation. For example, the directional derivative at orientation θ may be computed in terms of the x and y derivatives I_x and I_y as

$$\frac{\mathrm{d}I}{\mathrm{d}\vec{n}_\theta} = \begin{pmatrix} \cos\theta I_x \\ \sin\theta I_y \end{pmatrix}$$

For non-steerable filters such as Gabor filters, the response must be computed at each orientation, leading to higher computational complexity.

steganography: Concealing of information in non-suspect "carrier" data. For example, encoding information in the low-order bits of a digital image.

step edge: 1) A discontinuity in image intensity (compare with fold edge). 2) An idealized model of a step-change in intensity. This plot of intensity I versus X position shows an intensity step edge discontinuity in intensity I at a:

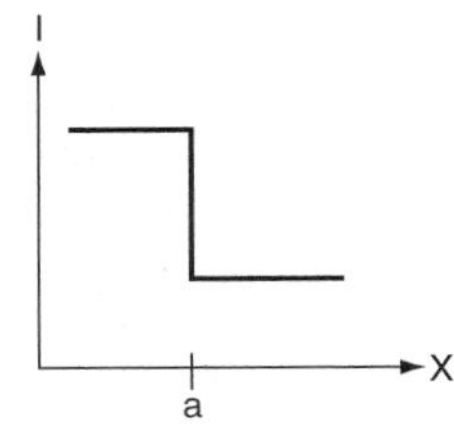

steradian: The unit of solid angle.

stereo: General term for a class of problems in which multiple images of the same scene are used to recover a 3D property such as surface shape, orientation or curvature. In binocular stereo, two images are taken from different viewpoints allowing the computation of 3D structure. In trifocal, trinocular and multiple-view stereo, three or more images are available. In photometric stereo, the viewpoint is the same, but lighting conditions are varied in order to compute surface orientation.

stereo camera calibration: The computation of intrinsic and extrinsic camera parameters for a pair of cameras. Important extrinsic variables are relative orientation: the rotation and translation relating the two cameras. Achieved in several ways: 1) conventional calibration of each camera independently; 2) computation of the essential matrix or fundamental matrix relating the pair, from which relative orientation may be computed along with one or two intrinsic parameters; 3) for a rigid stereo rig, moving the rig and capturing multiple image pairs.

stereo correspondence problem: The key to recovering depth from stereo is to identify 2D image points that are projections of the same 3D scene point. Pairs of such image points are called *correspondences*. The *correspondence problem* is to determine which pairs of image image points are correspondences. Unfortunately, matching features or image neighborhoods is usually ambiguous, leading to both massive amounts of computation and many alternative solutions. To reduce the space of matches, corresponding points are usually required to satisfy some constraints, such as having similar orientation and contrast, local smoothness, uniqueness of match. A powerful constraint is the epipolar constraint: from a single view, an image point is constrained to lie on a 3D ray, whose projection onto the second image is an epipolar curve. For pinhole cameras, the epipolar curve is a line. This greatly reduces the space of potential matches.

stereo convergence: The angle α between the optical axes of two sensors in a stereo configuration:

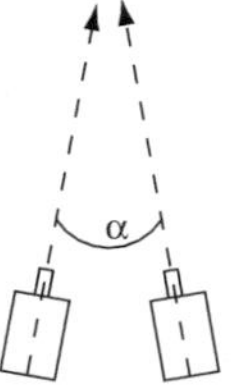

stereo fusion: The ability of the human vision system, when presented with a pair of stereo images, one to each eye independently, to form a consistent

3D interpretation of the scene, essentially solving the stereo correspondence problem. The fact that humans can perform fusion even on random dot stereograms means that high-level recognition is not required to solve all stereo correspondence problems.

stereo image rectification: For a pair of images taken by pinhole cameras, points in stereo correspondence lie on corresponding epipolar lines. Stereo image rectification resamples the 2D images to create two new images, with the same number of rows, so that points on corresponding epipolar lines lie on corresponding rows. This reduces computation for some stereo algorithms, although certain relative orientations (*e.g.*, translation along the optical axis) make rectification difficult to achieve.

stereo matching: See stereo correspondence problem.

stereo triangulation: Determining the 3D position of a point given its 2D positions in each of two images taken by cameras in known positions. In the noise-free case, each 2D point defines a 3D ray by back projection, and the 3D point is at the intersection of the two rays. With noisy data, the optimal triangulation is computed by finding the 3D point that maximizes the probability that the two imaged points are noisy projections thereof. Also used for the analogous problem in multiple views.

stereo vision: The ability to determine three dimensional structure using two eyes. See also stereo.

stimulus: 1) Any object or event that a computer vision system may detect. 2) The perceived radiant energy itself.

stochastic gradient: An optimization algorithm for minimizing a convex cost function.

stochastic completion field: A strategy for algorithmic discovery of illusory contours.

stochastic process: A process whose next state depends probabilistically on its current state.

stratification: A class of solutions to self-calibration in which a projective reconstruction is first converted to an affine reconstruction (by computing the plane at infinity) and then to a Euclidean reconstruction.

streaming video: Video presented as a sequence of images or frames. An algorithm processing such video cannot easily select a particular frame.

stripe ranging: See structured light triangulation.

strobe duration: The time for which a strobe light is illuminated.

strobed light: A light that is illuminated for a very short period, generally at high intensity.

structural pattern recognition: Pattern recognition where classification is achieved using high-level rules or patterns, often specified by a human designer. See also syntactic pattern recognition.

structural texture: A texture that is formed by the regular repetition of a primitive structure, for example an image of bricks or windows.

structure and motion recovery: The simultaneous computation of 3D scene structure and 3D camera positions from a sequence of images of a scene. Common strategies depend on tracking of 2D image entities (*e.g.*, interest points or edges) through multiple views and thus obtaining constraints on the 3D entities (*e.g.*, points and lines) and camera motion. Constraints are embodied in entities such as the fundamental matrix and trifocal tensor that may be estimated from image data alone, and then allow computation of the 3D camera positions. Recovery is up to certain equivalence classes of scenes, where any member of the class may generate the observed data, such as projective or affine reconstructions.

structure factorization: See motion factorization.

structure from motion: Recovery of the 3D shape of a set of scene points from their motion. For a more modern treatment, see structure and motion recovery.

structure from optical flow: Recovery of camera motion by computing optical flow constrained by the infinitesimal motion fundamental matrix. The small motion approximation replaces the rotation matrix $\mathbf{R}$ by $\mathbf{I} - [\vec{\omega}]_\times$ where $\vec{\omega}$ is the axis of rotation, the unique vector such that $\mathbf{R}\vec{\omega} = \vec{\omega}$.

structure matching: See recognition by components.

structured light: A class of techniques where carefully engineered illumination is employed to simplify computation of scene properties. Common examples include structured light triangulation, and moiré fringe sensing.

structured light source calibration: The special case of calibration in a structured light system where the position of the light source is determined.

structured light triangulation: Recovery of 3D structure by computing the intersection of

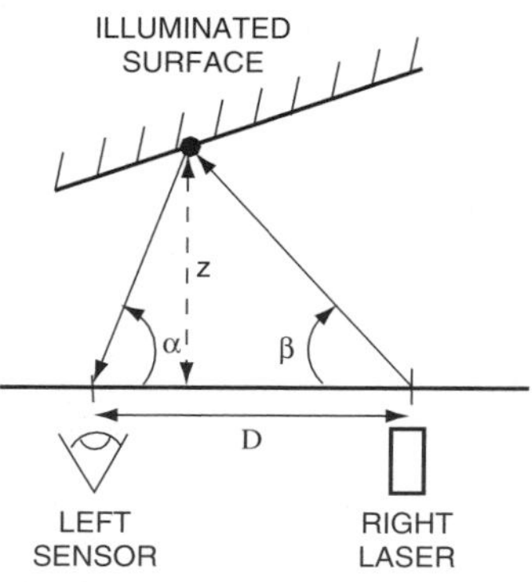

a ray (or plane or other light shape) of light with the ray determined by the image of the illuminated scene surface:

structured model: See hierarchical model.

structuring element: The basic neighborhood structure of morphological image processing. The structuring element is an image, typically small, that defines a shape pattern. Morphological operations on a source image combine the structuring element with the source image in various ways.

subband coding: A means of coding a discrete signal for transmission. The signal is passed through a set of bandpass filters, and each channel is quantized separately. The sampling rate of the individual channels is set such that, before quantization, the sum of the number of per-channel samples is the same as the number of samples of the original system. By varying the quantization for different bands, the number of samples may be reduced with small losses in signal quality.

subcomponent: An object part used in a hierarchical model.

subcomponent decomposition: Representation of a complete object part by a collection of smaller objects in a hierarchical model.

subgraph isomorphism: Equivalence of a pair of subgraphs of two given graphs. Given graphs A and B, the subgraph isomorphism problem is to enumerate all pairs of subgraphs (a, b) where: $a \subset A$; $b \subset B$; a is isomorphic to b; and some given predicate $p(a, b)$ is true. Appropriate modifications of the problem allow the solution of many graph problems including determining shortest paths and finding maximal cliques. A given graph has a number of subgraphs exponential in the number of vertices and the general problem is NP-complete. This example shows subgraph isomorphism with the matching graph being A:b-C:a-B:c:

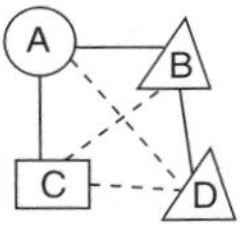

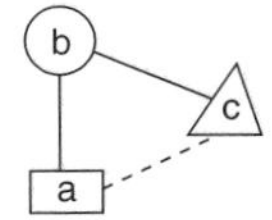

subjective contour: An edge perceived by humans in an image due to Gestalt completion, particularly when no image evidence is present.

In this example, the triangle that appears to float above the black discs is bounded partially by a subjective contour.

subpixel edge detection: Estimation of the location of an image edge by subpixel interpolation of the gradient operator response, to give a position more accurately than an integer pixel value.

subpixel interpolation: A class of techniques that essentially interpolate the position of local maxima in images to positions at a resolution smaller than integer pixel coordinates. Examples include subpixel edge detection and interest point detection. A rule of thumb is 0.1 pixel accuracy is often possible. If the input is an image $z(x, y)$ containing the response of some kernel to a source image, a typical approach might be as follows.

1. Identify a local maximum where $z(x, y) \geq z(a, b)$ where $(a, b) \in neighborhood(x, y)$.
2. Fit the quadratic surface $z = ai^2 + bij + cj^2 + di + ej + f$ to the set of samples $(i, j, z(x+i, y+j))$ in a neighborhood about (x, y).
3. Compute the position of the local maximum of the quadratic surface

$$\begin{pmatrix} i \\ j \end{pmatrix} = -\begin{pmatrix} 2a & b \\ b & 2c \end{pmatrix}^{-1} \begin{pmatrix} d \\ e \end{pmatrix}$$

4. If $-\frac{1}{2} < \{i, j\} < \frac{1}{2}$ then report a maximum at subpixel location $(x+i, y+j)$.

Similar strategies apply when computing the subpixel location of edges.

subsampling: Reducing the size of an image by producing a new image whose pixel values are more widely sampling the original image (*e.g.*, every third pixel). Interpolation can produce more accurate samples To avoid aliasing, spatial frequencies higher than the Nyquist limit of the coarse grid should be removed by low pass filtering the image. Also known as downsampling.

subspace learning: A subspace method where the subspace is learned from a number of observations.

subspace method: A general term describing methods that convert a vector space into a lower dimensional subspace, *e.g.*, projecting a set of N dimensional vectors onto their first two principal components to produce a 2D subspace.

subtractive color: The way in which color appears due to the attenuation/absorption of frequencies of light by materials (*e.g.*, we perceive that something is red it is because it is attenuating/absorbing all wavelengths other than those corresponding to red). See also additive color.

superellipse: A class of 2D curves, including the ellipses and Lamé curves as special cases. The general form of the superellipse is

$$\left(\frac{x}{a}\right)^{\alpha} + \left(\frac{y}{b}\right)^{\beta} = 1$$

although several alternative forms exist. Fitting superellipses to data is difficult due to the strongly nonlinear dependence of the shape on the parameters α and β.

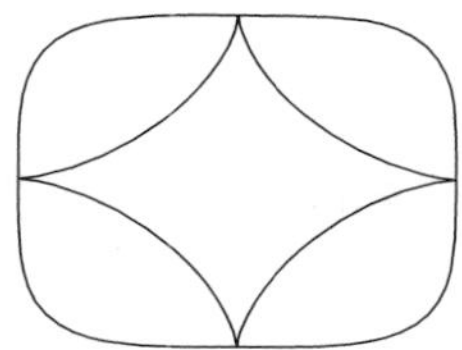

The above shows examples of two superellipses. The convex superellipse has $\alpha = \beta = 3$, the concave example has $\alpha = \beta = \frac{1}{2}$.

supergrid: An image representation that is larger than the original image and represents explicitly both the image points and the crack edges between them.

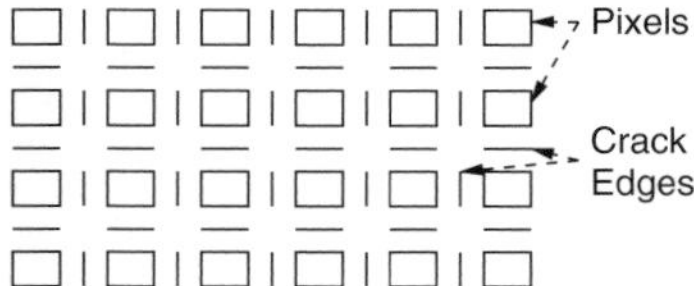

superpixel: A superpixel is a pixel in a high resolution image. An anti-aliasing computer graphics technique produces lower resolution image data by a weighted sum of the superpixels.

superquadric: A 3D generalization of the superellipse, the solution set of

$$\left(\frac{x}{a}\right)^{\alpha} + \left(\frac{y}{b}\right)^{\beta} + \left(\frac{z}{c}\right)^{\gamma} = 1$$

As with superellipses, fitting to 3D data is non-trivial, although some success has been achieved. Two examples of superquadrics, both with $\gamma = 2$:

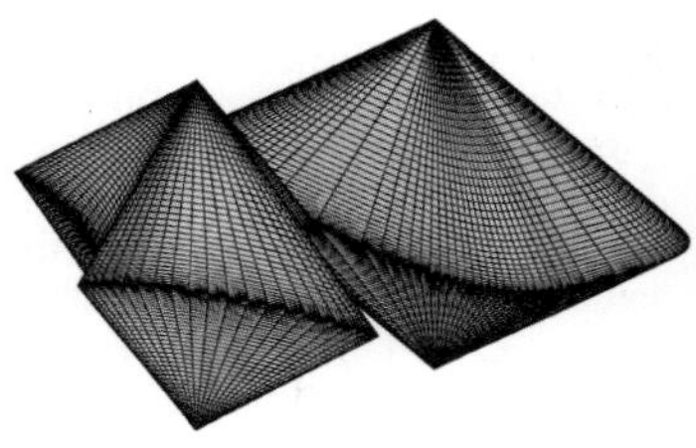

The convex superquadric has $\alpha = \beta = 3$, the concave example has $\alpha = \beta = \frac{1}{2}$.

superresolution: Generation of a high-resolution image from a collection of low-resolution images of the same object taken from different viewpoints. The key to successful superresolution is in the accurate estimation of the registration between viewpoints.

supervised classification: See classification.

supervised learning: A method for training a neural network where the network is presented (in a training phase) with a series of patterns and their correct classification. See also unsupervised learning.

support vector machine: A statistical classifier assigning labels l to points $\vec{x}$ in $\mathbb{R}^n$. The support vector machine has two defining characteristics. Firstly, the classifier places the decision surface that separates points $\vec{x}_i$ and $\vec{x}_j$ that have different labels $l_i \neq l_j$ in such a way as to maximize the margin between them. Roughly speaking, the decision surface is as far as possible from any $\vec{x}$. Secondly, the classifier operates not on the raw feature vectors $\vec{x}$, but on high dimensional projections $\vec{f}(\vec{x}) : \mathbb{R}^n \mapsto \mathbb{R}^N$, $N > n$. However, because the classifier only ever requires dot products such as $\vec{f}(\vec{x}) \cdot \vec{f}(\vec{y})$, we never form $\vec{f}$ explicitly, but specify instead the kernel function $K(\vec{x}, \vec{y}) = \vec{f}(\vec{x}) \cdot \vec{f}(\vec{y})$. Wherever the dot product between high-dimensional vectors is required, the kernel function is used instead.

support vector regression: A range of techniques for function estimation that attempts to determine a function to model data while ensuring that the function does not deviate from any data sample by more than a certain amount. See also support vector machine.

surface: A surface in general parlance is a 2D shape that is located in 3D. Mathematically, it is a 2D subset of $\mathbb{R}^3$ that is almost everywhere locally topologically equivalent to the open unit ball in $\mathbb{R}^2$. This means that a cloud of points is not a surface, but the surface may have cusps or boundaries. A *parameterization* of the surface is a function from $\mathbb{R}^2$ to $\mathbb{R}^3$ that defines the 3D surface point $\vec{x}(u, v)$ as a function of 2D parameters (u, v). Restricting (u, v) to subsets of $\mathbb{R}^2$ yields a subset of the surface. The surface is the set S of points on it, defined over a *domain* D:

$$S = \{\vec{x}(u, v) | (u, v) \in D \subset \mathbb{R}^2\}$$

surface area: Given a parametric surface $S = \{\vec{x}(u, v) | (u, v) \in D \subset \mathbb{R}^2\}$, with unit tangent vectors $\vec{x}_u(u, v)$ and $\vec{x}_v(u, v)$, the area of the surface is

$$\int_S |\vec{x}_u(u, v) \times \vec{x}_v(u, v)| \mathrm{d}u\mathrm{d}v$$

surface boundary representation: A method of defining surface models in computer graphics. It defines a 3D object as a collection of surfaces with boundaries. The model topology states which surfaces are connected, and which boundaries are shared between patches.

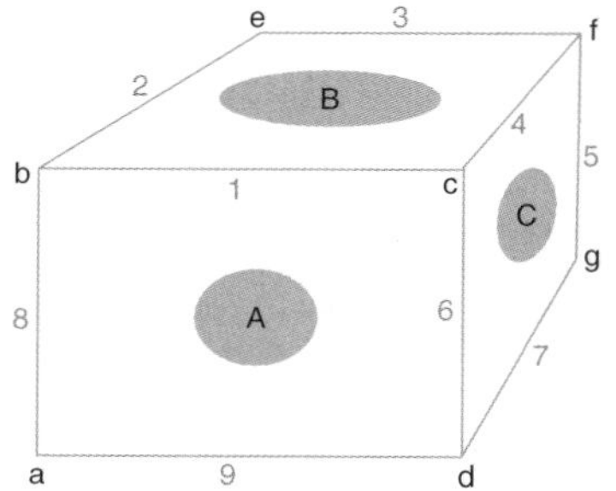

The B-rep model of these three faces comprises: 1) the faces A,B,C along with the parameters of their 3D surfaces; the edges 1–9 with 3D curve descriptions; and vertices a–g; 2) connectivities of these entities, for example face B is bounded by curves 1–4, curve 1 is bounded by vertices b and c.

surface class: Koenderink's classification of local surface shape into classes based on two functions of the principal curvatures:

- The *shape index* $S = \frac{2}{\pi}\tan^{-1}\frac{\kappa_1+\kappa_2}{\kappa_1-\kappa_2}$
- The *curvedness* $C = \sqrt{\frac{1}{2}(\kappa_1+\kappa_2)}$

where κ_1 and κ_2 are the principal curvatures. The surface classes are planar ($C = 0$), hyperboloid ($|S| < \frac{3}{8}$) or ellipsoid ($|S| > \frac{5}{8}$) and cylinder ($\frac{3}{8} < |S| < \frac{5}{8}$), subdivided into concave ($S < 0$) and convex ($S > 0$). Alternative classification systems exist based on the mean and Gaussian curvature or the principal curvatures. The former distinguishes more classes of hyperboloid surfaces.

surface continuity: Mathematically, surface continuity is defined at a single point parameterized by (u, v) on the surface $\{\vec{x}(u, v) | (u, v) \in D \subset \mathbb{R}^2\}$. The surface is continuous *at that point* if infinitesimal motions in any direction away from (u, v) can never cause a sudden change in the value of $\vec{x}$. The surface is *everywhere continuous*, or just *continuous* if it is continuous at all points in D.

surface curvature: Surface curvature measures the *shape* of a 3D surface (the characteristics of the surface that are constant if the surface is rotated or translated in 3D space). The shape is specified by the surface's principal curvatures at each point. To compute the principal curvatures, we need two pieces of machinery, called the *first* and *second fundamental forms*. In the differential geometry of surfaces, the first fundamental form encapsulates arc-length of curves in a surface. If the surface is defined in *parametric* form by a smooth function $\vec{x}(u, v)$, the surface's tangent vectors at (u, v) are given by the partial derivatives $\vec{x}_u(u, v)$ and $\vec{x}_v(u, v)$. From these, we define the dot products $E(u, v) = \vec{x}_u \cdot \vec{x}_u$, $F(u, v) = \vec{x}_u \cdot \vec{x}_v$, $G(u, v) = \vec{x}_v \cdot \vec{x}_v$. Then arclength along a curve in the surface is given by the first fundamental form $ds^2 = Edu^2 + 2Fdudv + Gdv^2$. The *matrix* of the first fundamental form is the 2×2 matrix

$$\mathbf{I} = \begin{pmatrix} E & F \\ F & G \end{pmatrix}$$

The second fundamental form encapsulates the curvature information. The second partial derivatives are $\vec{x}_{uu}(u, v)$ etc. The surface normal at (u, v) is the unit vector $\vec{n}(u, v)$ along $\vec{x}_u \times \vec{x}_v$. Then the matrix of the second fundamental form at (u, v) is the 2×2 matrix

$$\mathbf{II} = \begin{pmatrix} \vec{x}_{uu} \cdot \vec{n} & \vec{x}_{uv} \cdot \vec{n} \\ \vec{x}_{vu} \cdot \vec{n} & \vec{x}_{vv} \cdot \vec{n} \end{pmatrix}.$$

If $\vec{d} = (du, dv)^\top$ is a direction in the tangent space (so its 3D direction is $\vec{t}(\vec{d}) = du\vec{x}_u + dv\vec{x}_v$), then the *normal curvature* in the direction $\vec{d}$ is given by $\kappa(\vec{d}) = \frac{\vec{d}^\top \mathbf{II} \vec{d}}{\vec{d}^\top \mathbf{I} \vec{d}}$. The minima and maxima of κ as $\vec{d}$ varies at a point (u, v) are the principal curvatures at the point, given by the generalized eigenvalues of $\mathbf{II}\vec{z} = \kappa \mathbf{I}\vec{z}$, *i.e.*, the solutions to the quadratic equation in κ given by $\det(\mathbf{II} - \kappa \mathbf{I}) = 0$.

surface discontinuity: A discontinuity is a point at which the surface, or its normal vector, is not continuous. These are often fold edges, where the surface normal has a large change in direction. See also surface continuity.

surface fitting: A family of parametric surfaces $\vec{x}_\theta(u, v)$ is parameterized by a vector of parameters θ. For example, the family of 3D spheres is parameterized by four parameters: three for the center, one for the radius. Given a set of n sampled data points $\{\vec{p}_1, .., \vec{p}_n\}$, the task of surface fitting is to find the parameters θ of the surface that best fits the given data. Common interpretations of "best fit" include finding the surface for which the sum of Euclidean distances from the points to the surface is smallest, or that maximize the probability that the data points could be noisy samples from the surface. General techniques include least squares fitting or nonlinear optimization over the surface parameters.

surface interpolation: Generating a continuous surface from sparse data such as 3D points. For example, given a set of n sampled data points $S = \{\vec{p}_1, .., \vec{p}_n\}$, one might wish to generate other points in $\mathbb{R}^3$ that lie on a smooth surface that passes through all the points in S. Techniques include radial basis functions, splines, natural neighbor interpolation.

surface matching: Identifying corresponding points on two 3D surfaces, often as a precursor to surface registration.

surface mesh: A surface boundary representation in which the faces are typically planar and the edges are straight lines. Such representations are often associated with efficient data structures (*e.g.*, winged edge, quad edge) that allow fast computation of various geometric and topological properties. Hardware acceleration of polygon rendering is a feature of many computers.

surface normal: The direction perpendicular to a surface. For a parametric surface $\vec{x}(u, v)$, the normal is the unit vector parallel to $\frac{\partial \vec{x}}{\partial u} \times \frac{\partial \vec{x}}{\partial v}$. For an implicit surface $F(\vec{x}) = 0$, the normal is the unit vector parallel to $\nabla F = [\frac{\partial F}{\partial x}, \frac{\partial F}{\partial y}, \frac{\partial F}{\partial z}]$. The figure shows the surface normal as defined by the small neighborhood at the point X:

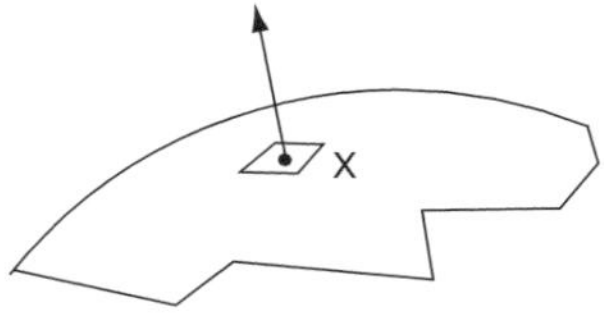

surface orientation: The convention that decides whether the surface normal or its negation points outside the space bounded by the surface.

surface patch: A surface whose domain is finite.

surface reconstruction: The problem of building a surface mesh or B-rep model from unorganized point data.

surface reflectance: A description of the manner in which a surface interacts with light. See reflectance.

surface roughness characterization: An inspection application where estimates of the roughness of a surface are made, *e.g.*, when inspecting spray-painted surfaces.

surface segmentation: Division of a surface into simpler patches. Given a surface defined over a domain D, determine a partition $D = \{D_{1..n}\}$ on which some goodness criteria are well satisfied. For example, it might be required that the maximal distance of a point of each D_i from the best-fit quadric surface is below a threshold. See also range data segmentation.

surface shape classification: The use of curvature information of a surface to classify each point on the surface as locally ellipsoidal, hyperbolic, cylindrical or planar. See also surface class. For example, given a parametric surface $\vec{x}(u, v)$, the classification function $c(u, v)$ is a mapping from the domain of (u, v) to a set of discrete class labels.

surface shape discontinuity: A discontinuity in the value of a surface shape classification over a surface. For example, a discontinuity in the classification function $c(u, v)$. Another example occurs at the fold edge at point X:

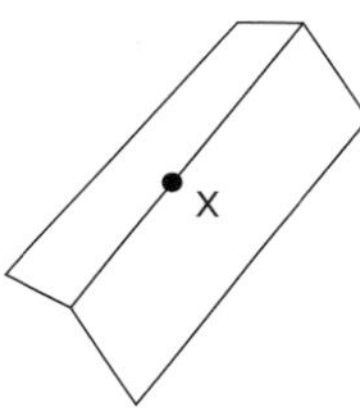

surface tracking: Identification of the same surface through the frames of a video sequence.

surface triangulation: See surface mesh.

surveillance: An application area of vision concerned with the monitoring of activities in a scene. Typically this will involve at least background modeling and human motion analysis.

SUSAN corner finder: A popular interest point detector developed by Smith and Brady. Combines the smoothing and central difference stages of a derivative-based operator into a single center–surround comparison.

SVD: See singular value decomposition.

SVM: See support vector machine.

swept object representation: A volumetric representation scheme in which 3D objects are formed by sweeping a 2D cross section along an axis or trajectory. A brick can be formed by sweeping a rectangle. Some schemes, like the geon or generalized cylinder representation, allow changes to the size of the cross section and curved trajectories. A cone is defined here by sweeping a circle along a straight axis with a linearly decreasing radius:

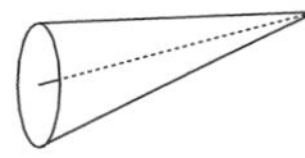

symbolic: Inference or computation expressed in terms of a set of *symbols* rather than a signal. Where a digital signal is a discrete representation of a continuous function, symbols are inherently discrete. For example, an image (signal) is converted to a list of the names of people who appear in it (symbols).

symbolic object representation: Representation of an object by lists of symbolic terms like "plane", "quadric", "corner", or "face", etc., rather than the points or pixels of the shape itself. The representation may include the shape and position of the objects, too.

symmetric axis transform (SAT): A transformation that locates all points on the skeleton of a region by identifying those points that are the locus of centers of bitangent circles. See also medial axis skeletonization. In the following example the medial axis is derived from a binary segmentation of a moving subject.

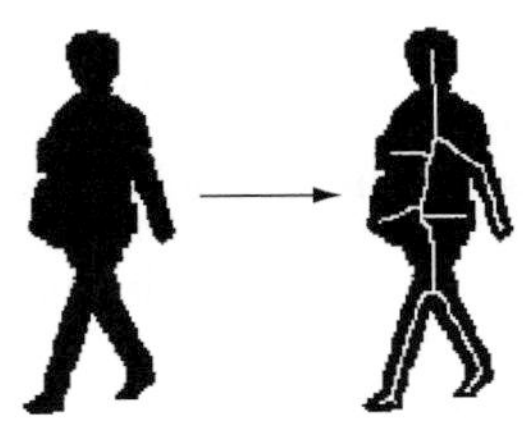

symmetry: A shape that remains invariant under at least one non-identity transformation from some pre-specified *transformation group* is symmetric. For example the set of points comprising an ellipse is the same after the ellipse is subjected to the Euclidean transformation of rotation by 180° about its center. The image of the outline of a surface of revolution under perspective projection is invariant under a certain homography, so the silhouette exhibits a projective symmetry. Affine symmetries are sometimes known as skew symmetries and symmetry induced by reflection about a line are called bilateral symmetries.

symmetry detection: A class of algorithms that search for symmetry in imaged curves, surfaces and point sets.

symmetry line: The axis of a bilateral symmetry. The solid line rectangle has two dashed line symmetry lines:

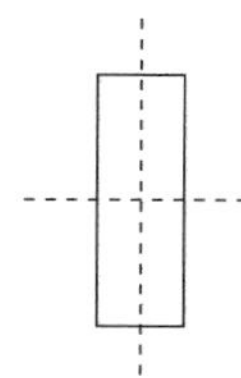

symmetry plane: The axis of a bilateral symmetry in 3D. The dashed lines show three symmetry planes of this cube:

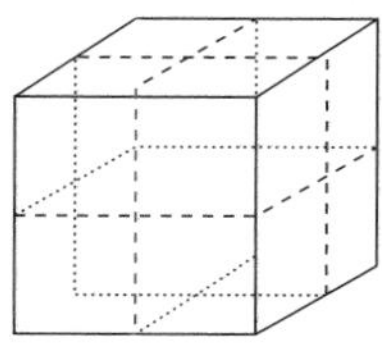

sync pulse: Abbreviation of "synchronization pulse". Any electrical signal that allows two or more electronic devices to share a common time frame. Commonly used to synchronize the capture instants of two cameras in a stereo image capture system.

syntactic pattern recognition: Object identification by converting an image of the object into a sequence or array of symbols and using grammar parsing techniques to match the sequence of symbols to grammar rules in a database.

syntactic texture description: Description of texture in terms of grammars of local shapes or image patches and transformation rules. Good for modeling synthetic artificial textures.

synthetic aperture radar (SAR): An imaging device that transmits long-wavelength (in comparison to visible light) radio waves from airborne or space platforms and builds a 2D image of the intensities of the returned reflections. Clouds are transparent at these (centimeter) wavelengths, and the active transmission means that images may be taken at night. The images are captured as a sequence of low-resolution ("small aperture") 1D slices as the platform translates across the target area, with a final high-resolution ("synthetic [large] aperture") image recoverable via a Fourier transform after all slices have been captured. The time-of-flight of the returned signal determines the distance from the transmitter and therefore, assuming a planar (or known geometry) surface, the pixel location in the cross-path direction.

systolic array: A class of parallel computer in which processors are arranged in a directed graph. The processors synchronously receive data from one set of neighbors (*e.g.*, North and West in a rectangular array), perform a computation, and transmit the computed quantity to another set of neighbors (*e.g.*, South and East).

T

tabu search: A heuristic search technique that seeks to avoid cycles by forbidding or penalizing moves taking the search to previously visited solution spaces (hence "tabu").

tangent angle function: Given a curve $(x(t), y(t))$, the function $\theta(t) = \tan^{-1} \frac{\dot{y}(t)}{\dot{x}(t)}$.

tangent plane: The plane passing through a point on a surface that is perpendicular to the surface normal.

tangential distortion (lens): A particular lens aberration created, among others, by lens decentering, usually modeled only in high-accuracy calibration systems.

target image: The image resulting from an image processing operation.

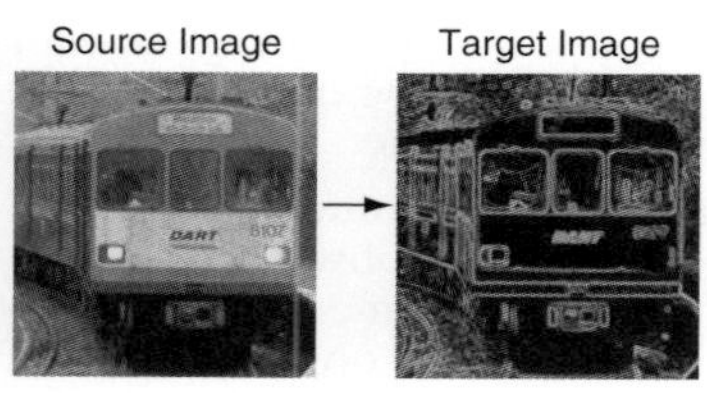

target recognition: See automatic target recognition.

task parallelism: Parallel processing achieved by the concurrent execution of relatively large subsets of a computer program. A large subset might be defined as one whose run time is of the order of tens of milliseconds. The parallel tasks need not be identical, *e.g.*, from a binary image, one task may compute a moment while another computes the perimeter.

tee junction: An intersection between line segments (possibly representing edges) where a straight line meets and terminates somewhere along another line segment. See also blocks world. Tee junctions

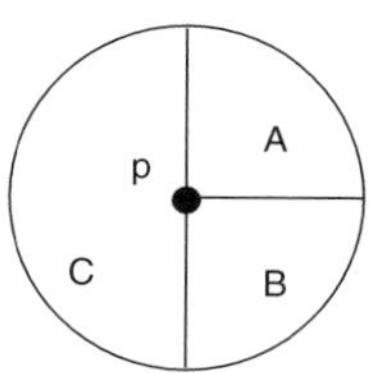

can give useful depth-ordering cues. Here we can hypothesize that surface c lies in front of the surfaces A and B, given the tee junction at p.

telecentric optics: A lens system arranged such that moving the image plane along the optical axis does not change the magnification or image position of imaged world points. One embodiment is to place an aperture in front of the lens so that when an object is imaged off the focal plane of the lens, the center of the (blurred) object is the ray through the center of the aperture, rather than the center of the lens. Placing the aperture at the lens's front focal plane will ensure these rays are parallel after the lens.

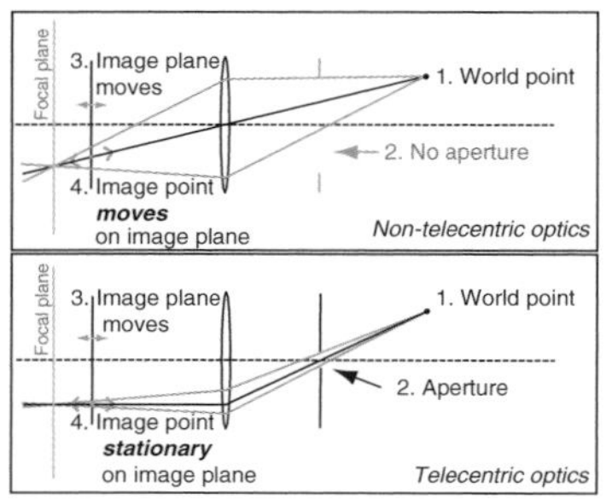

telepresence: Interaction with objects at a location remote from the user via vision or robotic devices. Examples include slaving of remote cameras to the motion of a head-mounted display worn by the user, transmission of audio from the remote location, use of local controls to operate remote machinery, and haptic (*i.e.*, touch) feedback from the remote to the local environment.

template matching: A strategy for location of an object in an image. The *template*, a 2D image of the object, is compared with all windows of the same size as the template in the image to be searched. Windows where the difference with the template (as computed by, *e.g.*, normalized cross-correlation or sum of squared differences (SSD)) is within a threshold are reported as instances of the object. Interesting as a brute-force matching strategy. To obtain invariance to scale, rotation or other transformations, the template must be subjected explicitly to the transformations.

temporal averaging: Any procedure for noise reduction in which a signal that is known to be static over time is sampled at different times and the results averaged.

temporal representation: A model representation that encodes the dynamics of how an object's shape or position can vary over time.

temporal resolution: The frequency of observations with respect to time (*e.g.*, one per second) as opposed to the spatial resolution.

temporal stereo: 1) Stereo achieved through movement of the camera rather than use of two separate cameras. 2) Integration of multiple stereo views of a dynamic scene to produce a better estimate of each view.

temporal tracking: See tracking.

tensor product surface: A parametric representation for a curved surface commonly used in computer modeling and graphics applications, The surface shape is defined by the product of two polynomial (usually cubic) curves in the independent surface coordinates.

terrain analysis: Analysis and interpretation of data representing the shape of the planet surface. Typical data structures are digital elevation maps or triangulated irregular networks (TINs).

tessellated viewsphere: A division of the viewsphere into distinct subsets of (approximately) equal areas. Often used as a data structure for representation of functions of the form $f(\vec{n})$ where $\vec{n}$ is a unit normal vector in $\mathbb{R}^3$. Typically constructed by subdivision of the viewsphere into a polygon mesh such as an icosahedron:

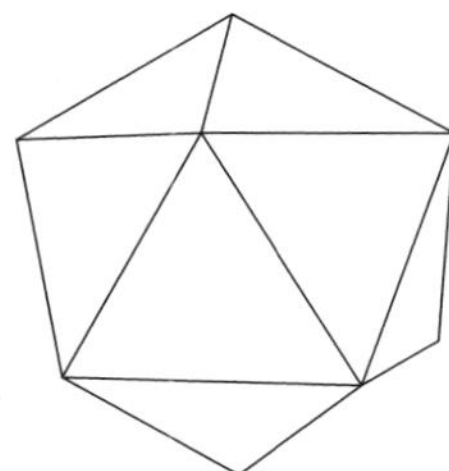

test set: The set used to verify a classifier or other algorithm. The test set contains only examples not included in the training set.

tetrahedral spatial decomposition: A method of decomposing 3D space into packed tetrahedrons instead of the more commonly used rectangular voxel decomposition. A tetrahedral decomposition allows a recursive subdivision of a tetrahedron into eight smaller tetrahedra. This figure illustrates the decomposition with one of the eight smaller volumes shaded:

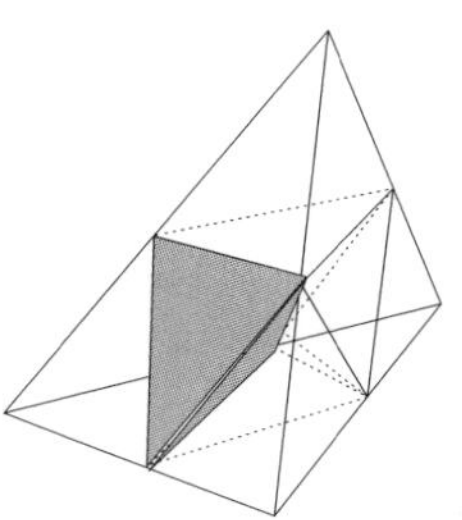

texel: See texture element.

texon: See texture element.

textel: See texture element.

texton: Julesz' 1981 definition of the units in which texture might be perceived. In the texton-based view, a texture is a regular assembly of textons.

texture: The phenomenon by which uniformity is perceived in regular (etymologically, "woven") patterns of (possibly irregular) elements. In computer vision, texture usually refers to the patterns in the appearance or reflectance on a surface. The texture may be regular, *i.e.*, satisfy some texture grammar or may be statistical texture *i.e.*, the distribution of pixel values may vary over the image. Texture could also refer to variations in the local shape on a surface, *e.g.*, its degree of roughness. See also shape texture.

texture-based image retrieval: Content-based image retrieval that uses texture as its classification criterion.

texture boundary: The boundary between adjacent regions in texture segmentation. The boundary perceived by humans between two regions of different textures. This figure shows the boundary between three regions of different color and shape texture:

texture classification: Assignment of an image (or a window of an image) to one of a set of texture classes. The texture classes are typically defined by presentation of training images representing each class by a human. The basis of texture segmentation.

texture descriptor: A vector valued function computed on an image subwindow that is designed to produce similar outputs when applied to different subwindows of the same texture. The size of the image subwindow controls the scale of the detector. If the response at a pixel position (x, y) is computed as the maximum over several scales, an additional scale output $s(x, y)$ is available. See also texture primitive.

texture direction: The texture gradient or a 90° rotation thereof.

texture element (texel): A small geometric pattern that is repeated frequently on some surface resulting in a texture.

texture energy measure: A single-valued texture descriptor with strong response in textured regions. A texture descriptor may be formed by combining the results of several texture energy measures into a vector.

texture enhancement: A procedure analogous to edge-preserving smoothing in which texture boundaries rather than edges are to be preserved.

texture field grouping: See texture segmentation.

texture field segmentation: See texture segmentation.

texture grammar: Grammar used to describe textures as

instances of simpler patterns with a given spatial relationship (including other textures defined previously in this way). A sentence from this grammar would be a syntactic texture description.

texture gradient: The gradient of a single scalar output $\nabla s(x, y)$ of a texture descriptor. A common example is the scale output, for homogeneous texture, whose texture gradient can be used to compute the foreshortening direction.

texture mapping: In computer graphics, rendering of a polygonal surface where the surface color at each output screen pixel is obtained by interpolating values from an image, called the texture map. The source image pixel location is computed using correspondences between the polygon's vertex coordinates and texture coordinates on the texture map.

texture matching: Matching of regions based on texture descriptions.

texture model: The theoretical basis for a class of texture descriptor. For example, autocorrelation of linear filter responses, statistical texture descriptions, or syntactic texture descriptions.

texture orientation: See texture gradient.

texture primitive: A basic unit of texture (*e.g.*, a small pattern that is repeated) as used in syntactic texture descriptions.

texture recognition: See texture classification.

texture region extraction: See texture field segmentation.

texture representation: See texture model.

texture segmentation: Segmentation of an image into patches of coherent texture. This figure shows a region segmented into three regions based on color and shape texture:

texture synthesis: The generation of synthetic images of textured scenes. More particularly, the generation of images that appear perceptually to share the texture of a set of training examples of a texture.

Theil–Sen estimator: A method for robust estimation of curve fits. A family of curves is parameterized by parameters $a_{1..p}$, and is to be fit to data $\vec{x}_{1..n}$. If q is the smallest number of points that uniquely define $a_{1..p}$, then the Theil–Sen estimate of the optimal parameters $\hat{a}_{1..p}$ are the parameters that have the median error measure of all the q-point estimates. For example, for line fitting, the number of parameters (slope and intercept, say) is $p = 2$, and the number of points required to give a fit is also $q = 2$. Thus the Theil–Sen estimate of the slope gives the median error of the $\binom{n}{q}$ two-point slope estimates. The Theil–Sen estimator is not statistically efficient, nor does it have a particularly high breakdown point, in contrast to such estimators as RANSAC and least median of squares.

thermal noise: In CCD cameras, additional electrons released by thermal vibration in the substrate that are counted with those released by incident photons. Thus, the gray scale values are corrupted by an additive Poisson noise process.

thickening operator: Thickening is a morphological operation that is used to grow selected regions of foreground pixels in binary images, somewhat like dilation or closing. It has several applications, including determining the approximate convex hull of a shape, and determining the skeleton by zone of influence. Thickening is normally only applied to binary images, and it produces another binary image as output. This is an example of thickening six times in the horizontal direction:

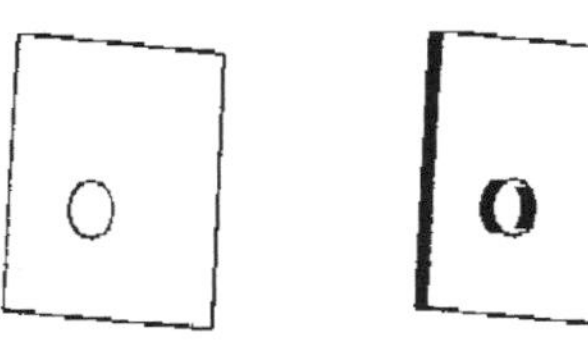

thin plate model: A model of surface smoothness used in the variational approach. The internal energy (or bending energy) of a thin plate represented as a parametric surface $(x, y, f(x, y))$ is given by $f_{xx}^2 + 2f_{xy}^2 + f_{yy}^2$.

thinning operator: Thinning is a morphological operation that is used to remove selected foreground pixels from binary images, somewhat like erosion or opening. It can be used for several applications, but is particularly useful for skeletonization and to tidy up the output of edge detectors by reducing all lines to single pixel thickness. Thinning is normally only applied to binary images and produces another binary image as output. This is a diagram illustrating the thinning of a region:

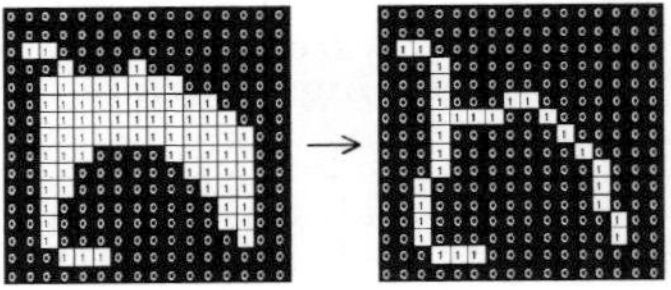

three view geometry: See trinocular stereo.

three dimensional imaging: Any of a class of techniques that obtain three dimensional information using imaging equipment. 1) 3D volumetric imaging: obtaining measurements of scene properties at all points in a 3D space, including the insides of objects. This is used for inspection, but more commonly for medical imaging. Techniques include nuclear magnetic resonance, computerized tomography, positron emission tomography and single photon emission computed tomography. 2) 3D surface imaging: obtaining surface information embedded in a 3D space. Active techniques generally include a source of structured light (or other electromagnetic or sonar radiation), and a sensor such as a camera or microphone. Either triangulation or time of flight computations allow the distance from the sensor system to be computed. Common technologies include laser scanning, texture projection systems, and moiré fringe methods. Passive 3D imaging depends only on external (and hence unstructured) illumination sources. Examples of such systems are stereo reconstruction and shape from focus techniques.

threshold selection: The automatic choice of threshold values for conversion of a scalar signal (such as a gray scale image) to binary. Often (*e.g.*, Otsu's 1979 method) proceeds by analysis of the histogram of the sample values. Different assumptions about the underlying distributions yield different strategies.

thresholding: Quantization into two values. For example, conversion of a scalar signal (such as a gray scale image) to binary. This figure shows an input image and its threshold output:

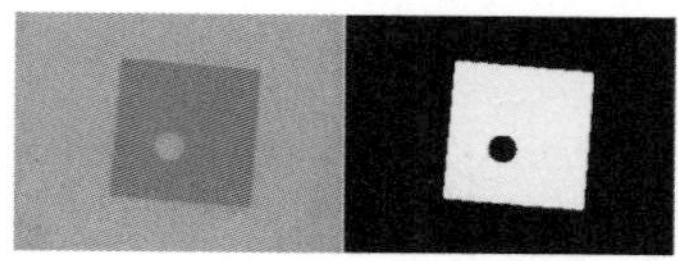

thresholding with hysteresis: Thresholding of a time-varying scalar signal where the threshold value is a function of previous signal and threshold values. For example a thermostat control based on temperature receives a signal $s(t)$, and generates an output signal $b(t)$ of the form

$$b(t) = \begin{cases} s(t) > \tau_{\text{cold}} & \text{if}\, b(t-1) = 0 \\ s(t) < \tau_{\text{hot}} & \text{if}\, b(t-1) = 1 \end{cases}$$

where the value at time t depends on the previous decision $b(t-1)$. In computer vision, often associated with the edge following stage of the Canny edge detector.

TIFF: Tagged Image File Format.

tilt: The *tilt direction* of a 3D surface patch as observed in a 2D image is parallel to the projection of the 3D surface normal into the image. If the 3D surface is represented as a depth map $z(x,y)$ in image coordinates, then the tilt direction at (x,y) is the unit vector parallel to $(\frac{\partial z}{\partial x}, \frac{\partial z}{\partial y})$. The *tilt angle* may be defined as $\tan^{-1}(\frac{\partial z}{\partial y} / \frac{\partial z}{\partial x})$.

time derivative: A technique for computing how an image

sequence changes over time. Typically used as part of shape from motion.

time to collision: See time to contact.

time to contact: From a sequence of images $I(t)$, computation of the value of t at which, assuming constant motion, an image object will intersect the plane parallel to the image plane that contains the camera center. It can be computed even in the absence of metric information about the imaging system—*i.e.*, in an uncalibrated setting.

time to impact: See time to contact.

time-of-flight range sensor: A sensor that computes distance to target points by emitting electromagnetic (or other) radiation and measuring the time between emitting the pulse and observing the reflection of the pulse.

tolerance band algorithm: An algorithm for incremental segmentation of a curve into straight line elements. Assume that the current straight line segment defines two parallel boundaries of a tolerance zone at a pre-selected distance from the line segment. When a new curve point leaves the tolerance zone the current line segment is ended and a new segment is started. A tolerance band is illustrated.

TOLERANCE ZONE

EXIT POINT

tolerance interval: An interval within which a stated proportion of some population will lie.

Tomasi–Kanade factorization: A maximum-likelihood solution to structure and motion recovery in the situation where points in a static scene are observed by affine cameras and the observed (x, y) positions are corrupted by Gaussian noise. The method depends on the observation that if m points are observed over n views, the $2n \times m$ measurement matrix containing all the observations (after certain transformations have been performed) is of rank 3. The closest rank-3 approximation of the matrix is reliably obtained via singular value decomposition, after which the 3D points and camera positions are easily extracted, up to an affine ambiguity.

tomography: A technique for the reconstruction of a 3D volumetric dataset based on a number of 2D slices. The most common examples occur in medical imaging (*e.g.*, nuclear magnetic resonance, positron emission tomography).

top-down: A reasoning approach that searches for evidence for high-level hypotheses in the data. For example, a hypothesize-and-test algorithm might have a strategy for making good guesses as to the position of circles in an image and then compare the hypothesized circles to edges in the image, choosing those that have good support. Another example is a human body recognizer that employs body part recognizers (*e.g.*, heads, legs, torso) that, in turn, either directly use image data or recognize even smaller subparts.

top hat operator: A morphological operator used to remove structures from images. The top-hat filtering of image I by structuring element S is the difference $I - \text{open}(I, S)$, where $\text{open}(I, S)$ is the morphological opening of I by S.

topological representation: Any representation that encodes connectedness of elements. For example, in a surface boundary representation comprising faces, edges and vertices, the *topology* of the representation is the list of face–edge and edge–vertex connections, which is independent of the *geometry* (or spatial positions and sizes) of the representation. In this case, the fundamental relation is "bounded by", so a face is bounded-by one or more edges, and an edge is bounded-by zero or more vertices.

topology: 1) Properties of point sets (such as surfaces) that are unchanged by continuous reparameterizations (homeomorphisms) of space. 2) The connectedness of objects in discrete geometry (see topological representation). One speaks of the topology of a network, meaning the set of connections within the network, or equivalently the set of neighborhood relationships that describe the network.

torsion: A concept in the differential geometry of curves formally representing the intuitive notion of the local twisting of a 3D curve as you move along the curve. The torsion $\tau(t)$ of a 3D space curve $\vec{c}(t)$ is the scalar

$$-\vec{n}(t) \cdot \frac{\mathrm{d}}{\mathrm{d}t}\vec{b}(t) = \frac{\left[\dot{\vec{c}}(t), \ddot{\vec{c}}(t), \dddot{\vec{c}}(t)\right]}{\|\ddot{\vec{c}}(t)\|^2}$$

where $\vec{n}(t)$ is the curve normal and $\vec{b}(t)$ the binormal. The notation $[\vec{x}, \vec{y}, \vec{z}]$ denotes the scalar triple product $\vec{x} \cdot (\vec{y} \times \vec{z})$.

torus: 1) The volume swept by moving a sphere along a circle in 3D. 2) The surface of such a volume.

total variation: A class of *regularizer* in the variational approach. The total variation regularizer of function $f(\vec{x}) : \mathbb{R}^n \mapsto \mathbb{R}$ is of the form $R(f) = \int_\Omega |\nabla f(\vec{x})| d\vec{x}$ where Ω is (a subset of) the domain of f.

tracking: A means of estimating the parameters of a dynamic system. A dynamic system is characterized by a set of parameters (*e.g.*, feature point positions, target object positions, human joint angles) evolving over time, of which we have *measurements* (*e.g.*, photographs of the human) obtained at successive time instants. The task of tracking is to maintain an estimate of the probability distribution over the model parameters, given the measurements, as well as *a priori* models of how the parameters change over time. Common algorithms for tracking include the Kalman filter and particle filters. Tracking may be viewed as a class of algorithms that operate on sequences of inputs, using assumptions about the coherence of successive inputs to improve performance of the algorithm. Often the task of the algorithm may be cast as estimation of a state vector—a set of parameters such as the joint angles of a human body—at successive time instants t. The state vector $\vec{x}(t)$ is to be estimated using a set of sensors that yield observations, $\vec{z}(t)$, such as the 2D positions of bright spots attached to a human. In the absence of temporal coherence assumptions, $\vec{x}$ must be estimated at each time step solely using the information in $\vec{z}(t)$. With coherence assumptions, the system uses the set of all observations so far $\{\vec{z}(\tau), \tau < t\}$ to compute the estimate at time t. In practice, the estimate of the state is represented as a probability density over all possible values, and the current estimate uses only the previous state estimate $\vec{x}(t-1)$ and the current measurements $\vec{z}(t)$ to estimate $\vec{x}(t)$.

traffic analysis: Analysis of video data of automobile traffic, *e.g.*, to identify number plates, detect accidents, detect congestion, compute throughput, etc.

training set: The set of labeled examples used to learn the parameters of a classifier. In order to build an effective classifier, the training set should be representative of the examples that will be encountered in the eventual domain of application.

trajectory: The path that a moving point makes over time. It could also be the path that a whole object takes if less precision of usage is desired.

trajectory estimation: Determination of the 3D trajectory of an object observed in a set of 2D images.

transformation: A mapping of data in one space (such as an image) into another space (*e.g.*, all image processing operations are transformations).

translation: A transformation of Euclidean space that can be represented in the form $\vec{x} \mapsto T(\vec{x}) \equiv \vec{x} \mapsto \vec{x} + \vec{t}$. In projective space, a transformation that leaves the plane at infinity pointwise invariant.

translation invariant: A property that keeps the same value even if the data, scene or the image from which the data comes is translated. The distance between two points is translation invariant.

translucency: The transmission of light through a diffusing interface such as frosted glass. Light entering a translucent material has multiple possible exit directions.

transmittance: Transmittance is the ratio of the ("outgoing") power transmitted by a transparent object to the incident ("incoming") power.

transparency: The property of a surface to be traversed by radiation (*e.g.*, by visible light), so that objects on the other side can be seen. A non-transparent surface is called opaque.

tree classifiers: A classifier that applies a sequence of binary tests to input points $\vec{x}$ in order to determine the label l of the class to which it belongs.

tree search method: A class of algorithms to optimize a function defined on tuples of values taken from a finite set. The tree describes the set of all such tuples, and the order in which tuples are explored is defined by the particular search algorithm. Examples are depth-first, breadth-first, A^* and best-first search. Applications include the interpretation tree.

triangulated models: See surface mesh.

triangulation: See Delaunay triangulation, surface triangulation, stereo triangulation, structured light triangulation.

trifocal tensor: The geometric entity that relates the images of 3D points observed in three perspective 2D views. Algebraically represented as a $3 \times 3 \times 3$ array of values T_i^{jk}. If a single 3D point projects to x, x', x'' in the first, second, and third views respectively, it must obey the nine equations

$$x^i (x'^j \epsilon_{j\alpha r})(x''^k \epsilon_{k\beta s}) T_i^{\alpha\beta} = 0_{rs}$$

for r and s varying from 1 to 3. In the above, ϵ is the epsilon-tensor for which

$$\epsilon_{ijk} = \begin{cases} 1 & ijk \text{ an even permutation of } 123 \\ 0 & \text{two of } i, j, k \text{ equal} \\ -1 & ijk \text{ an odd permutation of } 123 \end{cases}$$

As this equation is linear in the elements of T, it can be used to estimate them given enough 2D point correspondences x, x', x''. As not all $3 \times 3 \times 3$ arrays represent realizable camera configurations, estimation must also incorporate several nonlinear constraints on the tensor elements.

trilinear constraint: The geometric constraint on three views of a point (*i.e.*, the intersection of three epipolar lines). This is similar to the epipolar constraint which is applied in the two view scenario.

trilinear tensor: Another name for the trifocal tensor.

trilinearity: An equation in a set of three variables in which holding two of the variables fixed yields a linear equation in the remaining one. For example $xyz = 0$ is trilinear in x, y and z, while $x^2 = y$ is not, as holding y fixed yields a quadratic in x.

trinocular stereo: A multiview stereo process that uses three cameras.

tristimulus theory of color perception: The human visual system has three types of cones, with three different spectral response curves, so that the perception of any incident light is represented as three intensities, roughly corresponding to long (maximum about 558– 580 nm), medium (531–545 nm) and short (410–450 nm) wavelengths.

tristimulus values: The relative amounts of the three primary colors that need to be combined to match a given color.

true negative: A hypothesis which is false that has been corrected rejected.

true positive: A hypothesis which is true that has been corrected accepted.

truncated median filter: An approximation to mode filtering when image neighborhoods are small. The filter sharpens blurred image edges as well as reducing noise}. The algorithm truncates the local distribution on the mean side of the median and then recomputes the median of the new distribution. The algorithm can iterate and, under normal circumstances, converges approximately to the mode even if the observed distribution has very few samples with no obvious peak.

tube camera: See tube sensor.

tube sensor: A tube sensor converts light to a video signal using a vacuum tube with a photoconductive window. Once the only type of light sensor, the tube camera is now largely superseded by the CCD, but remains useful for some high dynamic range imaging. The image orthicon tube or "immy" is remembered in the name of the US Academy of Television Arts and Sciences Emmy awards.

twist: A 3D rotation representation component that specifies a rotation about the vector defined by the azimuth and elevation. This figure shows the pitch rotation direction:

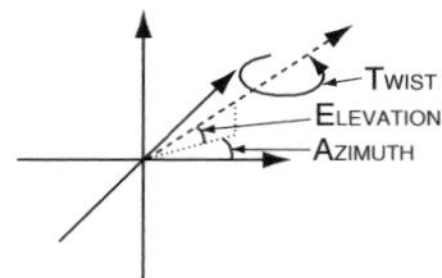

twisted cubic: The curve $(1, t, t^2, t^3)$ in projective 3-space, or any projective transformation thereof. The general form is thus

$$\begin{pmatrix} x_1 \\ x_2 \\ x_3 \\ x_4 \end{pmatrix} = \begin{pmatrix} a_{11} & a_{12} & a_{13} & a_{14} \\ a_{21} & a_{22} & a_{23} & a_{24} \\ a_{31} & a_{32} & a_{33} & a_{34} \\ a_{41} & a_{42} & a_{43} & a_{44} \end{pmatrix} \begin{pmatrix} 1 \\ t \\ t^2 \\ t^3 \end{pmatrix}$$

The projection of a twisted cubic into a 2D image is a rational cubic spline.

two view geometry: See binocular stereo.

type I error: A hypothesis which is true that has been rejected.

type II error: A hypothesis which is false that has been accepted.

U

ultrasonic imaging: Creation of images by the transmission and recording of reflected ultrasonic pulses. A phased array of transmitters emits a set of pulses, and then records the returning pulse intensities. By varying the relative timings of the pulses, the returned intensities can be made to correspond to locations in space, allowing measurements to be taken from within ultrasonic-transparent materials (including the human body, excluding air and bone).

ultrasound sequence registration: Registration of overlapping ultrasound images.

ultraviolet: Description of electromagnetic radiation with wavelengths between about 300–420 nm (near ultraviolet) and 40–300 nm (far ultraviolet). The short wavelengths make it useful for fine-scale examination of surfaces. Ordinary glass is opaque to UV radiation, quartz glass is transparent. Often used to excite fluorescent materials.

umbilic: A point on a surface where the curvature is the same in every direction. Every point on a sphere is an umbilic point.

umbra: The completely dark area of a shadow caused by a particular light source (*i.e.*, where no light falls from the light source).

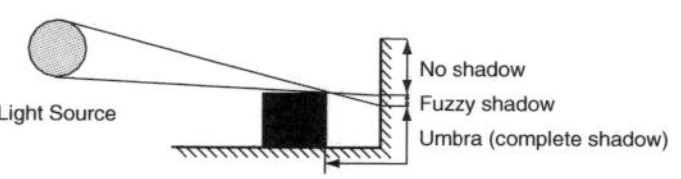

uncalibrated approach: See uncalibrated vision.

uncalibrated stereo: Stereo reconstruction performed without precalibration of the cameras. Particularly, given a pair of images taken by unknown cameras, the fundamental matrix is computed from point correspondences,

Dictionary of Computer Vision and Image Processing R.B. Fisher, K. Dawson-Howe, A. Fitzgibbon, C. Robertson and E. Trucco

after which the images may be rectified and conventional calibrated stereo may proceed. The results of uncalibrated stereo are 3D points in a projective coordinate system, rather than the Euclidean coordinate system that a calibrated setup admits.

uncalibrated vision: The class of vision techniques that require no quantitative information about the camera used in capturing the images on which they operate. For example, techniques that can be applied on archive footage. In particular, applied to geometric problems such as stereo reconstruction that traditionally required that the images be from a camera system upon which calibration measurements had been previously made. Uncalibrated approaches include those, such as uncalibrated stereo where the traditional calibration step is replaced by procedures that can use image features directly, and others, such as time-to-contact computations that can be expressed in ways that factor out the calibration parameters. In general, uncalibrated systems will have degrees of freedom that cannot be measured, such as overall scale, or projective ambiguity.

uncertainty representation: A strategy for representation of the probability density of a variable as used in a vision algorithm. In a similar manner, an interval can be used to represent a range of possible values.

under-segmented: Describing the output of a segmentation algorithm. Given an image where a desired segmentation result is known, the algorithm under-segments if regions output by the algorithm are generally the union of many desired regions. This image should be segmented into three regions but it was under-segmented into two regions:

uniform distribution: A probability distribution in which a variable can take any value in the given range with equal probability.

uniform illumination: An idealized configuration in which the arrangement of lighting within a scene is such that each point receives the same amount of light energy. In computer vision, sometimes uniform illumination has a different meaning: that each point in an image of the scene (or a part thereof such as the background) has similar imaged intensity.

uniform noise: Additive corruption of a sampled signal. If the signal's samples are s_i then the corrupted signal is $\tilde{s}_i = s_i + n_i$

where the n_i are uniformly randomly drawn from a specified range $[\alpha, \beta]$.

uniqueness stereo constraint: When performing stereo matching or stereo reconstruction, matching can be simplified by assuming that points in one image correspond to only one point in other images. This is generally true, except at object boundaries and other places where pixels are not completely opaque.

unit ball: An N dimensional sphere of radius one.

unit quaternion: A quaternion is a 4-vector $\vec{q} \in \mathbb{R}^4$. Quaternions of unit length can be used to parameterize 3D rotation matrices. Given a quaternion with components (q_0, q_1, q_2, q_3) the corresponding rotation matrix **R** is (letting $S = q_0^2 - q_1^2 - q_2^2 - q_3^2$):

$$\begin{bmatrix} S+2q_1^2 & 2q_1q_2+2q_0q_3 & 2q_3q_1-2q_0q_2 \\ 2q_1q_2-2q_0q_3 & S+2q_2^2 & 2q_2q_3+2q_0q_1 \\ 2q_3q_1+2q_0q_2 & 2q_2q_3-2q_0q_1 & S+2q_3^2 \end{bmatrix}$$

The identity rotation is given by the quaternion $(1, 0, 0, 0)$. The rotation axis is the unit vector parallel to (q_1, q_2, q_3).

unit vector: A vector of length one.

unitary transform: A reversible transformation (*e.g.*, the discrete Fourier transform). **U** is a unitary matrix where $\mathbf{U}^*\mathbf{U} = \mathbf{I}$, $\mathbf{U}^*$ is the adjoint matrix and **I** is the identity matrix.

unrectified: When a stereo camera pair has not been rectified.

unsharp operator: An image enhancement operator that sharpens edges by adding a high pass filtered version of an image to itself. The high pass filter is implemented by subtracting a smoothed version of the image yielding

$$I_{\text{unsharp}} = I + \alpha(I - I_{\text{smooth}})$$

This shows an input image and its unsharped output:

unsupervised classification: See clustering.

unsupervised learning: A method for training a neural network or other classifier where the network learns to recognize patterns (in a training set) automatically. See also supervised learning.

updating eigenspace: Algorithms for the incremental updating of eigenspace representations. These algorithms facilitate approaches such as active learning.

USB camera: A camera conforming to the USB (Universal Serial Bus) standard.

V

validation: Testing whether or not some hypothesis is true. See also hypothesize and verify.

valley: A dark elongated object in a gray scale image, so called because it corresponds to a valley in the image viewed as a 3D surface or elevation map of intensity *versus* image position.

valley detection: An image processing operator (see also bar detector) that enhances linear features rather than light-to-dark edges. See also valley.

value quantization: When a continuous number is encoded as a finite number of integer values. A common example of this occurs when a voltage or current is encoded as integers in the range 0–255.

vanishing line: The 2D line that is the image of the intersection of a 3D plane with the plane at infinity. The horizon line in an image is the image of the intersection of the ground plane with the plane at infinity, just as a pair of railway lines meeting in a vanishing point is the intersection of two parallel lines and the plane at infinity. This sketch shows the vanishing line for the ground plane with a road and railroad:

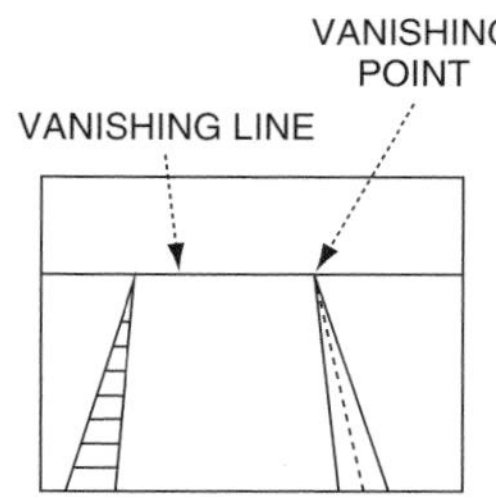

vanishing point: The *image of* the point at infinity where two parallel 3D lines meet. A pair of parallel 3D lines are represented as $\vec{a}+\lambda\vec{n}$ and $\vec{b}+\lambda\vec{n}$. The vanishing point is the image of the 3D direction $\binom{\vec{n}}{0}$. This sketch shows the vanishing points for a road and railroad:

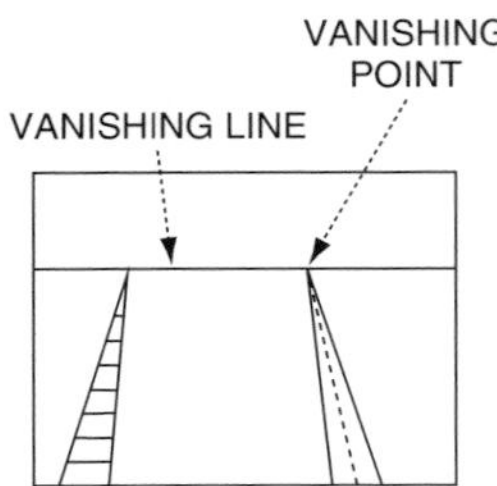

variable focus: A camera system with a lens system that allows zoom to be changed under user or program control. An image sequence in which focal length varies through the sequence.

variational approach: Signal processing expressed as a problem of variational calculus. The input signal is a function $I(t)$ on the interval $t \in [-1, 1]$. The processed signal is a function P defined on the same interval, that minimizes an *energy functional* $E(P)$ of the form

$$E(P) = \int_{-1}^{1} f(P(t), \dot{P}(t), I(t))\mathrm{d}t.$$

The calculus of variations shows that the minimizing P is the solution to the associated Euler–Lagrange equation

$$\frac{\partial f}{\partial P} = \frac{\mathrm{d}}{\mathrm{d}t}\frac{\partial f}{\partial \dot{P}}$$

In computer vision, the functional is often of the form

$$E = \int \mathrm{truth}(P, I) + \lambda\ \mathrm{beauty}(P)$$

where the "truth" term measures fidelity to the data and the "beauty" term is a *regularizer*. These can be seen in a specific example: smoothing. In the conventional approach, smoothing might be considered the result of an *algorithm*: convolve the image with a Gaussian kernel. In the variational approach, the smoothed signal P is the signal that best trades off smoothness, measured as the square of the second derivative $\int (\dot{P}(t))^2 \mathrm{d}t$, and fidelity to the data, measured as the squared difference between the input and the output $\int (P(t) - I(t))^2 \mathrm{d}t$, with the balance chosen by a parameter λ:

$$E(P) = \int (P(t) - I(t))^2 + \lambda(\dot{P}(t))^2 \mathrm{d}t$$

variational method: See variational approach.

variational problems: See variational approach.

vector field: A multi-valued function $\vec{f} : \mathbb{R}^n \mapsto \mathbb{R}^m$. For example, the 2D-to-2D function $\vec{f}(x, y) = (y, \sin \pi x)$ illustrated below. An RGB image

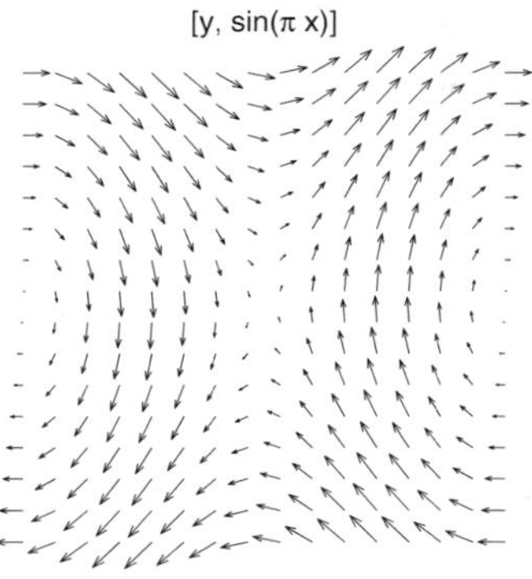

$I(x, y) = (r(x, y), g(x, y), b(x, y))$ is an example of a 2D-to-3D vector field.

vector quantization: Representation of a set of vectors by associating each possible vector with one of a small set of "codebook" vectors. For example, each pixel in an RGB image has 256^3 possible values, but one might expect that a particular image uses only a small subset of these values. If a 256-element colormap is computed, and each RGB value is represented by the nearest RGB vector in the colormap, the RGB space has been *quantized* into 256 elements.

vehicle detection: An example of the object recognition problem where the task is to identify vehicles in video imagery.

vehicle license/number plate analysis: When a visual system locates the license plate in a video image and then recognizes the characters.

vehicle tracking: An example of the tracking problem applied to images of vehicles.

velocity: Rate of change of position. Generally, for a curve $\vec{x}(t) \in \mathbb{R}^n$ the velocity is the n-vector $\frac{\partial \vec{x}(t)}{\partial t}$

velocity field: The image velocity of each point in an image. See also optical flow field.

velocity moment: A moment that integrates information about region velocity as well as position and shape distribution. Let $m_{pq}^{[i]}$ be the pq^{th} central moment of a binary region in the ith image. Then the Cartesian velocity moments are defined as $v_{pqrs} = \sum_i (\bar{x}_i - \bar{x}_{i-1})^r (\bar{y}_i - \bar{y}_{i-1})^s m_{pq}^{[i]}$, where $(\bar{x}_i, \bar{y}_i)$ is the center of mass in the ith image.

velocity smoothness constraint: Changes in the magnitude or direction of an image's velocity field occur smoothly.

vergence: 1) The angle between the optical axes in a stereo system, when the two cameras fixate on the same scene point. 2) The difference between the pan angle settings of the two cameras.

vergence maintenance: The action of a control loop which ensures that the optical centers of two cameras—whose positions are under program control—are looking at the same scene point.

verification: In the context of object recognition, a class of algorithms aiming to test the validity of various hypotheses (models) explaining the data. Back projection is such a technique, typically used with geometric models. See also object verification.

vertex: A point at the end of a line (edge) segment. Often vertices are common to two or more line segments.

video: 1) Generic term for a set of images taken at successive instants with small time intervals between them. 2) The analogue signal emitted by a video camera. Each frame of video corresponds to about 40 ms of electrical signal that encodes the start of each scan line, the image encoding of each video scan line, and synchronization information. 3) A video recording.

video annotation: The association of symbolic objects, such as text descriptions or index terms with frames of video.

video camera: A camera that records a sequence of images over time.

video coding: The conversion of video to a digital bitstream. The source may be analogue or digital. Generally, coding also compresses or reduces the bitrate of the video data.

video compression: Video coding with the specific aim of reducing the number of bits required to represent a video sequence. Examples include MPEG, H.263, and DIVX.

video indexing: Video annotation with the aim of allowing queries of the form "At what frame did event x occur?" or "Does object x appear?".

video rate system: A real time system that operates at the frame rate of the ambient video standard. Typically 25 or 30 frames per second, 50 or 60 fields per second.

video restoration: Application of image restoration to video, often making use of the temporal coherence of video, or correcting for video-specific degradations.

video segmentation: Application of segmentation to video, 1) with the requirement that the segmentation exhibit the temporal coherence in the original footage and 2) to split the video sequence into different groups of consecutive frames, *e.g.*, when there is a change of scene.

video sequence: See video.

video transmission format: A description of the precise form of the analog video signal coding conventions in terms of duration of components such as number of lines, number of pixels, front porch, sync and blanking.

vidicon: A type of tube camera, successor of the image orthicon tube.

view based object recognition: Recognition of 3D objects using multiple 2D images of the objects rather than a 3D model.

view combination: A class of techniques combining prototype views linearly to form appearance models. See also appearance model, eigenspace based recognition, prototype, representation.

view volume: The infinite volume of 3D space bounded by the camera's center of projection and the edges of the viewable area on the image plane. The volume might also be bounded near and far by other planes because of focusing and depth of field constraints. This figure illustrates the view volume:

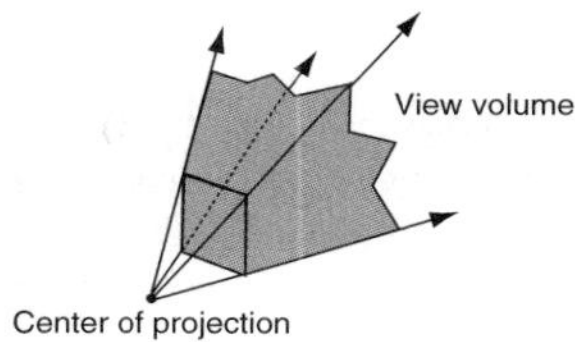

viewer centered representation: A representation of the 3D world that an observer (*e.g.*, robot or human) maintains. In the viewer centered version, the global coordinate system is maintained on the observer, and the representation of the world changes as the observer moves. Compare object centered representation.

viewing space: The set of all possible locations from which an object or scene could be viewed. Typically these locations are grouped to give a set of typical or characteristic views of the object. If orthographic projection is used, then the full 3D space of views can be simplified to a viewsphere.

viewpoint: The position and orientation of the camera when an image was captured. The viewpoint may be expressed in absolute coordinates or relative to some arbitrary coordinate system, in which case the relative position of the camera and the scene (or other cameras) is the relevant quantity.

viewpoint consistency constraint: Lowe's term for the concept that a 3D model matched to a set of 2D line segments must admit at least one 3D camera position that projects the 3D model to those lines. Essentially, the 3D and 2D data must allow pose estimation.

viewpoint dependent representations: See viewer centered representation.

viewpoint planning: Deciding where an active vision system will look next, in order to maximize the likelihood of achieving some preset goal. A common example is computing the location of a range sensor in several successive positions in order to gain a complete 3D model of a target object. After n pictures have been captured, the viewpoint planning problem is to choose the position of picture $n+1$ in order to maximize the amount of new data acquired, while ensuring that the new position will allow the new data to be registered to the n existing images.

viewsphere: The set of camera positions from which an object can be observed. If the camera is orthographic, the viewsphere is parameterized by the 2D set of points on the 3D unit sphere. At the camera position corresponding to a particular point on the viewsphere, all images of the object due to camera rotation are related by a 2D-to-2D image transformation, *i.e.,* no parallax effects occur. See aspect graph. The placement of a camera on the viewsphere is illustrated here:

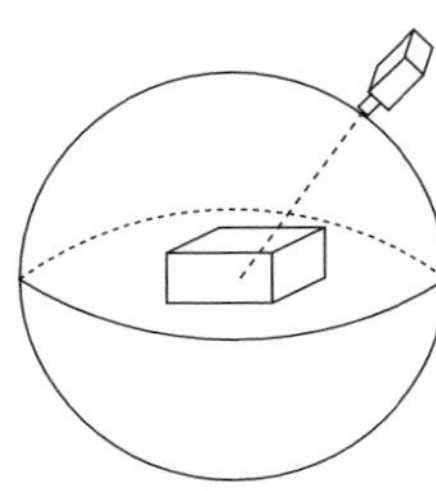

vignetting: Darkening of the corners of an image relative to the image center, which is related to the degree to which the points are off the optical axis.

virtual bronchoscopy: Creation of virtual views of the pulmonary system based on *e.g.,* magnetic resonance imaging as a replacement for endoscope imaging.

virtual endoscopy: Simulation of a traditional endoscopy procedure using virtual reality representation of physiological data such as that obtained by an X-ray CAT-scan or magnetic resonance imaging.

virtual reality: The use of computer graphics and other interaction tools to confer on a user the sensation of being in, and interacting with, an alternative environment. This includes simulation of visual, aural, and haptic cues. Common ways in which the visual environment is displayed are: rendering a 3D model of the world into a head-mounted display whose viewpoint is tracked in 3D so that the user's head movements generate images corresponding to their viewpoint; placing the user in a computer augmented virtual environment (CAVE), where as much as possible of the user's field of view can be manipulated by the controlling computer.

virtual view: Visualization of a model from a particular viewpoint.

viscous model: A deformable model based on the concept of a viscous fluid (*i.e.,* a fluid with a relatively high resistance to flow).

visible light: Description of electromagnetic radiation with wavelengths between about 400 nm (blue) and 700 nm (red), corresponding to the range to which the rods and cones of the human eye are sensitive.

visibility: Whether or not a particular feature is visible from a camera position.

visibility class: The set of points where exactly the same portion of an object or scene is visible. For example, when viewing the corner of a cube, an observer can move about in about one-eighth of the full viewing space before entering a new visibility class.

visibility locus: All camera positions from which a particular feature is visible.

VISIONS: The early scene understanding system of Hanson and Riseman.

visual attention: The process by which low level feature detection directs high level scene analysis and object recognition strategies. In humans, the results of the process are evident in the pattern of fixations and saccades in normal observation of the world.

visual cortex: A part of the brain dedicated to the processing of visual information.

visual hull: A space carving method for approximating shape from multiple images. The method finds the silhouette contours of a given object in each image. The region of space defined by each camera and the associated image contour imposes a constraint on the shape of the target object. The visual hull is the intersection of all such constraints. As more views are taken, the approximation becomes better. See the shaded areas in

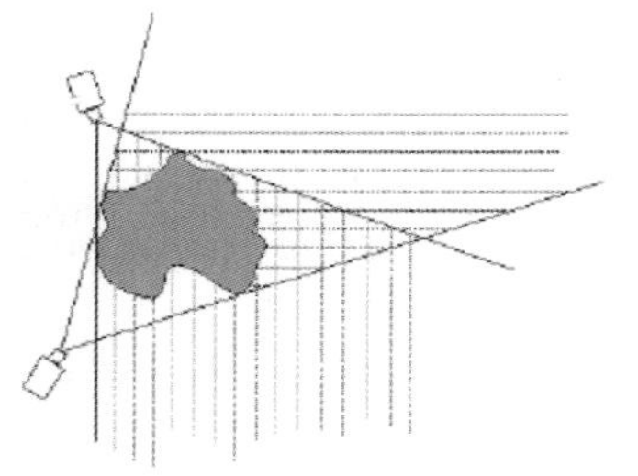

visual illusion: The perception of a scene, object or motion not corresponding to the world actually causing the image or sequence. Illusions are caused, in general, by the combination of special arrangements of the visual stimuli, viewing conditions, and responses of the human vision system. Well-known examples include the Ames room (two persons are seen as having very different heights in a seemingly normal room) and the Ponzo illusion:

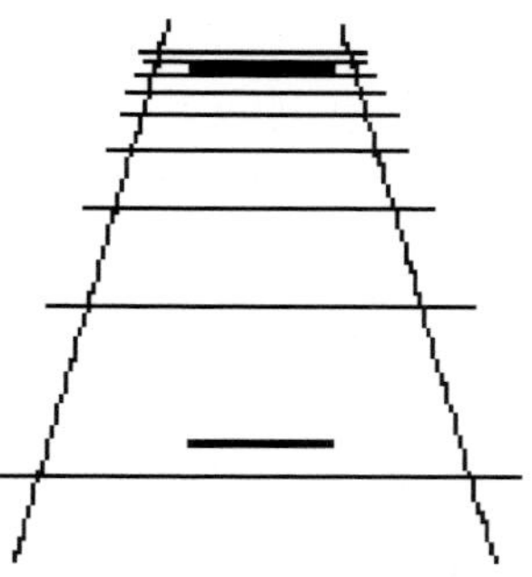

Here two equal segments seem to be different lengths as interpreted as 3D projections). The well-known ambiguous figure–background drawings of the Gestalt psychology (see Gestalt), like the famous chalice–faces pattern, are a related subject.

visual industrial inspection: The use of computer vision techniques in order to effect quality control or to control processes in an industrial setting.

visual inspection: A general term for analyzing a visual image to inspect some item, such as might be used for quality control on a production line.

visual learning: The problem of learning visual models from sets of images (examples), or in general knowledge that can be used to carry out vision tasks. An area of the vast field of automated learning. Important applications employing visual learning include face recognition and image database indexing. See also unsupervised learning, supervised learning.

visual localization: The problem of estimating the location of a target in space given one or more images of it. Solutions differ according to several factors including the number of input images (one, as in model based pose estimation, multiple discrete images, as in stereo vision, or video sequences, as in motion analysis), the *a priori* knowledge assumed (*i.e.*, camera calibration available or not, full perspective or simplified projection model, geometric model of target available or not).

visual navigation: The problem of navigating (steering) a robot through an environment using visual data, typically video sequences. It is possible, under diverse assumptions, to determine the distance from obstacles, the time-to-contact, and the shape and identity of the objects in view. Both video and range sensors have been used, including acoustic sensors (see sonar). See also visual servoing, visual localization.

visual routine: Ullman's 1984 term for a subcomponent of a visual system that performs a specific task, analogous to a behavior in robotics.

visual salience: A (numerical) assessment of the degree to which pixels or areas of a scene attract visual attention. The principle of Gestalt organization.

visual search: The task of searching an image for a particular prespecified object. Often used as a an experimental tool in psychophysics.

visual servoing: Robot control via motions that make *the image of*, *e.g.*, the robot end effector coincide with the image of the target position. Typically, the system has little or no *a priori* knowledge of the camera locations, their relation to the robot, or the robot kinematics. These parameters are learned as the robot moves. Visual servoing allows the calibration to change during robot operation. Such systems can adapt well to anomalous conditions, such as an arm bending under a load or motor slippage, or where calibration may not provide sufficient precision to allow the desired actions to be reliably produced purely from the modeled robot kinematics and dynamics. Because only image measurements are available, the inverse kinematic problem may be harder than in conventional servoing.

visual surveillance: Surveillance dependent only on the use of electromagnetic sensors.

volume: 1) A region of 3D space. A subset of $\mathbb{R}^3$. A (possibly infinite) 3D point set. 2) The space bounded by a closed surface.

volume detection: The detection of volume-shaped entities in 3D data sets, such as might be produced by an nuclear magnetic resonance scanner.

volume matching: Identification of correspondence between objects or subsets of objects defined using a volumetric representation.

volume skeletons: The skeletons of 3D point sets, by extension of the definitions for 2D curves or regions.

volumetric image: A voxmap or 3D array of points where each entry typically represents some measure of material density or other property in 3D space. Common examples include computerized tomography and nuclear magnetic resonance data.

volumetric reconstruction: Any of several techniques that derive a volumetric representation from image data. Examples include X-ray tomography, space carving and visual hull computation.

volumetric representation: A data structure by means of which a subset of 3D space is represented digitally. Examples include voxmap, octree and the space bounded by surface representations.

Voronoi cell: See Voronoi diagram.

Voronoi diagram: Given n points $\vec{x}_{1..n}$, the Voronoi diagram of the point set is a partition of space into n regions or cells $R_{1..n}$. Every point p in cell R_i is closer to point $\vec{x}_i$ than to any other $\vec{x}$. The hyperplanes separating the Voronoi regions are the perpendicular bisectors of the edges in the Delaunay triangulation of the point set. The Voronoi diagram of these four points are the four cells surrounding them:

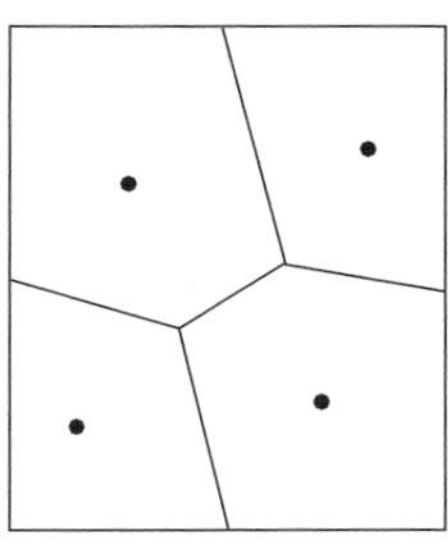

voxel: From "volume element" by analogy with "pixel". A region of 3D space, named by analogy with pixel. Usually voxels are axis-aligned rectangular solids or cubes. A component of the voxmap representation for 3D volumes. A voxel, like a pixel, may have associated attributes such as color, occupancy, or the density of some measurement at that point.

voxmap: A volumetric represen tation that describes a 3D volume by dividing space into a regular grid of voxels, arranged as a 3D array $v(i, j, k)$. For a boolean voxmap, cell (i, j, k) intersects the volume iff $v(i, j, k) = 1$. The advantages of the representation are that it can represent arbitrarily complex topologies and is fast to look up. The major disadvantage is the large memory usage, addressed by the octree representation.

VRML: Virtual Reality Markup Language. A means of defining 3D geometric models intended for Internet delivery.

W

walkthrough: A classification of the infinite number of paths between two points into one of nine equivalence classes of the eight relative directions between the points plus the ninth having no movement. Point B is in equivalence class 2 relative to A:

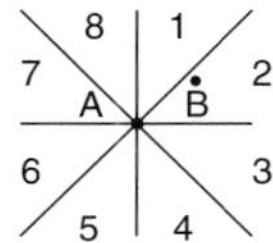

Walsh function: The Walsh functions of order n are a particular set of square waves $W(n, k) : [0, 2^n) \mapsto \{-1, 1\}$ for k from 1 to 2^n. They are orthogonal, and the product of Walsh functions is a Walsh function. The square waves transition only at integer lattice points so each function can be specified by the vector of values it takes on the points $\{\frac{1}{2}, 1\frac{1}{2}, \ldots, 2^n - \frac{1}{2}\}$. The collection of these values for a given order n is the Hadamard matrix $\mathbf{H}_{2^n}$ of order 2^n. The two functions of order 1 are the rows of

$$H_2 = \begin{bmatrix} 1 & 1 \\ 1 & -1 \end{bmatrix}$$

and the four of order 2 (depicted below) are

$$\mathbf{H}_4 = \begin{bmatrix} \mathbf{H}_2 & \mathbf{H}_2 \\ \mathbf{H}_2 & -\mathbf{H}_2 \end{bmatrix}$$

$$= \begin{bmatrix} 1 & 1 & 1 & 1 \\ 1 & -1 & 1 & -1 \\ 1 & 1 & -1 & -1 \\ 1 & -1 & -1 & 1 \end{bmatrix}$$

In general, the functions of order $n+1$ are generated by the relation

$$\mathbf{H}_{2^{n+1}} = \begin{bmatrix} \mathbf{H}_{2^n} & \mathbf{H}_{2^n} \\ \mathbf{H}_{2^n} & -\mathbf{H}_{2^n} \end{bmatrix}$$

and this recurrence is the basis of the fast Walsh transform. The four Walsh functions of order 2 are:

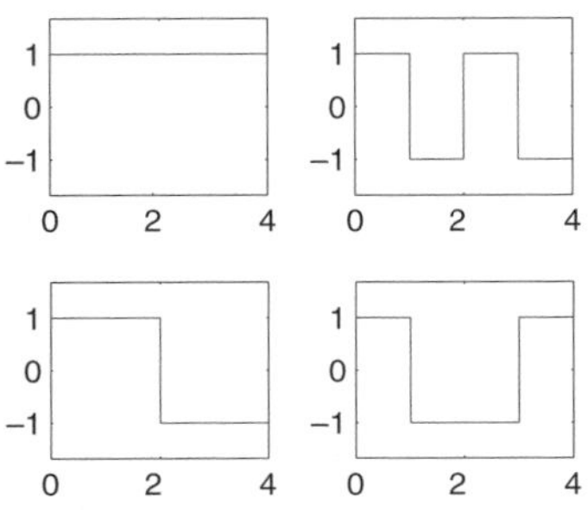

Walsh transform: Expression of a 2^n-element vector v in terms of a basis of order-n Walsh functions; the multiplication by the corresponding Hadamard matrix. The Walsh transform has applications in image coding, logic design and the study of genetic algorithms.

Waltz line labeling: A scheme for the interpretation of line images of polyhedra in blocks-world images. Each image line is labeled to indicate what class of scene edge gave rise to it: concave, convex, occluding, crack or shadow. By including the constraints supplied by junction labeling in a constraint satisfaction problem, Waltz demonstrated that collections of lines whose labels were locally ambiguous could be globally disambiguated. This is a simple example of Waltz line labeling showing concave edges (−), convex edges (+) and occluding edges (>).

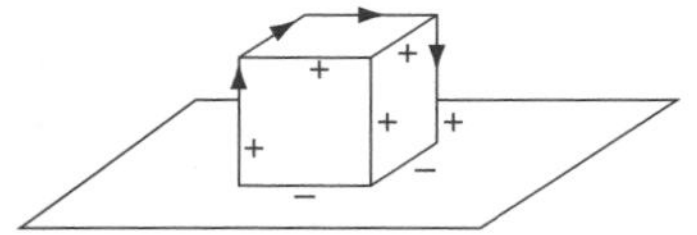

warping: Transformation of an image by reparameterization of the 2D plane. Given an image $I(\vec{x})$ and a 2D-to-2D mapping $w : \vec{x} \mapsto \vec{x}'$, the warped image $W(\vec{x})$ is $I(w(\vec{x}))$. Warping functions w are often designed so that certain control points $\vec{p}_{1..n}$ in the source image are mapped to specified locations $\vec{p}'_{1..n}$ in the destination image. See also image morphing.

The original image $I(\vec{x})$; Warping function represented by arrows joining points $\vec{x}$ to $w^{-1}(\vec{x})$; Warped image $W(\vec{x})$.

watermark: See digital watermarking.

watershed segmentation: Image segmentation by means of the watershed transform. A typical implementation proceeds thus: 1. Detect edges; 2. Compute the distance transform D of the edges; 3. Compute watershed regions in $-D$.

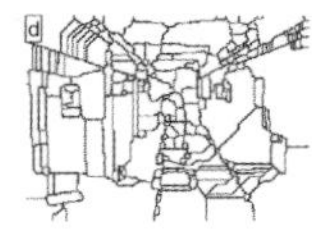

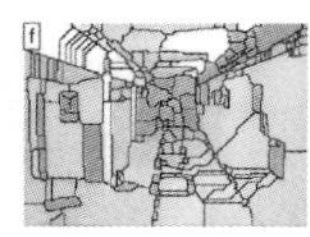

(a) Original image; (b) Canny edges; (c) Distance transform; (d) Region boundaries of watershed transform of (c); (e) Mean color in watershed regions; (f) Regions overlaid on image.

watershed transform: A tool for morphological image segmentation. The watershed transform views the image as an elevation map, with each local minimum in the map given a unique integer label. The watershed *transform* of the image assigns to each *non-minimum* pixel, p, the label of the minimum to which a drop of water would fall if placed at p. Points on "ridges" or watersheds of the elevation map, that could fall into one of two minima are called watershed points and the set of pixels surrounding each minimum that share its label are called watershed regions. Efficient algorithms exist for the computation of the watershed transform. Above is an image with minima superimposed, the same image viewed as a 3D elevation map and the watershed transform of the image, where different minima have different colored regions and watershed pixels are shown in white. One particular watershed is indicated by arrows.

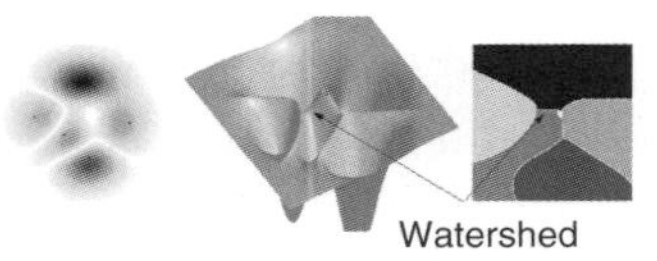

wavelength: The wavelength of a wave is the distance between successive peaks. Denoted λ, it is the wave's speed divided by the frequency. Electromagnetic waves, particularly visible light, are often important in computer vision, with wavelengths of the order of 400–700 nm.

wavelet: A function $\phi(x)$ that has certain properties that mean it can be used to derive a set of basis functions in terms of which other functions can be approximated. Comparing to the Fourier transform basis functions, note that they can be viewed as a set of scalings and translations of $f(x) = sin(\pi x)$, for example $cos(3\pi x) = sin(3\pi x + \frac{\pi}{2}) = f(\frac{6x+1}{2})$. Similarly, a wavelet basis is made from a mother wavelet $\phi(x)$ by translating and scaling: each basis function $\phi_{jk}(x)$ is of the form $\phi_{jk}(x) = \text{const} \cdot \phi(2^{-j}x - k)$. The conditions on ϕ ensure that different basis functions (*i.e.*, with different j and k) are orthonormal. There are several popular choices (*e.g.*, by Haar and Daubechies) for ϕ, that trade off various desirable properties, such as compactness in space and time, and ability to approximate certain classes of functions.

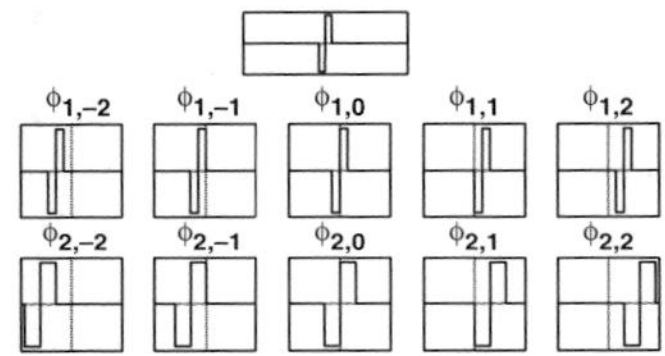

The mother Haar wavelet and some of the derived wavelets $\phi_{j,k}$.

wavelet descriptor: Description of a shape in terms of the coefficients of a wavelet decomposition of the original signal, in a manner similar to Fourier shape descriptors for 2D curves. See also wavelet transform.

wavelet transform: Representation of a signal in terms of a basis of wavelets. Similar to the Fourier transform, but as the wavelet basis is a two-parameter family of functions ϕ_{jk}, the wavelet transform of a d-D signal is an $(d+1)$-D function. However, the number of distinct values needed to represent the transform of a discrete signal of length n is just $O(n)$. The wavelet transform has similar applications to the Fourier transform, but the wavelet basis offers advantages when representing natural signals such as images.

weak perspective: An approximation of viewing geometry between the pinhole or full perspective camera and the orthographic imaging model. The projection of a homogeneous 3D point $\vec{X} = (X, Y, Z, 1)^\top$ is given by the formula

$$\begin{pmatrix} x \\ y \end{pmatrix} = \begin{pmatrix} p_{11} & p_{12} & p_{13} & p_{14} \\ p_{21} & p_{22} & p_{23} & p_{24} \end{pmatrix} \vec{X}$$

for the affine camera, but with the additional constraint that the vectors (p_{11}, p_{12}, p_{13}) and (p_{21}, p_{22}, p_{23}) are scaled rows of a rotation matrix, *i.e.*,

$$p_{11}p_{21} + p_{12}p_{22} + p_{13}p_{23} = 0$$

weakly calibrated stereo: Any two-view stereo algorithm for which the only calibration information needed is the fundamental matrix between the cameras is said to be *weakly calibrated*. In the general, multi-view, case, means the camera calibration is known up to a projective ambiguity. Weakly calibrated systems cannot determine Euclidean properties such as absolute scale but will return results that are projectively equivalent to the Euclidean reconstructions.

Weber's Law: If a difference can be just perceived between two stimuli of values I and $I + \delta I$ then it should be possible to perceive a difference between two stimuli with different values J and $J + \delta J$ where $\frac{\delta I}{I} \leq \frac{\delta J}{J}$.

weighted least squares: A least square error estimation process in which the data elements also have a weight associated. The weights might specify the confidence or quality of the data item. The use of weights can help make the estimation more robust.

weighted walkthrough: A discrete measure of the relative position of two regions. The measure is a histogram of the walkthrough relative positions of every pair of points selected from the two regions.

weld seam tracking: Using visual feedback to control a robot welding device, so it maintains the weld along the desired seam.

white balance: A system of color correction to deal with differing light conditions, in order for white objects to appear white.

white noise: A noise process in which the noise power at all frequencies is equal (as compared to pink noise). When considering spatially distributed noise, white noise means that there is distortion at all spatial frequencies (*i.e.*, large distortions as well as small).

whitening filter: See noise-whitening filter.

wide angle lens: A lens with a field of view greater than about 45°. Wide angle lenses allow more information to be collected in a single image, but often suffer a loss of resolution, particularly at the periphery of the image. Wide angle lenses are also more likely to require correction for nonlinear lens distortion.

wide baseline stereo: The stereo correspondence problem (sense 1) in the particular case when the two images for which correspondence is to be determined are significantly different because the cameras are separated by a long baseline. In particular, a 2D window around a point in one image is expected to look significantly different in the second image due to foreshortening, occlusion, and lighting effects.

wide field-of-view: Where the optics is designed to capture light rays forming large angles (say 60° or more) with the optical axis. See also wide angle lens, panoramic image mosaic, panoramic image stereo, plenoptic function representation.

width function: Given a 2D shape (closed subset of the plane) $S \subset \mathbb{R}^2$, the width function $w(\theta)$ is the width of the shape as a function of orientation. Specifically, the projection $P(\theta) := \{x \cos\theta + y \sin\theta \mid (x, y) \in S\}$, and $w(\theta) := \max P(\theta) - \min P(\theta)$.

Wiener filter: A regularized inverse convolution filter. Given a signal g that is known to be the convolution of an unknown signal f and a known corrupting signal k, it is desired to undo the effect of k and recover f. If (F, G, K) are the respective Fourier transforms of (f, g, k), then $G = F \cdot K$, so the inverse filter can recover $F = G \div K$. In practice, however, G

is corrupted by noise, so that when an element of K is less than the average noise level, the noise is amplified. Wiener's filter combats this tendency by adding an estimate of the noise to the divisor. Because the divisor is complex, a real formulation is as follows:

$$F = \frac{G}{K} = \frac{GK^*}{KK^*} = \frac{GK^*}{|K|^2}$$

and adding the frequency domain noise estimate N, we obtain the Wiener reconstruction of F given G and K:

$$F = \frac{GK^*}{|K|^2 + N}$$

windowing: Looking at a small portion of a signal or image through a "window". For example, given the vector $\vec{x} = \{x_1, \ldots, x_{100}\}$, one might look at the window of 11 values centered around 50, $\{x_{45..55}\}$. Often used in order to restrict some computation such as the Fourier transform to a small part of the image. In general, windowing is described by a *windowing function*, which is multiplied by the signal to give the windowed signal. For example, a signal $f(\vec{x}) : \mathbb{R}^n \mapsto \mathbb{R}$ and windowing function $w(\sigma; \vec{x})$ are given, where σ controls the scale or width of w. Then the windowed signal is

$$f_w(\vec{x}) = f(\vec{x})w(\sigma; \vec{x} - \vec{c})$$

where $\vec{c}$ is the center of the window. The Bartlett $(1 - \frac{|\vec{x}|}{\sigma})$, Hanning $(\frac{1}{2} + \frac{1}{2}\cos\frac{\pi|\vec{x}|}{\sigma})$, and Gaussian $(\exp(-\frac{|\vec{x}|^2}{\sigma^2}))$ windowing functions in 2D are shown here:

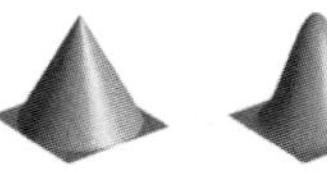
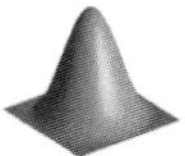
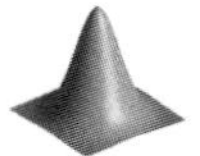

winged edge representation: A graph representation for polyhedra in which the nodes represent vertices, edges and faces. Faces point to bounding edge nodes, that point to vertices, that point back to connecting edges, that point to adjacent faces. The winged edge term comes from the fact that edges have four links that connect to the previous and successor edges around each of the two faces that contain the given edge, as seen here:

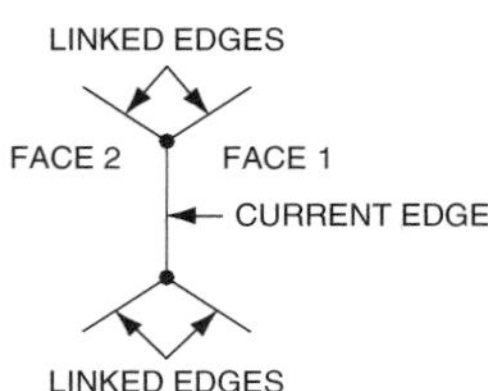

winner-takes-all: A strategy whereby only the best candidate (*e.g.*, algorithm, solution) is chosen, and any other is abandoned. Commonly found in the neural network and learning literature.

wire frame representation: A representation of 3D geometry in terms of vertices and edges linking the vertices. It does not include descriptions of the surface between the edges, and in particular, does not include information for hidden line removal. This is a wire frame model of a cube:

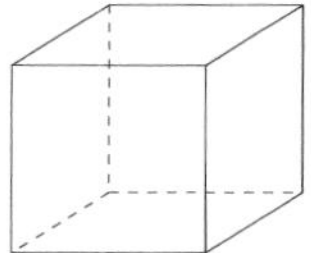

world coordinates: A coordinate system useful for placing objects in a scene. Usually this is a 3D coordinate system with some arbitrarily placed origin (*e.g.*, at a corner of a room). This contrasts with object centered representations, viewer centered representations or camera coordinates. world coordinates

X

X-ray: Electromagnetic radiation of shorter wavelength than ultraviolet light, *i.e.*, less than about 4–40 nm. Very short X-rays are called gamma rays. Useful for medical imaging because of their power to penetrate most materials, and for other areas such as lithography because of the short wavelength.

X-ray CAT/CT: Computed axial tomography or computer-assisted tomography. A technique for dense 3D imaging of the interior of a material, particularly the human body. Characterized by use of an X-ray source and imaging system that rotate round the object being scanned.

xnor operator: A combination of two binary images A, B where each pixel (i,j) in A xnor B is 0 if exactly one of $A(i,j)$ and $B(i,j)$ is 1. The output is the complement of the xor operator. The rightmost image is the xnor of the two left images:

xor operator: A combination of two binary images A, B where each pixel (i,j) in A xor B is 1 if exactly one of $A(i,j)$ and $B(i,j)$ is 1. The output is the complement of the xnor operator. The rightmost image is the xor of the two left images:

Y

YARF: Yet Another Road Follower. A Carnegie-Mellon University autonomous driving system.

yaw: A 3D rotation representation component (along with pitch and roll) often used for cameras or moving observers. The yaw component specifies a rotation about a vertical axis to give a side-to-side change in orientation. This figure shows the yaw rotation direction:

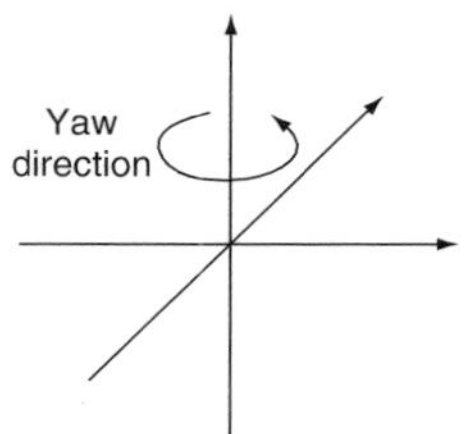

YCrCb: See YUV where U = Cr and V = Cb.

YIQ: Color space used in NTSC television. Separates Luminance (Y) and two color signals: In-phase (roughly orange/blue), and Quadrature (roughly purple/green). Conversion to YIQ from RGB is by $[Y, I, Q]' = \mathbf{M}[R, G, B]'$ where

$$\mathbf{M} = \begin{bmatrix} 0.299 & 0.596 & 0.212 \\ 0.587 & -0.275 & -0.523 \\ 0.114 & -0.321 & 0.311 \end{bmatrix}$$

YUV: A color representation system in which each point is represented by luminance (Y) and two chrominance channels (U which is Red minus Y, and V which is Blue minus Y).

Dictionary of Computer Vision and Image Processing R.B. Fisher, K. Dawson-Howe, A. Fitzgibbon, C. Robertson and E. Trucco

Z

Zernike moment: The dot product of an image with one of the Zernike polynomials. The Zernike polynomial

$$U_n^m(\rho, \phi) = R_n^m(\rho)e^{im\phi}$$

is defined in polar coordinates (ρ, ϕ) on the plane, only within the unit disk. When projecting an image, data outside the unit disk are generally ignored. The real and imaginary parts are called the *even* and *odd* polynomials respectively. The *radial* function $R_n^m(t)$ is given by

$$\sum_{l=0}^{(n-m)/2} (-1)^l \frac{t^{n-2l}(n-l)!}{l!\left(\frac{n+m}{2}-l\right)!\left(\frac{n-m}{2}-l\right)!}$$

The Zernike polynomials have a history in optics, as basis functions for modeling nonlinear lens distortion. Below, the leftmost column shows the real and imaginary parts of $e^{im\phi}$ for $m = 1$. Columns 2–4 show the real and imaginary parts of Zernike polynomials U_1^1, U_3^1, and U_2^2:

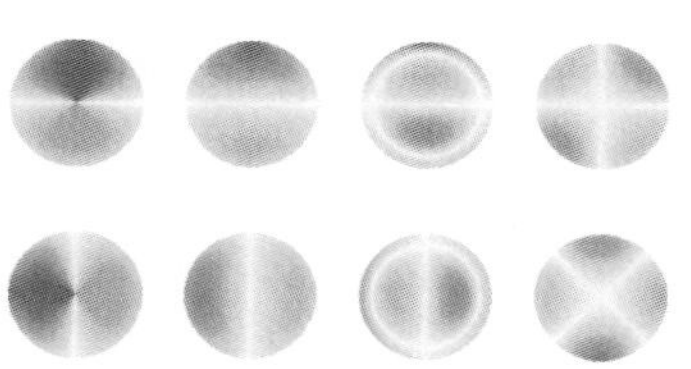

zero crossing operator: A class of feature detector that, rather than detecting maxima in the first derivative, detects zero crossings in the second derivative. An advantage of finding zero crossings rather than maxima is that the edges always form closed curves, so that regions are clearly delineated. A disadvantage is that noise is enhanced, so the image must be carefully smoothed before the second derivative is computed. A common kernel that combines smoothing and second derivative computation is the Laplacian of Gaussian.

zero crossings of the Laplacian of a Gaussian: See zero crossing operator.

zipcode analysis: See postal code analysis.

zoom: 1) To change the effective focal length of a camera in order to increase magnification of the center of the field of view. 2) Used in referring to the current focal-length setting of a zoom lens.

zoom lens: A lens that allows the effective focal length (or "zoom") to be varied after manufacture. Zoom lenses may be manipulated manually or electrically.

Zucker–Hummel operator: A convolution kernel for surface detection in volumetric images. There is one $3 \times 3 \times 3$ kernel for each of the three derivatives. For example, if $v(x, y, z)$ is the volume image, $\frac{\partial v}{\partial z}$ is computed as the convolution of the kernel $c = [-S, 0, S]$ where S is the 2D smoothing kernel

$$S = \begin{bmatrix} a & b & a \\ b & 1 & b \\ a & b & a \end{bmatrix}$$

and $a = 1/\sqrt{3}$ and $b = 1/\sqrt{2}$. Specifically, the kernel $Dz(i, j, k) = S(i, j)c(k)$, and the kernels for $\frac{\partial v}{\partial x}$ and $\frac{\partial v}{\partial y}$ are permutations of Dz given by $Dx(i, j, k) = Dz(j, k, i)$ and $Dy(i, j, k) = Dz(k, i, j)$.

Zuniga–Haralick operator: A corner detection operator that is based on the coefficients of a cubic polynomial approximating the local neighborhood.